LAND, AIR & OCEAN

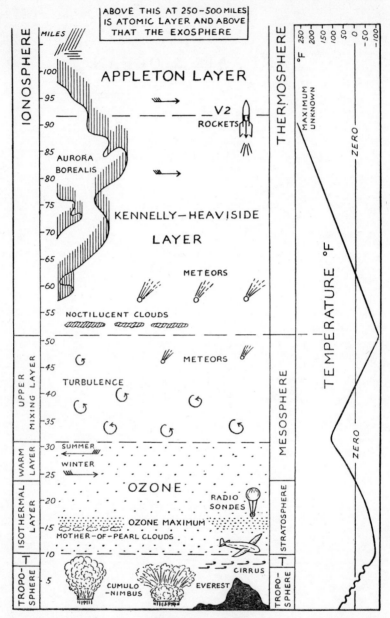

The Atmosphere

T denotes tropopause at its height in the tropics. In and near the tropopause, the Jet Stream, a narrow belt of upper westerly winds, blows strongly in circumpolar and in subtropical latitudes. Artificial rockets now ascend far beyond the upper limits of the Earth's atmosphere

LAND, AIR & OCEAN

by

R. P. BECKINSALE
M.A. (LOND.), M.A., D.PHIL. (OXON.)
*Senior Lecturer in Geography,
University of Oxford*

GERALD DUCKWORTH & CO. LTD.
3 HENRIETTA STREET LONDON W.C.2

First published 1943
Reprinted 1945, 1946, 1949
Second edition, revised, enlarged and re-set 1956
Third edition, revised 1960
Fourth edition, revised and enlarged 1966

LIBRARY
FLORIDA STATE UNIVERSITY
TALLAHASSEE, FLORIDA

Printed in Great Britain by
The Camelot Press Ltd., London and Southampton

PREFACE

THIS book caters for the general public who are interested in 'the heaven above, the earth beneath or in the water under the earth' though it has been written primarily for students at universities and training colleges. It aims at giving within a small compass an up-to-date and unified account of the physical processes by land, sea, and air which more directly affect mankind. The matter is meant to be intelligible rather than academic and selective rather than encyclopædic.

Many of the recent advances in physical geography, particularly in the study of climate, land-forms, and the oceans have been incorporated. The re-setting of the volume has allowed a considerable revision and enlargement. Many sections have been re-written and extended and three new chapters—on river régimes; lakes; and climate near the ground—have been added. Throughout, however, an attempt has been made to keep aspects in their proper proportion and to elucidate problems which the author has found from experience to puzzle the average student.

The brief summary on human geography given at the end of each chapter is intended merely as a suggestion for further study or as a pointer to the close relations between the two main branches of geography. It is hoped that the physical aspects described here will have an intrinsic appeal and that they will ensure that the reader finds a continual interest in an environment, of which, for our survival, we can hardly know too much.

The lists of references for each chapter contain the books and periodicals most consulted in the preparation of the text. These either give more intimate detail or approach a topic from another angle. It was felt that they might help the student in his further progress and the librarian in his quest for relevant volumes.

It is a pleasure to thank those persons who assisted me in the preparation of the first edition of this book, especially to the following, from whom I received many valuable suggestions:

Prof. A. A. Miller of the University of Reading; Mr. A. F. Martin, M.A. of the University of Oxford; and Mr. W. V. Lewis, M.Sc., of the University of Cambridge.

In the new edition I am greatly indebted for helpful advice to Mr. W. V. Lewis; Dr. A. W. Brewer of the Clarendon Laboratory,

Oxford; and Dr. Y. F. Tuan. It is also a great pleasure to express my thanks to Miss E. M. Potter who drew most of the diagrams and to my wife who helped with the text and compiled the index.

Acknowledgements are due to the following publishers and authors who kindly permitted the use of diagrams and plates:

Messrs. Edward Arnold, London, and Prof. L. J. Wills for three diagrams from *The Physiographical Evolution of Britain*; Messrs. Macmillan, London, and Dr. W. B. Wright for four diagrams from *The Quaternary Ice Age*; Messrs. Victor Gollancz and Prof. A. E. Trueman for two diagrams from *The Scenery of England and Wales*, since reprinted as *Geology and Scenery in England and Wales* (Penguin), 1949; the Oxford University Press and Dr. G. W. Tyrrell for two diagrams from *Volcanoes*; Dr. L. C. King and Oliver & Boyd, Edinburgh for one diagram from *South African Scenery*; and Dr. C. E. P. Brooks and the National Smoke Abatement Society for one diagram from his pamphlet, 'Atmospheric Pollution in Great Britain'. The *Geological Survey* photographs are reproduced by permission of H.M. Stationery Office. Acknowledgements for other photographs are given below the plates.

April, 1955   R. P. BECKINSALE

This Third Edition has given an opportunity to make minor amendments and to include some of the numerous findings on Earth facts and sciences since 1955. In response to many requests, the Book Lists have been strengthened as well as being brought up-to-date in regard to the latest editions of book references. It is a pleasure to thank Mr. M. J. Webb, M.A., for kindly helping to recast the section on the Indian Monsoon.

March, 1960   R. P. BECKINSALE

This Fourth Edition has allowed a full revision of all parts, including the Book Lists, and the addition of several new topics and of two new chapters on Soils and Vegetation. I am greatly indebted for skilful professional help to Dr. P. H. T. Beckett of the Soil Science Laboratory, University of Oxford; Dr. D. R. Harris of University College, London; and R. D. Beckinsale of University College, Oxford. Acknowledgements for use of the many new diagrams and plates are given below them.

October, 1965   R. P. BECKINSALE

CONTENTS

Section One

THE TERRESTRIAL GLOBE

CHAP.   PAGE

1. THE EARTH AS A GLOBE 21

   Shape and Size, p. 21; Great Circle Routes, p. 21; Revolution of
   the earth, p. 22; Rotation of the Earth, p. 23; Longitude, p. 23;
   Time, p. 23; Effect of Tilt of Earth's Axis, p. 24; The Seasons,
   p. 24; Deflection of Winds, p. 25; North-south Winds, p. 25;
   East-west Winds, p. 25; Minor Effects of Rotation and Revolution
   of Earth, p. 27; Precession of the Equinoxes, p. 27.

2. THE THREE ELEMENTS OF THE GLOBE . . . 28

   The Atmosphere, p. 28; Its Composition, p. 28; Height, p. 28;
   Density, p. 28; Twilight, p. 29; Hydrosphere and Lithosphere,
   p. 30; Distribution and Extent, p. 30; Relief of the Land and of
   the Ocean Floor, p. 31; Modern Conceptions of the Earth's
   Interior, p. 32; Nature of Earth's Interior, p. 32; Relations
   between Lithosphere and Interior, p. 33; Marine Transgressions,
   p. 34; The Theory of Continental Drift, p. 35.

Section Two

THE ATMOSPHERE

3. THE TEMPERATURE OF THE ATMOSPHERE . . . 41

   Insolation, p. 41; At Upper Surface of Atmosphere; p. 41; At
   Earth's Surface, p. 42; Heating Effect of the Earth's Surface, p. 43;
   Vertical Decrease of Temperature, p. 44; Horizontal Distribution
   of Temperature, p. 46; Effect of Relief, p. 46; Effect of Land and
   Water Masses, p. 47; Local Factors, p. 48.

4. THE MOISTURE OF THE ATMOSPHERE . . . 49

   Evaporation, p. 49; Humidity, p. 49; Absolute, p. 49; Relative,
   p. 50; Causes of Precipitation, p. 51; Major Forms of Precipitation,
   p. 55; Clouds, p. 55; Rain, p. 56; Snow, p. 56; Hail, p. 57;
   The Artificial Stimulation of Rain, p. 57.

5. GENERAL CIRCULATION OF THE ATMOSPHERE . . . 60

   Causes of High and Low Pressure, p. 60; General Latitudinal
   Distribution of Pressure, p. 60; Winds and Pressure Gradients,
   p. 62; Chief Wind Systems of the World, p. 64; Trade Winds,
   p. 64; Westerlies, p. 66; Polar Winds, p. 66; Factors Modifying
   the General Wind Systems, p. 67; Migration of Pressure Belts,
   p. 67; The Sudan Type of Climate, p. 68; Dry Season, p. 69;
   Wet Season, p. 69; Mediterranean Climates, p. 72; Summer
   p. 72; Winter, p. 73; Effect of Differential Heating of Land and
   Sea, p. 74; Land and Sea Breezes, p. 74; Monsoons, p. 75;

CHAP.   PAGE

The Indian Monsoon, p. 75; The Cool Weather Season, p. 75;
Hot Weather Season, p. 76; The Rainy Season, p. 78; The
Coming of the Monsoon, p. 78; The Retreat of the Monsoon,
p. 78; Air-Mass concept applied to the Monsoon, p. 79.

6. THE MAJOR CONTROLS OF CLIMATE   .   .   .   82

Effect of Latitude, p. 83; Equatorial and Polar Climates, p. 83;
Influence of Distance from the Sea, p. 88; Continental Climates,
p. 88; Importance of the Character of Prevailing Winds, p. 92;
Hot Deserts, p. 92.

7. THE INFLUENCE OF TOPOGRAPHY ON CLIMATE   .   .   98

Influence on Insolation and Temperature, p. 98; Insolation, p. 98;
Effects of Shade and Exposure, p. 99; Influence of Shape of Land-
forms, p. 100; Inversion of Temperature, p. 101; Influence on
Winds, p. 102; Formation of Local Winds, p. 102; Föhn Winds,
p. 105; Shelter from Winds, p. 107; Influence on Precipitation,
p. 107; Influence on Vegetation and Human Life, p. 109.

8. THE CYCLONIC INFLUENCE   111

Cyclonic Belts of the World, p. 111; Depressions or 'Lows', p. 112;
Air-masses, p. 112; Frontal Zones, p. 114; Formation, p. 114;
Older Depressions, p. 117; Anticyclones or 'Highs', p. 118;
Characteristic Weather, p. 118; Formation, p. 119; Intermediate
Types of Pressure Systems, p. 120; Secondary Depression, p. 120;
V-shaped Depression, p. 121; Wedge, p. 121; Col, p. 121; The
Polar Front, p. 122; Cyclonic Paths, p. 124; Weather in Cyclonic
Belts, p. 124; The Human Response, p. 127; Jet Streams, p. 128.

Section Three

THE OCEANS

9. GENERAL CHARACTERISTICS AND CIRCULATION OF OCEAN
       WATERS   131

General Characteristics, p. 131; Salinity of Sea Water, p. 131;
Variations at Surface, p. 133; Temperature of Oceans and Seas,
p. 133; At Surface, p. 133; Beneath Surface, p. 134; Movements
of Sea Water, p. 135; Atlantic Ocean, p. 137; South Atlantic,
p. 137; North Atlantic, p. 139; Pacific, p. 141; Influence of Ocean
Currents on the Climate of Coastal Lands, p. 141; Temperature,
p. 141; Precipitation, p. 142; Effect of Currents on Marine Life,
p. 143.

10. WAVES AND TIDES   145

Nature and Movement of Waves, p. 145; Propagation, p. 145;
Water Particles in Ocean Waves, p. 145; Waves on a Coastline,
p. 147; Tides, p. 147; Their Causes; Attraction of the Moon and
Sun, p. 147; Influence of Continental Masses, p. 151; Tides on
the Continental Shelf, p. 153; Tidal Currents, p. 155; Effects of
Tides on Estuaries and Ports, p. 156.

CHAP.   PAGE

11. SUBMARINE RELIEF AND ISLANDS 157

The Continental Shelf, p. 157; Continental Islands, p. 160; The Deep Sea, p. 161; Atlantic Ocean, p. 161; Indian Ocean, p. 162; Pacific Ocean, p. 162; Ocean Deposits, p. 163; Pelagic Deposits, p. 165; Oceanic Islands, p. 167; Volcanic, p. 168; Coral, p. 168; Island Life, p. 175.

Section Four

THE LAND SURFACES

12. THE COMPOSITION OF THE SURFACE OF THE LITHOSPHERE 179

Sedimentary Rocks, p. 179; Characteristics of Structure, p. 181; Igneous and Metamorphic Rocks, p. 182; Economic Contrasts of the Rock Groups, p. 184.

13. CRUSTAL MOVEMENTS 185

Rapid Movements, p. 185; Earthquakes, p. 185; Gradual changes, p. 187; Relative Elevation of the Land, p. 187; Marine Sediments, p. 187; Beaches, p. 187; Warping, p. 188; Subsidence of the Land, p. 188; Submerged Forests, p. 188; Drowned Valleys, p. 188; Orogenic Movements, p. 189; Simple Folding, p. 189; Complex Folding, p. 190; Mountain-building Movements in Britain, p. 194; Faulting and Fracturing, p. 195; Fault Scarps, p. 196; Rift Valleys, p. 197; Summary, p. 199; Mountains and Mankind, p. 199.

14. VULCANISM 201

Volcanoes, p. 201; Distribution, p. 201; Nature, p. 202; Different Types of Volcanic Cones, p. 203; Topographic Importance of Volcanic Cones, p. 206; Fissure Eruptions, p. 208; Intrusions of Magma, p. 211; Signs of Decaying Vulcanism, p. 213; Effect of Vulcanism on Human Life, p. 213.

15. THE WORK OF THE ATMOSPHERE 215

Chemical Weathering; Oxidization; Carbonation; Hydration, p. 215; Physical Weathering, p. 216; Work of the Wind, p. 216; Arid Topographies, p. 216; Scenic Variations in Hot Deserts, p. 217; Weathering, p. 218; Sand Corrasion, p. 218; Sand Deposition and Transportation, p. 219; Loess, p. 221; Sand on Shore Lines, p. 222; Surface Water in Arid Lands, p. 222; Human Significance of the Work of Winds, p. 224.

16. THE WORK OF GROUND WATER 226

Nature of Ground Water, p. 226; Its Position, p. 226; Dry Valleys, p. 228; Mechanical Work of Ground Water, p. 230; Slumping and Creeping, p. 230; Landslips, p. 230; Chemical Action of Ground Water, p. 231; Solution, p. 231; Limestone Topographies, p. 231; Ground Water and Human Affairs, p. 236.

17. THE WORK OF RUNNING WATER 239

Work of Flowing Surface Water, p. 239; Erosion, p. 239; Transport, p. 240; Deposition, p. 241; Local Factors Influencing the

CHAP.   PAGE

Work of Rivers, p. 242; Relief, p. 242; Climate, p. 242; Geology,
p. 242; Shaping of River Valleys, p. 244; Deepening, p. 244;
Widening, p. 244; Shallowing, p. 246; Deltas, p. 248; Suggestions
on Human Geography, p. 251.

18. THE DEVELOPMENT OF RIVER-SYSTEMS   .   .   . 254

Headward Extension of Rivers, p. 254; Watershed Regression,
p. 255; River-capture, p. 257; Northumberland, p. 259; The
Lowther Hills, p. 260; The Weald, p. 261; The Cotswolds, p. 262;
River Development in Areas of Folded Rock, p. 264; Interruptions
in the Cycle of Erosion, p. 264; Rejuvenation, p. 265; River
Terraces, p. 269; River Courses largely Independent of Surface
Structure, p. 270; Superimposed Drainage, p. 270; Antecedent
Drainage, p. 272.

19. THE WORK OF SNOW AND ICE   274

Snow-line and Existing Ice-fields, p. 274; Glaciers and Moving
Ice, p. 275; Movement of Ice, p. 275; Work of Moving Ice, p. 276;
Erosion, p. 276; Transport, p. 277; Deposition, p. 278; Quaternary
Ice-sheets and Glaciers, p. 278; The Ice Age, p. 278; Extent of
Ice-sheets, p. 279; Changes made by Glacial Erosion, p. 280; Upon
Plateau Surfaces, p. 280; On Mountains and Valleys, p. 282;
U-shaped Valleys and Truncated Spurs, p. 283; Ribbon-shaped
Lakes, p. 284; Glacial Overflow Channels, p. 285; Changes made
by Glacial Deposition, p. 285; Moraines and Glacial Drift, p. 285;
Roches Moutonnées, p. 287; Drumlins and Eskers, p. 287; The
Moraine Zone and Outwash Plain, p. 290; Diversion of Pre-glacial
River Drainage, p. 290; Glaciation and Human Affairs, p. 292.

20. THE WORK OF THE SEA: COASTAL TOPOGRAPHIES   . 295

Erosive Action of Waves, p. 295; Marine Benches and Sea Cliffs,
p. 296; Influence of Geological Structure, p. 297; Constructive
Action of Waves and Currents, p. 299; Beaches, p. 299; Spits and
Bars, p. 299; Coastal Topographies, p. 304; Flat Coastlines, p. 304;
Mountainous Coastlines, p. 305; The Human Aspect, p. 311.

21. WATER ON THE LAND SURFACES: RIVER RÉGIMES   .   . 313

Drainage of the Land Surfaces, p. 313; Factors of River Discharge,
p. 313; Types of River Régime, p. 315; Equatorial, p. 315; Savanna
and Monsoon, p. 316; Mediterranean, p. 317; Rainy Temperate,
p. 319; Cool and Cold Continental, p. 320; Mountain, p. 320;
Human Significance, p. 321.

22. WATER ON THE LAND SURFACES: LAKES   .   .   . 324

Nature of Lakes, p. 324; Formation of Lakes, p. 326; Lakes in
Hollows Formed mainly by a Deposited Barrage, p. 326; Unevenly
Deposited Sheets of Glacial Debris, p. 326; Moraines, p. 326; Ice,
p. 326; Landslips and Screes, p. 327; Water-borne Debris, p. 329;
Marine Deposits, p. 330; Volcanic Deposits, p. 330; Organic
Deposits, p. 331; Lakes in Hollows Formed mainly by Erosion,
p. 332; Wind, p. 332; Ice, p. 332; Fluvial Action, p. 334; Lakes
in Basins Formed by Earth Movements and Volcanic Explosions,
p. 334; Folding, p. 334; Warping, p. 335; Fracturing, p. 335;
General Uplift of Coastal Areas, p. 337; Volcanic Explosions,
p. 337; Lakes and Human Affairs, p. 337.

Section Five

CLIMATE, SOILS AND VEGETATION

CHAP.   PAGE

23. THE INFLUENCE OF SURFACE COVER ON LOCAL WEATHER   343

The Need for Surface Recordings, p. 343; General influences,
p. 344; Shape, p. 344; Aspect or Orientation, p. 345; Insolation,
p. 345; Exposure to Winds, p. 347; Colour and Composition,
p. 349; Influence of Surface Cover, p. 350; Bare Soils, p. 350;
Grass, p. 351; Other Crops, p. 352; Woods and Forests, p. 353;
Water Bodies and Snow, p. 354; Built-up Areas, p. 355; Human
Significance, p. 358.

24. SOILS: THEIR FORMATION AND DISTRIBUTION   .   .   361

Definition, p. 361; Nature of Mineral Content, p. 361; Organic
Content, p. 364; Air and Water Content, p. 366; Factors in Soil
Formation: Climate, p. 372; Parent Material, p. 373; Flora and
Fauna, p. 375; Time-Scale and Past Land-Use, p. 377; Relief,
p. 378; World Classifications of Soils, p. 380; Tundra Soils, p. 384;
Podzols, p. 385; Grey-Brown Podzolics, p. 387; Red-Yellow Pod-
zolics, p. 387; Ferruginous and Ferralitic Tropical Soils, p. 388;
Chernozemic Soils, p. 391; Desertic Soils, p. 393; Intrazonal Soils:
Halomorphic, p. 394; Hydromorphic, p. 394; Calcimorphic,
p. 396; Azonal Soils, p. 396; Man and Soils, p. 397.

25. VEGETATION   400

Nature of Vegetation, p. 400; How Plants Spread, p. 402; Climax
Plant Formations, p. 403; Plant Successions, p. 404; Factors In-
fluencing Growth of Vegetation: Temperature and Light, p. 406;
Precipitation, p. 408; Soil and Slope, p. 412; Biotic Factors, p. 413;
Major Plant Formations: Tropical Rainforest, p. 415; Tropical
Semi-Evergreen and Deciduous Forest, p. 418; Tropical Thorn
Forest and Semi-Desert Scrub and Desert, p. 421; Middle-Latitude
Evergreen Forest, p. 422; Cool-Season Broad-Leaved Deciduous
Forest, p. 424; Intermixed Broad-Leaved Deciduous and Coni-
ferous Forest, p. 425; Boreal Needle-Leaved Coniferous Forest,
p. 425; Tundra, p. 426; Grasslands: Tropical Savanna, p. 428;
Middle-Latitude Grasslands, p. 431; Vegetation on Mountains,
p. 432; Conservation of Plants and Vegetation, p. 436.

INDEX   439

LIST OF DIAGRAMS

PAGE

The Atmosphere   *Frontispiece*

1. The Great Circle Route from London to Winnipeg   22
2. Graphical method of finding approximate declination of sun   24
3. Deflection of north-south winds in the northern hemisphere   26
4. Relative extent of Land and Ocean each 10° of latitude   30
5. The Earth, showing the approximate thickness of its component zones   33
6. The Nature of Floating Continents   34
7. Structural relations of the opposing sides of the Atlantic Ocean   37
8. Latitudinal distribution of Insolation on June 21st   41
9. Explanation of decrease of temperature with increase of latitude   43
10. Average decrease of temperature with increase of altitude over south-eastern England   45
11. Mountains as a break in isothermal layers: the Himalayas   46
12. Rate of cooling of ascending air before and after condensation   52
13. Relation of rainfall to relief: (a) the Lake District, (b) Sierra Nevada   53
14. Convection currents as a cause of wind   62
15. General atmospheric circulation   65
16. Seasonal migrations of the main pressure belts   67
17. Generalized world distribution of chief climatic types   70
18. Relation between rainfall and the zenithal sun   71
19. Possible cause of the 'burst' of the summer monsoon in India and Pakistan   77
20. World map of annual mean range of temperature   89
21. Mean temperatures during 'inversions'   103
22. Explanation of up-valley and down-valley winds   104
23. Atmospheric conditions during a föhn   105
24. Main air-mass sources and frontal zones of the northern hemisphere   113
25. Life-cycle of a depression   115
26. Weather sequence of an occluded depression   117

PAGE

27. Weather associated with an anticyclone or 'high' 119
28. Weather associated with a Secondary depression 121
29. Weather associated with (*a*) a V-shaped depression, (*b*) a wedge 122
30. Mean position of the polar front in winter 123
31. Main cyclonic tracks over Western Europe 125
31a. Jet streams of the northern hemisphere 128
32. Salinity of the surface of the Atlantic Ocean 132
33. Salinity and temperature of sea water at the Strait of Gibraltar 134
34. Seasonal swing of equatorial currents in the Atlantic 136
35. Seasonal change in direction of currents in the Indian Ocean 137
36. General distribution of ocean currents in the Atlantic 138
37. The Gulf Stream Drift 140
38. Motion of water particles in a free wave 146
39. Explanation of tidal pull 148
40. General distribution of tide-raising forces 149
41. Cause of spring and neap tides 151
42. Co-tidal lines in the Atlantic Ocean 153
43. Co-tidal lines around the British Isles 154
44. The orography of the North Sea bed 158
45. The submarine valley off the mouth of the Congo 159
46. Main structural relations of south-eastern England and the Continent 160
47. The Tonga and Kermadec Deeps 164
48. General vertical distribution of deep-sea deposits 165
49. Section across a coral reef 169
50. Funafuti, a typical atoll 171
51. Theories on the formation of coral islands 173
52. Normal sequence of deposition on a shoreline 180
53. Effect of dip on topography: (*a*) Leckhampton Hill, (*b*) Hog's Back 182
54. Effect of joints on weathering of granite: Great Mountain Tor, Dartmoor 183
55. Simple symmetrical folds: south-western Ireland 189
56. Asymmetrical fold in chalk of the Isle of Wight 190
57. Complex folds. Recumbent folding leading to overfolds and formation of nappes 192

		PAGE
58.	Inversions of relief: (*a*) Snowdonia, (*b*) the Weald	194
59.	Main types of faulting: (*a*) Vale of Eden, Cumberland; (*b*) Great Basin, Oregon	196
60.	Ingleborough and Giggleswick Scar viewed from the south-east	197
61.	The Rhine rift valley	197
62.	The rift valley of East Africa	198
63.	The island of Hawaii	205
64.	The north-east quadrant of Etna	206
65.	Volcanic plugs; Dumgoyn and Dumfoyn, Scotland	208
66.	Basalt lava plateau of the Deccan, India	209
67.	Chief Tertiary volcanic regions of the British Isles	210
68.	Main types of igneous intrusion and extrusion	211
69.	The Whin Sill	212
70.	Desert dunes	220
71.	Structural nature of artesian wells: (*a*) London Basin, (*b*) Queensland	227
72.	Position of water-table and of springs	228
73.	Diminution of surface streams in limestone country: river Churn	229
74.	Cycle of erosion in a karst land	234
75.	Formation of gorge in a limestone escarpment	235
76.	Areas of maximum turbulence in a mature stream	240
77.	Upstream recession of waterfalls: High Force, Teesdale	243
78.	Normal evolution of the cross-section of a valley	244
79.	Widening of a valley-floor	245
80.	Downstream migration of meanders	246
81.	Delta of the Nile	250
82.	Delta of the Ebro	251
83.	Lowering and shifting of a divide	256
84.	Main drainage of the Lowther Hills	256
85.	Development of streams on tilted strata of unequal hardness.	258
86.	Headward erosion of a spring	259
87.	Rivers of Northumberland	260
88.	Drainage of the Weald	261
89.	Drainage of the Cotswold Hills	263
90.	Profile and cross-sections of the valley of the North Tyne	265
91.	Incised meanders	266

PAGE

92. Abandoned hanging meander of the Herefordshire Wye at Newland   267
93. Valley of the Dee at Llangollen   268
94. River terraces: (*a*) the upper Thames, (*b*) the lower Thames   270
95. Course of the Wye at Symond's Yat   271
96. Longitudinal profile of a glacier valley   277
97. Extent of main Quaternary ice-sheets in the northern hemisphere   280
98. Direction of ice-flow and limits of glaciation in the British Isles   281
99. Valley filled with glacial débris: the Cam at Littlebury   282
100. Formation of cirques   283
101. Origin of U-shaped valleys and truncated spurs   284
102. General sequence of deposition at the snout of an ice-tongue or valley-glacier   286
103. Contour map of drumlin-topography in the neighbourhood of Ballintra, Co. Donegal   288
104. Portion of an esker near Tyrell's Pass, central plain of Ireland   289
105. The main eskers west of Tyrell's Pass, central plain of Ireland   289
106. Glacial Lake Pickering and the river Derwent   291
107. Marine benches   296
108. Influence of geological structure on a coastline: Torquay Headland and Tor Bay   298
109. Loe Bar, near Helston, Cornwall   300
110. Chesil Bank and the Isle of Portland   301
111. Hurst Castle Spit   302
112. Orfordness   303
113. The Dalmatian coast near Zara   305
114. The Rias of Galicia, north-western Spain   307
115. Kenmare River, a ria in south-western Ireland   307
116. Sogne Fiord and Hardanger Fiord, Norway   309
117. Stages in the erosion of a fault coastline   310
118. Equatorial river régimes   315
119. Hydrology of the Nile basin   317
120. River régimes of Italy   318
121. Rainy temperate river régimes   319

16 LAND, AIR AND OCEAN

PAGE

122. River régimes controlled by melting of snow and ice 320
123. Sources and use of Nile water at Aswan 322
124. Glacier dam across Shyok valley 328
125. Lake dammed by river-alluvium 329
126. Lava flow damming a river 330
127. Crater lakes in the Azores 331
128. Sudd-dammed lakes in the Sudan 332
129. A typical glacial trough lake: Coniston Water 333
130. Drainage systems of western Uganda 336
131. Regulating effect of Lake Garda 338
132. Soil and air temperatures: Poona and Munich 343
133. Insolation on horizontal surface and on slope of maximum sunniness at Kew 346
134. Mean daily totals of insolation on differently oriented surfaces at Potsdam 347
135. Influence of relief trend on surface winds 348
136. Mean monthly range of temperature just below surface and in screen on Salisbury Plain 350
137. Mean minima temperatures at Oxford 351
138. Influence of built-up areas and forests on wind-speed 353
139. Mean monthly maximum and minimum temperatures of urban and rural areas
140. Section across London from west to east to show how 356
atmospheric pollution affects winter sunshine 357
141. Simplified system of textural soil nomenclature 363
142. Physical-chemical structure of clay minerals 365
143. Idealized forms of water in the soil 367
144. Water molecule; and hydrated sodium ion 369
145. Leaching action of carbonic acid in soils 371
146. Development of rendzinas and terra rossa 375
147. Types of soil catenas 379
148. Profiles of podzol and grey-brown podzolic soils 386
149. Profiles of tropical ferralitic soils 390
150. Profiles of pedocals 392
151. Profiles of halomorphic soils 395
152. General limits of distribution of certain plants 402
153. A hydrosere in middle latitudes 405
154. The nutrient cycle of vegetation 410
155. Water régime of plants in different climates 411

PAGE

156. Cauliflory and buttress roots in tropical rainforest   416
157. Profile diagrams of tropical rainforest in Trinidad   417
158. Length of sunbaking period in and near the Tropics   420
159. Boreal forest–tundra transition in Labrador–Ungava   427
160. Main types of grassland east of the Rockies   431
161. Zonal vegetation on mountains from Arctic to Antarctic   433

LIST OF PLATES

FACING PAGE

 1. A Swiss alp upon the Silvretta massif   100
 2. The Rieder Alp, Valais, Switzerland   100
 3. A volcanic island. The Nuuanu-Pali road near Honolulu,
    the Hawaiian Islands   101
 4. Luxuriant growth of coral exposed at low tide on the outer
    edge of the Great Barrier Reef, Australia   101
 5. 25-foot (in foreground) and 100-foot raised beaches at
    Hilton of Cadboll, Ross-shire   192
 6. The drowned estuary of the River Salcombe, near Kings-
    bridge, Devon   192
 7. Young fold-mountains: the Alps in eastern Switzerland   193
 8. Old fold-mountains: the Cairngorms, Scotland   193
 9. A volcanic cone. Fuji Yama, Japan   208
10. Basalt sills at Pleaskin Head, near the Giant's Causeway,
    County Antrim   208
11. Moving dunes in Death Valley, southern California   209
12. Culbin Sands, Elgin, showing old land surface (on left)
    swept bare of advancing sand   209
13. Upper stretch of the gorge at Cheddar in the Mendips   224
14. Caves in limestone at Cheddar   224
15. A youthful stream. The Afon Dulas, Montgomery   225
16. A mature river. The Rheidol below Devil's Bridge,
    Cardigan   225
17. A lake delta. Buttermere and Crummock Water separated
    by the delta of the Sail Beck   240
18. Delta of the Vardar at the head of the Gulf of Salonika   240
19. Rejuvenation. Incised meander of the Rheidol Gorge,
    two miles north of Devil's Bridge, Cardigan   241
20. Incised meander of the Wye near Chepstow, Monmouth   241
21. South-western margin of the Greenland Ice-cap   288

B

FACING PAGE

22. Glacier and fiord south of Scoresby Sound, eastern
    Greenland   288
23. Plateau after glaciation by continental ice-sheet: the
    Laurentian Shield north of Sioux Lookout, Ontario   289
24. Mountains after glaciation: the English Lake District   289
25. Coastal erosion. The north cliffs near Camborne, Cornwall   304
26. Marine deposition. Chesil Bank with Portland Bill in the
    distance   304
27. Lake dammed by land-slips: Lac des Brenets   305
28. A volcanic crater lake: Lake Duluti in northern Tanzania   305
29. Lake dammed by up-fold: Lac de Joux   344
30. A rift-valley lake: Lake Manyara in the great East African
    rift-valley   344
31. Inversion fog in the Coxwold–Gilling gap, North Riding   345
32. Smoke pall over Manhattan, New York, on a calm, sunny
    day in winter   345

Between pages 424 and 425

33. Root system and soil particles with attached water-film
34. Root and root hairs of Italian rye-grass
35. Profiles of podzol and grey-brown podzolic soil
36. Profiles of brunizem and chernozem
37. Gully erosion in Tennessee valley before reclamation
38. Contour strip cropping in Dodge County, Minnesota
39. Tropical rainforest, New Britain
40. Lianes in tropical rainforest
41. Mangroves in Malaya
42. Teak forest, Burma
43. Trunk of baobab
44. Lower trunks of giant redwood
45. Semi-desert and desert scrub near Phoenix
46. Semi-desert and desert scrub in Gila Desert
47. Close boreal forest
48. Lichen-woodland in eastern Canada
49. Lichen and rock tundra, Ellesmere Island
50. Tundra, Baffin Island
51. Savanna woodland, Uganda
52. Dry bush savanna, Tanzania
53. Giant lobelia, Mt. Ruwenzori
54. Giant groundsels, Mt. Ruwenzori

Section One

THE TERRESTRIAL GLOBE

CHAPTER ONE

THE EARTH AS A GLOBE

Shape and Size

IN recent years photographs taken from the stratosphere have familiarized us with the fact that the horizon of the land is just as curved as that of the sea. So great is the curvature of the earth's surface that the visible horizon, or farthest point of any extensive flat area visible to a person standing upon it, is only about 3 miles.

Yet in dealing so frequently with maps of small areas it is easy for geographers to forget the importance of viewing the earth as a globe in the solar system. The shape of the earth differs slightly from that of a true sphere, the chief irregularity being a bulge at the Equator which makes it appear that there is a flattening at the poles. The actual variations from a true sphere are relatively insignificant, the polar axis being only 27 miles shorter than the diameter at the Equator (7,899 miles to 7,926 miles), and the shortest circumference only 42 miles less than the longest.

From the point of view of modern transport, distances on the globe's surface, such as the 25,000 miles round the Equator or the 12,500 miles round the latitude of 60°, are diminishing rapidly. If an all-sea route were possible, a modern liner at 25 knots could circumnavigate the Equator in about five weeks; if it were an all-land route, an express at 40 miles per hour could travel round it in less than a month; and a traveller by aeroplane that averaged 500 miles per hour could fly round it in about two days. Yet from the point of view of the movement of the physical phenomena that affect the earth's surface, these distances are enormous. A low-pressure system or cyclonic depression may take a week to cross the North Atlantic while the Gulf Stream Drift is warming our winters with heat accumulated near the Equator and West Indies a few years previously. It will be shown later that even these movements are rapid compared with the rate of certain physical changes, such as the erosion of land masses and the 'drifting' of the continents. Indeed, Nature usually works so slowly that it is difficult for man to appreciate the great length of time she needs for so many of her processes.

Great Circle Routes

The importance of viewing the earth as a globe has increased with advances in transport. The shortest distance between any two points on the earth's surface is along an arc of the Great Circle joining them. A Great Circle is any complete circumference of the earth whose diameter is practically equal to that of the Equator. For instance, all the meridians of longitude are Great Circles. The effect of Great Circle sailing is necessarily best seen in the routes between any ports lying in the same latitude, but at a considerable distance from the Equator. Thus, in endeavouring to follow the direct or Great Circle route between London and New York, ships in autumn travel

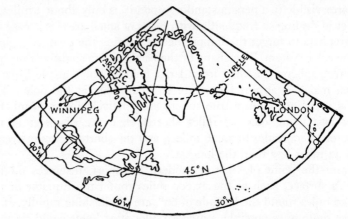

1. The Great Circle Route from London to Winnipeg

far northwards towards Greenland: the direct route from Cape Horn to Cape Town would lead ships far into the Antarctic ice, but the compromise route followed saves some 200 miles. Furthermore, the direct air-route from London to Winnipeg leads directly across Iceland and Greenland and passes well within the Arctic Circle, a fact which emphasizes the importance of recent Arctic expeditions (Fig. 1).

Revolution of the Earth

The earth moves round the sun in an ellipse of some 600 million miles, which it completes every 365¼ days. The sun is not in the centre of this ellipse and the difference between the length of the axes, some 3 million miles,[1] would appreciably affect our weather if the earth moved at a uniform speed along its orbit. As it is, owing to

[1] Perihelion in early January, 91,500,000 miles, Aphelion in early July, 94,500,000 miles.

the greater curvature at one end of the ellipse, the earth moves faster along its path in our winter, and this extra speed compensates for the closer proximity to the sun. A minor effect of the varying velocity of the earth in its orbit, coupled with the constant inclination of the earth's Equator to the plane of this orbit, is that time taken from the midday sun does not exactly correspond to mean solar time or a day of 24 hours. Sometimes the sun's noon falls before mean noon and sometimes after by as much as 16 minutes. Tables are readily available that give the correction, or Equation of Time, whereby the sun's time can immediately be corrected to mean time. The Equation of Time is essential for obtaining Greenwich or mean time from sun-dials. Further details are given on p. 27.

Rotation of the Earth

At the same time as the earth proceeds along its orbit, it rotates upon its own axis. This rotation can be demonstrated by the gradual changes in the direction of swing of a pendulum (for example, Foucault's) and by the deviation eastwards of a body dropped from a height, the deviation being about one inch for every 500 feet of fall in middle latitudes.

Longitude. The rotation of the earth is from west to east, and is the cause of day and night with all that these diurnal changes of temperature imply; such physical phenomena as land- and sea-breezes, mountain- and valley-winds and the expansion and contraction of rock surfaces come readily to mind. But the earth's rotation also supplies a means of fixing positions on the earth's surface, for while north to south or latitude positions are fixed by reference to the altitude of the sun,[1] east to west or longitude fixations are usually based on the time of the meridian passage of the sun compared with Greenwich time at that instant.

Time. In actual practice it is found that time in any country is usually fixed for convenience on a central line of longitude. Thus western Europe keeps Greenwich time and central Europe that of 15° E., or one hour fast on Greenwich. In large countries the territory is divided into standard time zones of about one hour each (15°). There are six such zones in Canada and eleven in Russian territory.

---

[1] For latitude, altitude of meridian sun or of circum-polar stars is read by sextant or theodolite and allowance made for declination. For longitude, Greenwich time is obtained from chronometer or wireless. Observers of meridian time must subtract or add Equation of Time to it before they compute its difference from Greenwich time.

Clocks on Siberian railway stations show both local and Moscow
time, and all clocks in the U.S.S.R. are permanently advanced one
hour in front of standard time.

Effect of Tilt of Earth's Axis: the Seasons. The rotation of
the earth would give equal days and nights all the year with no
seasonal changes were it not for the inclination of the earth's axis to
the plane of its orbit. Except at the equinoxes, this inclination leads
to unequal days and nights, and these cause the seasons. The variations in the latitudinal position of the overhead sun and in length of
daylight are shown diagrammatically in most atlases. In our latitudes
the longest and shortest length of daylight are roughly—

	21st June	*22nd December*
60° N.	18 hr. 27 min.	5 hr. 33 min.
50° N.	16 hr. 18 min.	7 hr. 42 min.

The apparent movement of the sun from overhead at the Tropic of
Cancer in June to overhead at the Tropic of Capricorn in December
is caused entirely by the plane of the Equator gradually changing its
position with regard to the sun as the tilted earth moves along its
orbit. This change is the same as the so-called *declination of the sun*
and must be known in order to compute the sun's altitude.

The formula for computing the altitude of the sun at noon on a
given day at any place is as follows:

$$(90 - \text{latitude of place}) \pm \text{declination of sun}.$$

The sun's declination, which is added
in the summer half-year and subtracted
in the winter, is given by the formula
$23\frac{1}{2}° \times sine$ of number of days from
nearest equinox.

An approximate graphical solution is
shown in Fig. 2. Assume that the year
consists of 4 periods of 90 days each
between Equinox and Solstice. Draw a
quadrant, each degree of arc representing
one day, and the radius representing $23\frac{1}{2}°$
latitude. Then, having determined the
number of days given date is from
nearest equinox, measure from the
vertical radius 1° arc for each day. Draw
a radius to the circumference; from it

2. Graphical method of
finding approximate de-
clination of sun

Given day is May 5 and declination is about 16° 30′.
Precise declination is 16° 2′

drop a perpendicular to the base, and intersection gives declination of the sun.

Effect of Rotation on Bodies moving on the Earth's Surface: Deflection of Winds, or Coriolis Force

Owing to the rotation of the earth, any point at rest on its surface at the Equator is rotating eastwards at about 1,050 m.p.h., and from 500 to 650 m.p.h. in our latitudes (50°– 60° N.). This rotational movement must necessarily affect the relative direction of bodies moving on and above the earth's surface. Ocean currents and north-south flowing rivers are affected, but the influence is greatest and most significant in the case of winds. Winds in the northern hemisphere are deflected to the right of their intended course and in the southern hemisphere to the left.

North-south Winds. In the case of a particle moving in a north or south direction the explanation of the deflection depends upon the following factors:

(a) the particle when at rest has a rotational velocity which it has acquired from the earth beneath it;

(b) the fact that if a body rotating at a uniform speed about a centre is pushed towards the centre, that body's velocity of rotation increases. Or, conversely, if the body is forced away from the centre its velocity decreases.

Thus if a particle A 'at rest' at the Equator is pushed due north towards B, both its greater initial velocity of rotation and its increasing velocity (due to the fact that it is moving towards the centre of rotation) will cause it to arrive in B's latitude well in front of B. In other words, the particle has been deflected to the right (Fig. 3a).

But if a particle 'at rest' at B is forced southwards towards the Equator, its slower initial speed of rotation and its decreasing rate of rotation (because it is moving away from the centre) combine to cause it to arrive in A's latitude behind A. In other words, it is deflected to the right (Fig. 3b). The same arguments will show how winds in the southern hemisphere are deflected to the left.

East-west Winds. In the case of particles moving due east or west the explanation is slightly different, for the impulse that causes these winds affects in the first place their velocity and not their distance from the axis of rotation. Now it happens that

(a) if the velocity of a rotating body is increased, its centrifugal force will also increase and it will tend to move away from the centre or axis of rotation;

(b) if the velocity decreases so does the centrifugal force, and the particle will tend then to move towards the axis of rotation.

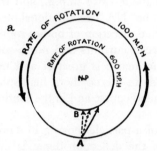 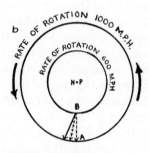

3. Deflection of north-south winds in the northern hemisphere:

(i) Arrow with broken line denotes direction if no deflection

(ii) Arrow with crosses represents deflection due to difference in initial velocity of rotation

(iii) Arrow with solid line represents actual deflection, which consists of (ii) and the gain (in *Fig. a*) or loss (*Fig. b*) of rotational force as a body moves towards or from the centre of rotation

Thus a particle forced eastwards will have its initial velocity of rotation increased, and owing to the increase in centrifugal force it will tend to move away from the centre towards the Equator. Therefore, in the northern hemisphere, a west wind will be deflected to the south of east or to the right. An east wind will also be deflected to the right, or to a north-westerly quarter, because its relative rate of rotation is diminished and it is drawn towards the pole. In the southern hemisphere the deflection will be to the left.

These arguments bring another point to our notice. At the Equator the centrifugal force acts at right angles to the surface, and so is directly opposed to the force of gravity. Therefore its effect merely diminishes slightly the weight of any article on the Equator, and in no way tends to cause that particle to move on the surface. This means that east or west winds blowing near to the Equator will suffer very little or no deflection. Hence air eddies or convergent winds only become common in latitudes outside the tropics where there is a great deflection in the intended direction of both east-west and north-south air currents.

Minor Effects of Rotation and Revolution of Earth

The revolution of the earth is necessarily such that its speed, or angular velocity, varies inversely as the square of the distance from the sun and that its radius vector (straight line from sun to earth) sweeps over equal areas in equal times. Thus the orbital speed lessens in aphelion because a longer radius sweeps over a smaller arc in a given time. Owing to these changes in velocity the summer half-year in the northern hemisphere is 8 days longer than the winter half-year and the reverse happens in the southern hemisphere. But differing distance from the sun also affects slightly the intensity of incoming solar radiation to the advantage of perihelion. Consequently, although each hemisphere receives the same total amount of solar radiation, to-day in the northern hemisphere each season is slightly moderated whereas in the southern hemisphere each season is slightly strengthened and seasonal differences increased.

Precession of the Equinoxes

The position of the equinoxes moves in a clockwise direction round the earth's orbit. This *precession*, of about 50 seconds of arc a year, is caused by the pull of the sun, moon and other planets on the earth's equatorial bulge and by the inclination of the equatorial plane to the plane of the sun's orbit (ecliptic) and of the moon's orbit. If the earth did not rotate, the gravitational action of the sun and moon would 'bring these planes into coincidence. But because of the gyroscopic effect of rotation, the action is at right angles to what one would otherwise expect.'[1] The earth's tilt remains at about $23\frac{1}{2}°$ and the position of the equinoxes moves completely round the orbit in about 25,800 years.

To-day the northern winter is in perihelion whereas 12,000 years ago it was in aphelion. Such changes have been used to account for Pleistocene ice-advances but although they could favour the growth of existing ice-sheets they could not in themselves cause an Ice Age.

LIST OF BOOKS

For Equation of Time and standard times, *Nautical Almanack; Whitaker's Almanack; The Air Almanac*
For daylight, and altitude of sun:
Harrison, L. C. *Sun, Earth, Time and Man*, 1960
Beckinsale, R. P. 'Altitude of the Zenithal Sun,' *Geog. Rev.*, 1945, pp. 596–600
Strahler, A. N. *Physical Geography*, 1960
Wylie, C. C. *Astronomy, Maps and Weather*, 1942
For Earth's deflective influence:
Proudman, J. *Dynamical Oceanography*, 1953 (mathematical)
Scorer, R. S. 'Vorticity', *Weather*, 1957, pp. 72–84.

[1] Munk, W. H. & Macdonald, J. F. *The Rotation of the Earth*, 1960, p. 7.

CHAPTER TWO

THE THREE ELEMENTS OF THE GLOBE

THE solid earth, of which the surface is partly covered by land and partly by water, is encircled by an envelope of gas known as the atmosphere.

The Atmosphere

Composition. Dry air is a remarkably uniform mixture of the gases shown in the following table:

	Percentage by volume or pressure
Nitrogen	78·08
Oxygen	20·95
Argon	0·93
Carbon Dioxide	0·03
Neon	·0018
Helium	·00053
Hydrogen ⎫ Krypton ⎪ Xenon ⎬ Ozone ⎭	slightest trace

Normal air, however, contains particles of water-vapour and dust to a varying degree, and these two constituents have a marked effect on physical phenomena occurring in the atmosphere.

Height. The height of the atmosphere exceeds 200 miles and may, according to the evidence of 'shooting' stars and of the Aurora Borealis, extend in an extremely rarefied form to over 600 miles. It was formerly thought that the proportion of the heavier gases diminished steadily with increase in altitude. To-day it is known that nitrogen and oxygen are also the chief constituents far into the upper thermosphere, or to about 400 miles.

Density. These assumptions, however, are less important than the rapid rarefying of the air with increase of height. High-level flying to-day is made possible only by the use of artificial oxygen, and oxygen apparatus may be carried on mountain expeditions beyond about 23,000 feet. Over two-fifths of the whole atmosphere lies below 16,000 feet, a fact which is usually stated as follows: at 3½ miles the atmospheric pressure decreases to one half, at 7 miles it decreases to one quarter, and at $10\frac{1}{2}$ miles to one-eighth of the pressure at sea-level.

Although pressure decreases so rapidly with increasing altitude, the amount of water-vapour in the air decreases even more rapidly. Probably, on an average, half the total water-vapour in the atmosphere lies below 7,500 feet, some three-quarters lies below 15,000 feet, and practically all lies below a height of 9 or 10 miles. The scarcity of atmospheric moisture above 15,000 feet does not, however, preclude the possibility of heavy snows on peaks such as Aconcagua and Everest, as moisture-laden currents from the surface layers are frequently forced to ascend into the layers of atmosphere above them.

From what has already been said, it follows that below a height not much exceeding that of the world's highest mountains lies a layer of air where water-vapour is relatively abundant. This same layer also contains most of the dust particles which help to scatter light and so to illuminate better the whole atmosphere. It happens that water-vapour and dust are the chief constituents of the atmosphere capable of abstracting heat from the sun's rays and of absorbing heat that is radiated into space by the solid part of the earth's surface. Consequently it is this lower layer of air which at night blankets in the heat gained by the earth from the sun during the preceding daylight, and which in day-time protects the earth from the sun's rays. Without this layer the earth's surface, especially in tropical latitudes, would be alternately torrid in day-time and frigid at night.

Twilight. Reflection from the dust and water-vapour in the atmosphere is the cause of twilight. Long before the sun has appeared above or after it has disappeared below the horizon these particles reflect downwards the sun's light so that a diffuse glow precedes its rising and follows its setting. The reflection begins or ends as the case may be, when the sun is 18° below the horizon and the whole period between then and sunrise and sunset respectively is called *astronomical twilight*. The illumination usually remains quite bright as long as the sun is 6° or less below the horizon and this more brilliant period of astronomical twilight is known as *civil twilight*.

Twilight varies in length with latitude in the same way and for the same reasons as daylight does. It depends basically on the altitude of the noon-day sun or on latitude. In the tropics, where the sun's rays fall almost vertically, civil twilight ends in about twenty minutes and astronomical twilight includes a further hour of fainter glow. In polar regions the low angle of the sun's rays gives a very long twilight in some seasons. Poleward of about 48° latitude at least for one day each year the sun's path never descends more than 18° below the horizon and there is no true night. At the poles twilight lasts continuously for two periods of seven weeks, one after the autumn equinox (September 23rd), and one before the spring equinox (March 21st). Consequently, the so-called 'polar night' prevails only from mid-November to late January. It will be seen, however, that in all latitudes, especially when local weather conditions are favourable, twilight is an atmospheric gift of considerable value to man.

Hydrosphere and Lithosphere

Distribution and Extent. The atmosphere and hydrosphere are intimately connected; atmospheric gases penetrate in solution to the

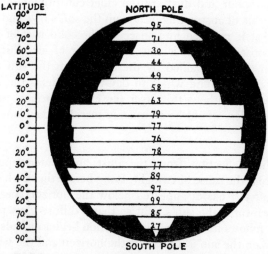

4. Relative Extent of Land and Ocean each 10°
of latitude

Numbers give percentage of area covered by sea

ocean depths and the atmosphere obtains its water-vapour by evaporation from the water surface. It is difficult to appreciate fully the vastness of the water expanses on the surface of the globe; they cover no less than 140 million square miles or 71 per cent. of its total area, leaving 57 million square miles or only 29 per cent. to be occupied by land surface. The approximate distribution of land and sea for each 10 degrees of latitude is shown in Figure 4.

It will be seen that the surface of the southern hemisphere is predominantly water (81 per cent. total area) while the world's great land masses lie mainly north of the Equator. The globe is almost encircled by sea in latitudes 40°–70° S., and is most nearly encircled by land in latitudes 60°–70° N. The usual division of the globe's surface into a land hemisphere, containing about 85 per cent. of the lithosphere and centring on London, and an ocean hemisphere centring upon New Zealand, is useful if it is noticed that even in the land hemisphere water covers rather more than half of the total area. The relatively small proportion of the globe's surface available for settlement is a vital factor in human affairs, especially as the world population of about 3,300 millions continues to increase.

Relief of the Land and Ocean Floor. The extreme inequalities of the earth's crust (Everest 29,030 feet to the Mariana Deep 36,204 feet) have a range of about $12\frac{1}{2}$ miles, which is insignificant $(\frac{1}{313})$ compared with the radius of the earth. The estimates in the following table emphasize the relative lowness of the land masses and the great depth of the oceans.

Depth or Height	LAND		SEA	
	Area Million sq. miles	Surface of whole Globe per cent.	Area	Surface of whole Globe per cent.
0–600 feet . . .	15	8	10	5
600–3,000 feet . .	26	13	7	3
3,000–6,000 feet . .	10	5	5	2
6,000–12,000 feet . .	4	2	27	15
12,000–18,000 feet . .	} 2	1	81	41
Over 18,000 feet . .			10	5
TOTAL .	57	29	140	71

It is estimated that just over 70 per cent. of the land surface is less than 3,000 feet above sea-level, while a similar proportion of ocean floor has a depth exceeding 10,000 feet. The mean height of Europe is about 1,100 feet, of Africa 2,200 feet, and of Asia 3,300 feet; but the oceans about them average between 11,000 feet (Atlantic) and over 13,000 feet (Pacific) in depth. When looked at from this point of view, the oceans cover the greatest relief features of the earth's surface.

The interactions of the hydrosphere and lithosphere are strongest upon the continental shelf where wave erosion, deposition, and relative changes in sea-level become of great significance. It is, however, the atmosphere that gives the oceans their waves and carries their climatic influence inland. These numerous relations between the three widely different elements of the globe, and especially the infinite variations they cause in physical phenomena occurring on the earth's surface, form the realm of physical geography.

Modern Conceptions of the Earth's Interior

The formation of the continents and oceans, is as yet, almost inexplicable. These primary features largely owe their existence to the manner of origin of the earth, and old theories on them, such as Lowthian-Green's tetrahedral hypothesis, are quite untenable.

Nature of Earth's Interior. The explanation of earth features of the second order, such as mountains, involves some discussion of the earth's interior. From astronomical computations it is known that the density of the whole earth is 5·53 grams per cc. whereas the lithosphere averages less than 3. Thus the interior substances exhibit a much greater density than the surface rocks. Seismology, the study of earthquake waves, provides the main key to the internal composition. Earthquakes generate three main types of waves: primary, or compressional-expansional (P), in which each particle vibrates to and fro in the direction of propagation; secondary, or transverse (S), in which each particle oscillates at right angles to the direction of propagation; and 'surface', or L waves, with a long wavelength. The velocity of P waves, the fastest, depends on the density and resistance of rocks to compression whereas the velocity of S waves is controlled by the density and rigidity (resistance to distortion) of rocks. As liquids offer no resistance to distortion, S waves cannot propagate through them.

The analysis of seismograph records show that the earth consists of a number of physically distinct concentric shells of which the most important are the inner and outer core at depths of 3,100 and 1,800 miles respectively; the mantle, which is about 1,800 miles thick; and the very thin outer crust (Fig. 5). The density of the earth's material increases gradually from 3·3 grams per cc. just below the crust to 5·66 at the base of the mantle, then jumps suddenly to 9·71 at the top of the core and increases steadily again to 11·76 at the bottom of the outer core and to between 14·5 and 18·0 at the earth's centre.

The composition of the mantle is, at the moment, a subject of great controversy. It may be analogous to that of some meteorites and certainly must be such that it can provide, by local and probably partial melting, the source of basaltic magmas. In the Hawaiian

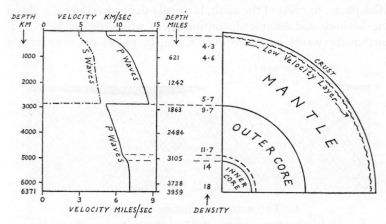

5. The Earth, showing the approximate thickness and density of its component zones

Left-hand diagram illustrates velocity of earthquake waves

islands, for example, the foci of earthquakes probably related to the origin and movements of magma occur at depths as great as 37 miles. The physical properties of the upper mantle vary slightly with depth. Extending from depths of about 37 miles to 155 miles is a layer characterized by seismic velocities significantly lower than in the surrounding mantle. This *low velocity layer* is close to its melting point and relatively plastic compared with the solid mantle enclosing it. It may be the source area for basaltic magmas and the seat of isostatic compensation. Earthquakes have not been recorded from the mantle below 450 miles (*see*, pp. 185–87).

The mantle-core boundary is a surface of very marked discontinuity across which S, or transverse, earthquake waves are not transmitted, indicating a junction between a solid and a fluid state. In contrast the boundary between the outer and the inner core is transitional and probably marks the change from a fluid to a solid

C

state. The problem of the composition of the core is far from solved. The traditional view is that it consists mainly of an iron-nickel alloy. The outer core may be the site of origin of most of the earth's magnetic field because circulations within a fluid encircling a solid inner core would function as a self-exciting dynamo. If this is not so, the generating source of the earth's magnetic field remains a mystery.

Relations Between the Lithosphere and Interior. The lithosphere, or crust of the earth, is usually defined as the layer above the Mohorovičić seismic discontinuity, at which the velocity of P earthquake waves jumps to over 8 km. (5 miles) per sec. from a

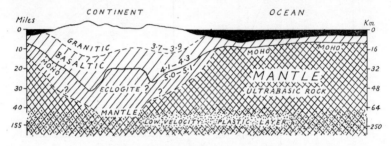

6. The nature of floating continents

Numbers in diagram denote average densities. Notice large gap in vertical scale

smaller value (often 6 to 7 km. per sec.) above it. This discontinuity was deduced from seismograph records of the great earthquake in 1909 near Zagreb in Croatia by Dr. Andrija Mohorovičić, and to-day is popularly called the Moho or M.

The depth to the M discontinuity varies almost inversely with the height of the surface relief, suggesting that the crust generally floats on the mantle in isostatic equilibrium. However, two types of crust, oceanic and continental, can be distinguished. Beneath true ocean basins the Moho lies at an average depth of only 10 or 11 km. Above it is a layer about 5 km. thick with a P wave velocity of 6·2 km. per sec., probably consisting of basalt and then a thin layer of unconsolidated sediments and the ocean itself (Fig. 6). In contrast, continental crust averages about 35 km. (22 miles) in thickness and lacks distinct layering. Probably it consists of what we observe at the surface, a complex of igneous, metamorphic and sedimentary rocks, although its overall composition, as indicated by the increase of seismic wave velocities with depth, becomes denser and more basaltic in its lower parts.

The Moho beneath the continents probably corresponds to a change from basalt to a denser metamorphic equivalent called eclogite and therefore differs markedly in this respect from the sub-oceanic Moho which rests directly on the mantle. At least two aspects of this difference are significant. First, to obtain actual rock samples from the mantle it will be best to drill under the ocean and since 1957 the Mohole project has been planning to reach the sub-oceanic mantle. Secondly, the Moho is so shallow that a great deal of the internal activity affecting major surface features takes place below it, probably in the plastic, low velocity layer of the upper mantle. The presence of a plastic layer at depths of 37 to 155 miles greatly favours the possibility of sub-crustal convection currents and of continental drift.

Marine Transgressions and Isostasy. The vast areas of sedimentary rocks of marine origin exposed on each continent prove that the continental blocks now above sea-level must in the past have been subjected to inundations by the sea. These marine transgressions and subsequent regressions must be a result of relative vertical movement between the continental blocks and sea-level, and, judging from the complicated succession of sedimentation and erosion revealed by geological strata in all continents, must have been of frequent occurrence. As discussed on pp. 187–89, some are due to changes in the total volume of sea water but most are due to the rising or sinking of land-masses and are associated with orogenesis and isostasy.

Isostasy is verified by gravity measurements which show that mountains have a low gravitational attraction and so must be underlain by roots of relatively low density going down to considerable depths, probably up to 37 miles. Thus, continental blocks float on the mantle in much the same way as icebergs float in the sea. The thinning of these blocks, for example by erosion, results in their uplift to restore isostatic equilibrium while extra surface weight, such as an ice cap, depresses them deeper into the mantle. During the Pleistocene period, a thick ice-sheet depressed the Scandinavian block and since the last ice retreat, or in the last 10,000 years, it has rebounded slowly leaving a tier of raised beaches to mark the isostatic recovery to date.

There are two principal theories regarding the detailed mechanism of isostatic compensation. First, that the compensation is entirely local, in which case the highest mountains would float on the deepest roots and the depth of the Moho would vary inversely or anti-pathetically with the height of the topography. Secondly, that some proportion of the load is borne by the elastic bending and depression of a much larger area of surface than covered by the load, and hence that the compensation is regional and that locally the depth of the Moho may not be closely related to surface height.

There is, however, more in the problem of changes in the relative height of land-masses than unloading and overloading of their surface. Thus the depression of mountain roots into the mantle could result from thickening of the continental crust by folding and overthrusting. Also, although, for example, large volcanoes could cause peripheral crustal depressions that would serve as basins of deposition and although new sediments must further depress any basin of deposition, the *primary cause* of the production of many such basins and indeed of the depression of continental blocks generally is highly problematic. A commonly invoked cause is that where slow convection currents in the upper mantle turn downwards they drag down the local crust, so producing a depression or geosyncline.

The Theory of Continental Drift. The idea that continents move laterally originated in the early seventeenth century but modern scientific discussion of it dates mainly from the writings of A. Wegener in 1912 and 1924. The general contention is that in Carboniferous times (350 million years ago) the continental masses of to-day were either conjoined or close together. Thus at that time Australia, Antarctica, peninsular India, Africa and Brazil formed a great southern block sometimes called Gondwanaland. This super-continent then gradually split up and portions of it drifted away to form separate land-masses. Simultaneously the magnetic poles gradually shifted their position.

The former unity was largely deduced from the amazing jig-saw fit geometrically and geologically of the opposite sides of the Atlantic and was favoured by similarities in vegetation. But difficulties arose because such evidence, although explicable in terms of movement apart of the continents, was not acceptable to many scientists as direct evidence of it. However, among the cogent evidence for the theory is the distribution of palaeoclimates (ancient climatic conditions). Strong geological effects of Permo-Carboniferous ice ages are recognized in eastern South America, Africa, India and southern Australia. Wegener suggested that these glaciations are all related to the south pole and that the continents in question were then clustered together about a south pole situated near the position of Durban to-day. There appears to be no other rational explanation of this remarkable distribution of ice-caps (Fig. 7). Furthermore if the continents are so grouped, the position of the Carboniferous equator lies through the world's major coalfields (indicative of tropical rain-forest) and these are flanked latitudinally by deposits of gypsum, salt and sandstone (indicative of dry desert conditions), just as to-day arid zones flank the humid equator.

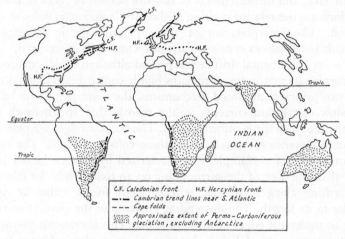

7. Structural relations of the opposing sides of the Atlantic Ocean

Maximum extent of Permo-Carboniferous glaciation is shown

Recent geological and geophysical investigations virtually prove continental drift. The horizontal movement on some faults is now known to be considerable. At the Great Glen fault which transects Scotland the block north of the fault has been displaced horizontally south-westward for 66 miles. Similarly, the Pacific floor off California is crossed by a series of east-west fault zones stretching for distances of 1,200 to 1,800 miles. Magnetic surveys suggest large horizontal displacements at these faults, for example of 736 miles at the Mendocino fault. Also, it is now certain that the Red Sea was formed by crustal separation and that its shores have undergone considerable horizontal displacement.

The evidence of palaeowinds and of palaeomagnetism also tends to substantiate continental drift. In some areas, barchan dunes preserved in aeolian sandstones show a dominant orientation, and so a dominant wind, over a wide tract. For example, Permian sandstones in Britain exhibit south-west trends indicating a location in the north-east Trade wind zone at the time of deposition.

Palaeomagnetism depends on the fact that in rocks such as basaltic lavas every crystal of ferro-magnetic material acquires a fixed magnetization when it cools. Thus this *remanent magnetization* has the orientation of the local magnetic field at the time of consolidation of the lava, and measurements of this orientation in rocks of known age indicate the relative position of the earth's magnetic poles at that period. The interpretation of these palaeomagnetic findings is difficult but becomes consistent if continental drift is accepted.

To-day, continental drift seems proved although evidence directly concerning its motive forces remains highly controversial in detail. Of the two principal proposed mechanisms, the more popular is based on slow convection currents in the mantle that rise beneath the oceans and push the continents apart. The oceanic ridges are believed to mark the sites of upwellings of these convective cells. The mid-Atlantic ridge, for example, is characterized by a much higher heat flow than the floor of the surrounding ocean basin while the faulting and rifting along it suggest tension from forces acting at right angles to its length. The descending currents of the convective cells may lie under the edges of the continents. This is certainly plausible for the west coast of South America where a plane of deep-focus earthquakes dipping under the continent and an ocean deep and a high mountain range parallel to the coast could all be attributed to the downturning of a convective cell which upwells beneath the mid-Atlantic ridge and whose horizontal flow passes beneath the American continent.

The other important theory is that the earth is expanding and causing the continents to be more widely separated from each other. This expansion may well have led to the progressive widening of the ocean basins and a steady emergence of the land-masses. It would also help to explain the slight slowing down of the earth's rotation during geological time. Such a retardation is supported by studies of corals, which contain about 360 daily growth lines a year to-day and about 400 daily growth lines annually in Devonian times, suggesting that the earth was then smaller.

Thus, in conclusion, there is some evidence both for the existence of convection currents in the mantle and for the expansion of the earth. Both could cause continental drift and perhaps both are involved.

LIST OF BOOKS

Concerning atmosphere: Massey, H. S. & Boyd, R. F. *The Upper Atmosphere*, 1958
Goody, R. M. *The Physics of the Stratosphere*, 1954
Malone, T. F. (ed.) *Compendium of Meteorology*, 1951
Ratcliffe, J. A. (ed.) *Physics of the Upper Atmosphere*, 1960
Craig, R. A. *The Upper Atmosphere*, 1965
Dobson, G. M. B. *Exploring the Atmosphere*, 1963
King-Hele, D. G. 'Satellites and the earth's outer atmosphere', *Q.J.R. Met. Soc.*, 1961, pp. 265–81

Concerning hydrosphere and lithosphere:
Wegener, A. *The Origin of Continents and Oceans*, 1924
du Toit, A. L. *Our Wandering Continents*, 1937 (reprinted 1957)
Jeffreys, H. *Earthquakes and Mountains*, 1950; *The Earth*, 1959
Gutenberg, B. *Physics of the Earth's Interior*, 1959
Coulomb, J. & Jobert, G. *The Physical Constitution of the Earth*, 1963
Anderson, D. L. 'The plastic layer of the Earth's mantle', *Scientif. Amer.*, July, 1962
Meinesz, F. A. V. *The Earth's Crust and Mantle*, 1964
Holmes, A. *Principles of Physical Geology*, 1965 (a magnificent revised edition of 1288 pp. which covers all topics mentioned here)

For continental drift *see* also:
King, L. C. *The Morphology of the Earth*, 1962
Runcorn, S. K. (ed.) *Continental Drift*, 1962; 'Geophysics and palaeoclimatology', *Q.J.R. Met. Soc.*, 1965, pp. 257–67
Bullard, E. C. 'Continental drift', *Q.J. Geol. Soc.*, 1964, pp. 1–33

Section Two

THE ATMOSPHERE

CHAPTER THREE

THE TEMPERATURE OF THE ATMOSPHERE

Insolation

THE earth's surface receives its heat directly from the sun's rays; the amount received from the internal heat of the earth is relatively insignificant and, its influence being spread uniformly over the globe except in volcanic regions, does not affect the world distribution of temperature.

The sun is a gaseous mass with a diameter some one hundred times that of the earth. Even its colder surface has a temperature of about 10,000° F. From its blazing mass radiant energy is emitted in all directions, and the terrestrial globe intercepts but a very minute fraction of it. The amount of radiant energy or of heat received from the sun in a given time is called insolation.

Insolation at the Upper Surface of the Atmosphere. Insolation at the upper limits of the air envelope depends on the angle of incidence of the sun's rays and on the length of daylight.[1] In equatorial latitudes the constant day of twelve hours means little seasonal

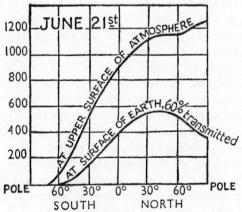

8. Latitudinal distribution of Insolation
on June 21st
(*After Angot*)
In kilowatt hours per square decameter

[1] Factors which, as already explained, depend on latitude and the tilt of the earth's axis.

variation in insolation, but even here the amount of heat received is slightly greater at the equinoxes than at the solstices (10:9). The length of the summer day increases polewards but the altitude of the sun, or incidence of its rays, diminishes. The noontime sun on mid-summer day has an altitude of only 47° at the polar circle, and of only 23½° at the pole. Yet here the sun never passes below the horizon, and the continuous daylight gives the upper atmosphere of polar regions more insolation than anywhere else on earth during the twenty-four hours in question. Thus the highest layers of atmosphere over Greenland receive more heat on 21st June than similar layers over the Sahara (Fig. 8). Similarly the high atmosphere above Antarctica on 22nd December receives considerably more heat than anywhere else in the world. From the great discrepancy between these statements and the temperatures actually experienced on the earth's surface, it will be recognized that the passage through the atmosphere has a great effect on the sun's rays.

Insolation at the Earth's Surface. Of the solar energy that strikes the upper atmosphere part is reflected back into space and lost, part is absorbed during the passage through the atmosphere, and part succeeds in reaching the surface of the earth.

The proportion of insolation that is absorbed by the air depends mainly on the angle of incidence of the heat rays and on the transparency of the atmosphere. It has been computed that when the length of the path of a ray is doubled, the passage reduces its heat to one quarter of the original amount. Thus the long oblique path of the sun's rays through the atmosphere in polar regions is one cause of the marked difference between their high upper-air insolation and their low surface temperatures. The reader will be familiar with Figure 9, which shows that heat rays near the poles pass through a greater thickness of atmosphere and also have to heat a greater area of surface than rays falling upon tropical regions.

The transparency of the atmosphere varies very considerably locally and even hourly. Water-vapour and dust are the chief constituents that abstract heat from the sun's rays. Water-vapour is most efficient when condensed in the form of cloud, but even when not condensed it will obstruct heat rays. At Montpellier in South France, some 70 per cent. of the insolation reaches the ground in December, and rather less than 50 per cent. in July, when the greater moisture content of the hotter air more than counterbalances the more direct path of the rays. This, however, does not mean that

Montpellier is cooler in summer, as then the hours of daylight are much longer ($15\frac{3}{4}$ hours, June; $8\frac{1}{4}$ hours, January). It will be seen

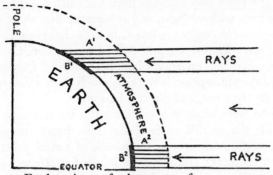

9. Explanation of decrease of temperature
with increase of latitude

Passage through atmosphere at A^1 is longer than at A^2; area of
land receiving heat at B^1 is greater than at B^2

that areas where the atmospheric moisture content is great or the cloud cover is fairly continuous will seldom experience particularly high temperatures.

The obstructional effect of dust, although very small compared with that of water-vapour, is appreciable; it becomes especially important over large cities (smoke) and in volcanic regions. Generally speaking, the atmosphere is most transparent over sub-tropical trade-wind deserts, as here the air is extremely dry and the occasional dust-storms are confined to the lowest layers of atmosphere. In middle latitudes, on an average, some 50 per cent. of the insolation is lost in the atmosphere even with a fairly clear sky; in polar regions the average amount that reaches the earth's surface is about 20 per cent. of the total. Hence, with regard to temperature, the atmosphere deprives the earth's surface of a considerable amount of its possible heat. This absorption of heat by the atmosphere at first sight appears to be a distinct gain to the air temperatures. In fact, however, the atmospheric dust and water particles radiate and lose their heat relatively quickly, and as will now be shown, the air would be considerably warmer if the heat normally absorbed by the atmosphere were allowed to penetrate to the earth's surface.

The Heating Effect of the Earth's Surface

The sun's rays that actually reach the earth are mainly absorbed by land or water and radiated back into the lower atmosphere. This *radiation* from the earth is the major factor in raising the air temperature. The main rays radiated from the earth are infra-red rays, which give heat without light, and it happens that air is much less transparent to these long-wave rays than to the light heat rays, mostly of medium wave-lengths, coming direct from the sun. Consequently air absorbs a higher proportion of the heat radiated from the earth's surface, and the atmosphere is warmed much more efficiently by the earth than by the passage of direct solar rays.

The lowest layer of air is also warmed in the day-time by actual contact with the earth's surface, the heat being transmitted by *conduction*, in the same way as the handle of a steel poker becomes warm when the other end is thrust into a fire. Hence, the lowest layers of air are warmed largely by radiation and partly by conduction from the earth's surface.

The third factor involved is *convection*. The heated bottom layer of the atmosphere expands, becomes lighter, and rises, its place being taken by denser, colder air. The process continues, and so the atmosphere is heated to a considerable height.

It is obvious that the temperature of the air at any time is the balance between the heat received, directly or indirectly, from the sun and that lost by radiation into space. At night, when the sun is below the horizon, the loss of heat exceeds the supply and the air temperature continues to fall, reaching its minimum just before sunrise. During daylight the heat received by the earth's surface so far exceeds the loss into space that the air temperature continues to rise until about 2 p.m. on land and 3 p.m. on the sea. The accumulative effect of the differences between day and night is reflected in the seasons.

Vertical Decrease of Temperature

From the facts stated above it follows that the air nearest the surface of the earth is warmest, and that temperature decreases with ascent. The decrease averages about 1° F. for every 330 feet of ascent, although this last figure hides differences ranging in the British Isles from about 410 feet in winter to about 270 feet in summer. The lower atmospheric layers, which are densest and contain most of the water-vapour and dust, act as a blanket and tend to prevent the loss of heat into space. The effect is well seen in Europe, where cloudy nights are normally relatively warm while clear nights are associated with the possibility of frost or cooler weather. In some Californian orchards this blanketing effect at night is achieved artificially by smoke from stoves, whereby the radiation of ground heat is obstructed. Where the dense lower layers of atmosphere are absent, as on mountains and high plateaux, the rarefied air cannot retain the heat radiated from the ground; the heat is lost rapidly into space and the temperature remains permanently low. At lower elevations the varying amounts of water-vapour and dust and the convection currents cause the thermal structure of the atmosphere to be very

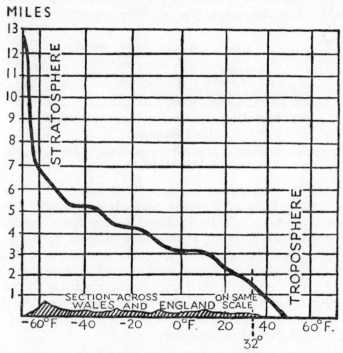

10. Average decrease of temperature with increase of altitude over south-eastern England
(*After Lempfert and Dines*)
Graph shows mean temperatures for the year

complex. Up to 7,000 feet (1½ miles) the vertical decrease of temperature is irregular and may occasionally be reversed. In the air envelope over south-eastern England, from 10,000 feet (2 miles) upwards there is a fairly steady decrease to about 30,000 feet (5½ miles).

There follows a relatively narrow zone where the temperature usually falls at a slightly faster rate. Then above a height of 6 or 7 miles is a zone where the temperature decrease is relatively insignificant. Already much is known about this isothermal layer (the stratosphere), the base of which increases in average height from about 7 miles in middle latitudes to 10 miles within the tropics and often to 12 miles at the Equator. Hence the 'ground-floor' of the atmosphere may be looked upon as consisting of a lower, denser layer (the troposphere), where weather phenomena abound, which is separated by a region of fairly sudden change (the tropo-pause) from the stratosphere (Fig. 10). Here temperature changes are

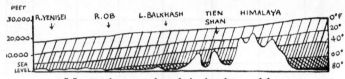

11. Mountains as a break in isothermal layers: the Himalayas

Diagrammatic section of relief and air temperature in July across Asia in longitude 80° E.

little, and as water-vapour is practically absent, 'weather' in the usual sense of the word is almost absent. It is no wonder that the stratosphere is regarded as the ideal level for long-distance air-flights. The stratosphere, however, extends only to about 25 miles, and above this height the temperature rises to form a warm zone at about 30 miles where the ozone absorbs most of the short-wave ultra-violet rays or 5 per cent. of the total sun's radiation entering the earth's atmosphere. Above this is probably a narrow frigid zone which gives way to a wide zone of great heat. The general scheme is summarized diagrammatically in the Frontispiece.

Horizontal Distribution of Temperature

Effect of Relief. Theoretically the surface temperatures of the earth should cause a hot tropical zone with no marked seasons, a temperate zone with more marked seasons and greater extremes, and a polar zone with extreme seasonal changes and no high temperatures. These zones, however, are greatly complicated by the effect of relief and the influence of the distribution of land and sea.

In world diagrams of mean monthly temperatures the isotherms are usually reduced to a sea-level, as otherwise they would largely correspond with the relief. Such maps only show the real temperature for places that are at or near sea-level. The true irregularities of temperature distribution upon the surface of any country can only be seen by reference to the actual temperature isotherms. The relation between altitude and temperature can also be shown as in Figure 11, where the relief forms an interruption or break in thermal layers of air. If temperatures were always regarded as isothermal layers, the separating effect of relief would not be so easily overlooked. It would then be readily seen, to take an extreme example, how the Himalayas tend to isolate the lower thermal layers of India from those of Asia to the north. It is a remarkable fact that an ascent of one mile causes about the same decrease of temperature as a poleward movement of 800 to 1,000 miles.

Effect of Land and Water Masses. The other major factor that upsets the theoretical distribution of heat on the earth's surface is the intricate distribution of land and sea.

Land heats and cools more rapidly than water. If similar areas of land and sea receive the same amount of heat in a given time, the temperature of the land would rise to nearly double that of the water. Yet the land would cool nearly twice as fast as the water. In actual practice these differences are brought about by the following factors:

Water reflects more heat rays than the land since land surfaces are usually poor reflectors, especially if the soils are black.

A water surface is cooled by evaporation; even on a spring day in England evaporation may cool the surface of a pond by 15° F.

Water is transparent to light and heat. This ensures that waterbodies are heated to a greater depth than land.

Water is mobile and consequently, owing to currents, a larger proportion of a water-body is affected. Daily changes of temperature have been observed down to about 3 feet on land and 15 feet on water; in middle latitudes, seasonal changes are not felt below 50–60 feet of the earth's surface, whereas corresponding figures for the Baltic Sea and Black Sea are 150–180 feet.

Largely owing to these different heating qualities of land and water, the latitudinal arrangement of sea-level isotherms is almost completely obscured except in the 'ocean-girdle' of the southern Pacific. The climates of the continental masses, partaking of the rapid heating and cooling of the land, tend to become extreme, whereas oceanic areas, with their usual cloud cover, tend to have no marked extremes, the sea cooling the air in summer and warming it in winter. As will be shown later, maritime regions also have this type of climate provided the prevailing winds blow from the sea.

The above discussion largely concerns world-wide examples, but the same principles generally hold good in a minor degree in any locality. Thus the Great Lakes of America temper the winter cold on their leeward shores; the eastern shores of Lakes Michigan and Superior have mean January temperatures that are 5° F. higher than those on the western shores, while their coldest nights are 13° F. less cold. The strip of country, about 30 miles wide, on the east shore of Lake Michigan that benefits from the westerly winds blowing off the lake has become a notable 'fruit belt.'

Local Factors. Even the nature of soils and of the surface cover cannot be ignored entirely. For instance, the amount of moisture in a soil greatly affects its heating qualities, damp clay soils being noticeably cold and sandy soils warm. In summer, the heat received tends to be more efficient than in winter when the soils are wetter. Or, to take another example, snow-cover makes such a good reflector and absorbs so much heat in melting, that in parts of England, spring (March–May) temperatures may be considerably lowered by snowfall and frequently spring may be colder than autumn. In ice-covered areas the loss by reflection is great, and much energy is used up in converting ice into water. Where the ice is so thick that the sun cannot melt it all, most of the sun's heat absorbed is used up in the melting process and the surface, and the air above it, remains near freezing-point. Hence under the brilliant sunshine of Alpine skies one can walk upon glistening ice-fields.

The climate of any place depends primarily on latitude and altitude, but position with regard to expanses of land and water is highly important. Local weather is the interaction of a score of minor influences upon these primary controls.

LIST OF BOOKS

Byers, H. R. *General Meteorology*, 1959
Kendrew, W. G. *Climatology*, 1957
H.M.S.O. *Handbook of Aviation Meteorology*, 1960
H.M.S.O. *Course in Elementary Meteorology*, 1962
Petterssen, S. *Introduction to Meteorology*, 1958
Geiger, R. *The Climate near the Ground*, 1950
Malone, T. F. (ed.) *Compendium of Meteorology*, 1951

For upper atmosphere *see* references on p. 39, especially Massey & Boyd; and Sawyer, J. S. 'Jet Stream features of the Earth's atmosphere', *Weather*, 1957, pp. 333-44

CHAPTER FOUR

THE MOISTURE OF THE ATMOSPHERE

Evaporation

WHEREVER unsaturated air is in contact with a damp surface, particles of water-vapour pass into the atmosphere. This evaporation process absorbs, as we have already noticed, a part of the energy of insolation especially over water-surfaces; its rate depends mainly on the temperature and moisture content of the air. Provided that moisture is available, evaporation will be greatest where temperatures are high and the air dry. Winds, too, assist evaporation by bringing supplies of less humid air against the moisture-yielding surface. London has an average annual evaporation equal to about 18 inches of rainfall. At the Equator, where the air is hotter and the soils are damper, the corresponding figure is about 40 inches. Over the oceans in trade-wind latitudes some 40 to 50 inches, or three times the annual rainfall, are lost each year. These figures are small compared with the mean annual evaporation over tropical deserts. In the Nile lands between Wadi Halfa and Khartoum the annual evaporation averages some 210 to nearly 250 inches, and nowhere in the 1,700 miles between Mongalla (5° N.) and Cairo (30° N.) does the Nile Valley experience a mean annual evaporation falling much below 90 inches. It is not surprising that rivers entering desert regions begin to diminish rapidly in volume away from their sources of water-supply.

Humidity

Absolute Humidity. The actual amount of water-vapour present in the air is called absolute humidity, and may be expressed as the pressure or barometric weight of the vapour. During a normal July afternoon in London the mean vapour pressure is nearly half an inch, which is only slightly less than that at Cairo and Athens. In January, London and Athens have mean vapour pressures of ·15 to ·2 inch, while that at Cairo is about ·25 inch. These pressures are small compared with those at Calcutta (about 1 inch in July and ·45 inch in January) and along the Equator, where Jakarta, for instance, has absolute humidities equal to ·8 inch pressure throughout the year. The absolute humidity can only be changed by loss of moisture

D

through condensation or by the addition of moisture by means of evaporation. Generally speaking, the higher the absolute humidity the heavier will be the rainfall when rains do occur.

Relative Humidity. The amount of water-vapour that air can hold is, however, limited for each degree of temperature. The following table shows the maximum vapour content of air at various temperatures:

Temperature °F.	30°	40°	50°	60°	70°	80°	90°	100°
Grains of water-vapour per cubic foot	1·9	2·9	4·1	5·7	8·0	10·9	14·7	19·7
Vapour pressure (inches)	·16	·25	·36	·52	·73	1·02	1·41	1·92

From the table it may be seen that at 50° F. a cubic foot of air holding 4·1 grains of moisture would have reached its saturation point.

The relative humidity of the air is the most important human aspect of this problem. The term denotes the proportion between the water-vapour actually in the air and the maximum amount that the air could hold at its present temperature. Thus a cubic foot of air at 60° F. containing 1·9 grains of water-vapour would have a relative humidity of 33 per cent. $\left(\text{see table:} \dfrac{1 \cdot 9}{5 \cdot 7} \times \dfrac{100}{1} \right)$, or it would contain one-third of its possible moisture content at that temperature. A decrease of temperature to 50° F. would increase the relative humidity to 46 per cent., and a further decrease to 30° F. would cause the air to be saturated (99 to 100 per cent.). Relative humidity, as measured by the wet-bulb thermometer, is of great physiological importance. It provides a means of measuring the feeling of humidity that the air imparts to the human body. The body cools itself and is stimulated by the evaporation of moisture from the lungs and skin; this proceeds most easily where the relative humidity is low. Retarded cooling results in a feeling of oppression, and prohibited cooling may result in a heat-stroke. Thus a dull warm day of a British summer (70° F. and relative humidity of 85 per cent.) would be more oppressive than a day in the Sahara with temperatures of 90° F. and a relative humidity of 20 per cent. At the Equator, when temperatures rise much above 80° F. and relative humidities exceed 90 per cent., life becomes almost unbearable for white settlers and the slightest breeze is welcome as it stimulates evaporation from the skin. The relation between heat and the relative moisture content of the air and the comfort of the human body is of vital importance in connection with white settlement in the tropics. From the point of view of physical geography, however, it should be noticed that what seems to us wet air in winter may often contain less water-vapour than a seemingly dry scorching atmosphere in summer. The cooling of air on a January day (say at 40° F.) will usually result in drizzle or steady rain, but the prolonged cooling of the hot air of a July afternoon will give intense rainfall.

Causes of Precipitation

If air is cooled it will eventually reach the temperature (the dewpoint) for which it contains the maximum possible moisture content. It is then said to be saturated, and prolonged decrease of temperature will cause moisture to be condensed as particles of water. Most cloud drops form about minute particles of hygroscopic substances, which consist chiefly of certain compounds of nitre and sulphur, and of salt from sea-spray. Dust and smoke particles, though more numerous, take little part in cloud formation. The condensed droplets may float as cloud or they may, in a manner mentioned later, become sufficiently large to fall to the earth.

The principal means whereby air is cooled beyond its dew-point is by ascent. In ascending, air is cooled in two ways:

First, by contact with colder layers of atmosphere; second, by expansion owing to diminishing pressure. The second is of great importance. Expansion due to decrease in pressure is always accompanied by a fall in temperature, the whole process being known as adiabatic or dynamical cooling.[1] In the denser lower layers of the atmosphere, where cooling by ascent is most rapid, the lapse or cooling rate of unsaturated air averages about 1° F. for every 185 feet of ascent, or 5·4° F. for every 1,000 feet (Fig. 12).

To understand the full significance of adiabatic cooling, the liberation of latent heat by condensation must be taken into account. It has already been shown that when water is changed into vapour, heat energy is used up from the surrounding air. Conversely, when vapour is condensed into water particles this latent heat is set free and helps to warm the surrounding air. Hence, when condensation occurs in rising air currents, the latent heat liberated partly compensates for the cooling due to expansion. As a result, saturated air is cooled some 3° F. for every 1,000 feet of ascent or at about half

[1] Air is adiabatically warmed by compression in descending in exactly the reverse manner.

the rate of unsaturated air. Consequently, ascending saturated currents tend to keep warmer and therefore less dense than the air about them. The liberation of latent heat feeds, as it were, the ascensional impulse and assists the air to rise until it is deprived of much of its moisture. This self-feeding nature of rising air currents is a very

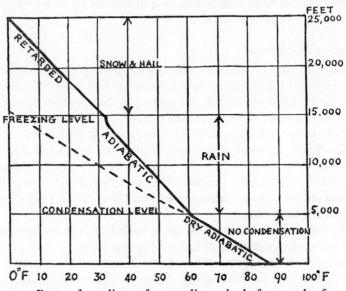

12. Rate of cooling of ascending air before and after condensation

important factor in causing heavy rain and largely explains the pulsatory character of convectional rainfall.

Air currents may be caused to rise by three main factors.

(i) **Convection.** Convection (or the warming, expansion, and ascent of surface layers) occurs on a large scale, especially where the lower air is abnormally hot as in equatorial latitudes. Convection currents of a more local kind occur chiefly during summer in the continental interiors of temperate latitudes and in the thunderstorms of the British Isles. The vortical ascent of air in small revolving storms, known as tornadoes, is largely an extremely localized form of convection, fed and maintained by intense rainfall.

(ii) **Obstruction by Relief.** Even where ocean winds meet a shoreline their surface layers are slightly impeded by friction with the land, and there is a tendency for the upper air to over-run the surface wind. This may result in a slight ridging-up of the air with the consequent formation of rain. Mountain barriers and hill features act in a more decisive way. Air currents are forced to ascend over the relief, and frequently orographical rainfall results. The effect of ascent is not limited to the actual relief feature, as the piling-up of air on the windward-side tends to give increased rainfall some distance from the foot of a mountain ridge. This is shown in the following table for stations at 65 feet above sea-level on the Gangetic plain.

Dacca	.	.	100 miles from Khasi Hills	78 in.
Bogra	.	.	60 miles from Khasi Hills	92 ins.
Mymensingh	.	30 miles from Khasi Hills	110 in.
Sylhet	.	.	20 miles from Khasi Hills	150 in.

The amount of precipitation caused by relief depends on the moisture content of the air and the height of the land obstruction. Heavy orographical rain is to be expected when strong warm ocean winds strike high mountains, in the same way as the Western Ghats obstruct the wet monsoon of India. Rainfall of this type increases up

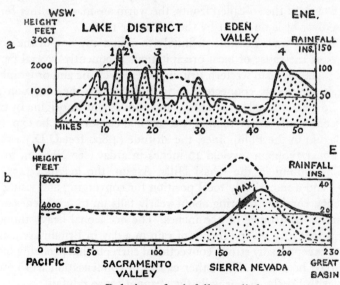

13. Relation of rainfall to relief

Broken line shows mean annual rainfall
(a) The Lake District. Rainfall maximum to lee of summit (*After Salter*)
     1. Great Gable   2. Styhead Pass   3. Helvellyn   4. Pennine Edge
(b) Sierra Nevada: San Francisco to Reno, California. Rainfall maximum
     below summit   (*After Bowman*)

to a certain height, beyond which it steadily decreases. The zone of maximum rainfall in Java is at 3,500 feet, and in the Sierra Nevada at 5,500 feet (Fig. 13). In the British Isles, where the zone of greatest precipitation is not reached, the wet air currents continue to rise beyond the mountain top and the heaviest rainfall may be to the leeward of the summit. Styhead Pass (1,070 feet O.D.) has the greatest rainfall in the Lake District, although the nearby peaks to the west are 1,000 feet higher, while in North Wales, the rainiest locality is at Llyn Lydau one mile east of the summit of Snowdon (Fig. 13a). On the other hand, it should be noticed that this effect is very localized, and that the leeward side of mountain ranges are frequently well-marked rain shadows. In Scotland, Ben Nevis has 170 inches a year and Nairn 25 inches, while in southern New Zealand, Hokitika, on the west coast has 117 inches a year and Christchurch only 25 inches.

(iii) **Convergence.** When air currents of different temperatures meet, the less dense, or warmer, current is forced to rise above the denser layers. This is what commonly happens in a cyclonic depression where, at the so-called fronts, the warm air-mass is either forced to rise over or is undercut by the colder air-mass.

It is not always possible definitely to assign any one of the above factors as the cause of local precipitation. Frequently several factors work together, as in western Britain, where cyclonic and orographical rainfall are closely connected, or in India where the monsoon rain may be convectional, orographic, and convergent in origin. In cases of exceptionally heavy rainfall, exceptional causes must be expected. At Baguio in the Philippines, the altitude (4,800 feet O.D.) assisted an intense typhoon to yield 46 inches in a day (see Note on p. 59). At Cherrapunji in the Khasi Hills, Assam, the height (4,455 feet O.D.) and a remarkably focal position for converging and rising air-currents, combine to bring about yearly falls up to 900 inches and a mean annual average of 450 inches. The occasional heavy thunder-storms that give several inches of rain in a day in Britain are usually due to exceptionally strong convectional currents that probably follow a marked heat-wave. The other causes of condensation will now be dealt with briefly, as normally they cannot cause rainfall.

Poleward Moving Winds. The movement of warm air into colder latitudes cannot in itself cause rain; it merely raises the relative humidity of the air and so increases the possibility of precipitation when other factors cause the air to rise.

Cooling by Contact with Cold Surfaces: Fog. At night the earth's surface cools the air above it by conduction, and some of the atmospheric moisture may be condensed as fog. As cold air shows no tendency to rise, the phenomenon is very shallow. Land fogs occur most frequently in autumn and winter in Britain, especially on calm nights, as the presence of strong wind would renew the surface layers of air and so retard their cooling. The particularly unpleasant fogs of cities are intensified by smoke and dust.

Some of the sea fogs off British coasts may be due to the meeting of slow-moving masses of air at different temperatures, but the major cause of coastal fog is the cooling of warm saturated air when it reaches a colder surface. Fogs are common over the cold ocean currents near Peru and Chile, California, and South-west Africa. The sea around Newfoundland is fog-bound for about seventy days a year, mainly because here the cold Labrador current meets the warmer waters of the Gulf Stream and the air above the currents partakes of their temperatures. In this region even the cold of large icebergs will occasionally form a fog-field that encompasses the berg to the great danger of shipping.

Major Forms of Precipitation

Clouds. Fog is really a cloud at ground-level, and consists, as does a cloud, of particles of water minute enough to be held in suspension in the atmosphere. Clouds, however, differ from fog in their modes of formation, in their varying heights, and characteristic shapes. The three main classes of cloud recognized are: *Cirrus*. The delicate, fibrous, feathery wisps of whitish cloud that may float high in the sky on a clear day. Cirrus consists of minute particles of ice formed at a height in Britain of five to six miles near the upward limit of cloud formation.

Cumulus. The massive billowy clouds of a cauliflower shape that are usually formed by convection currents and consequently are typical of the sky-cover of hot humid areas and of temperate zones in summer. The flattish bases of the clouds lie at about a mile above the earth, but the billowy rounded tops may protrude for several miles above their bases. Frequently the lower layers become dark and appear to be ragged round their edges; this characteristic is associated with rain or snow and is called *nimbus*, a term given to any rain cloud.

Stratus. A uniform layer of cloud with no distinguishable cloud shapes, that is so common on dull grey days in a British winter. The clouds may almost touch the tree tops or they may form horizontal layers as high as five miles above the earth. Stratus clouds are frequently connected with the junction and superimposition of horizontal air currents which have different temperatures and are travelling in different directions.

Rain. Any prolonged cooling of air containing moisture already condensed as cloud will frequently cause precipitation in the form of rain, snow, or hail.

The intensity of rain depends on the absolute humidity of the air and the rapidity of its ascent. Normally air ascends in a gradual incline, but strong convection currents or abrupt relief barriers may cause a sudden rise. Thunderstorms form an interesting example of rapid convectional ascent. It has been shown that raindrops in still air cannot fall faster than 24 feet a second, and similarly that an upward current of air exceeding 24 feet a second would prevent their descent. Moreover, there is a limit in size ($\frac{1}{4}$ inch in diameter) beyond which raindrops cannot pass without breaking up. In thunderstorms the strong convectional currents yield moisture that often cannot descend because the upward air movement is too strong. The drops go on coalescing and grow larger and larger until they exceed $\frac{1}{4}$ inch in diameter. They then split up, and at the splitting of each raindrop a small charge of electricity is released and is carried upwards into the cloud. Parts of the cloud eventually become so highly charged with electricity that lightning occurs. The upward ascent of the air is, however, pulsatory, and any lull in the strength of the ascending currents allows the water accumulated in the air above them to fall to the earth as a deluge. Thus thunderstorms usually give torrential rain of a spasmodic nature. It is not surprising that convectional ascent of this nature should occur mainly in the late afternoon and especially in equatorial lowlands.

Snow. When water-vapour condenses below the freezing-point the minute particles of ice unite to form flakes which, provided the lower atmosphere is not much warmer than 32° F., will fall to the earth as snow. On an average some 10 to 12 inches of new-fallen snow are equivalent to one inch of rain, a fact which in more temperate regions causes snowfall to be a greater hindrance to human movement than is rainfall. Many of the higher Alpine valleys experience winter snowfalls of over 25 feet, while on parts of the Sierra Nevada and Cascade Ranges of the western Rockies annual snowfalls of over 40 feet are common. Even in the higher lands of Britain snowfalls of

5 to 10 feet are not uncommon, and wind may drift the snow to double those depths. Spring is the snowy season of the British Isles, a fact which is frequently reflected in the relatively slow rise of temperature from February to April. Heavy snowfall is not to be expected in regions where temperatures are very low, such as below zero, for here the absolute humidity is correspondingly small. Consequently, there is a remarkable contrast between the small falls and persistent snow cover of polar regions and the heavy falls and relatively short periods of snow cover in more temperate lands.

Hail. A more unusual form of precipitation is hail which consists of frozen rain, and which formerly was thought to be formed by a succession of ascents and descents in pulsatory air-currents, each ascent adding a layer of ice to the hail-stone. To-day, it is considered that very small hailstones may in their earlier stages travel in this irregular way but that larger stones could not be supported by the updraft. It is thought that sizeable hailstones grow by the collection of supercooled water as they fall through the cloud. If, as in a thunderstorm, the updrafts are strong they merely serve to delay the speed of a stone's descent and so to increase its coatings of ice. In other words, the growth of the hailstone is the same as that of the accretion of ice on aircraft. The hailstones may reach four inches in diameter and weigh as much as two pounds. Destructive hail of this nature is known, for instance, on the South African plateau, but in the British Isles hailstorms are neither severe nor common enough to necessitate insurance against such damage. Ascending air currents in the upper parts of high cumulus clouds are often associated with thunder and hail in temperate latitudes. In Burgundy in France fairly successful efforts have been made recently to avert destruction to the ripening grapes by hailstones. When cumulus clouds seemed likely to form hail they were subjected to intensive cannon-fire which tended to cause precipitation prematurely as rain rather than as hail.

The Artificial Stimulation of Rain

Recent advances in the knowledge of atmospheric moisture, and especially of the formation of rain, provide man's main hope of influencing climate. Clouds, as we have seen, consist of minute droplets of moisture, of which about a million would need to coalesce to form an ordinary rain-drop or a drop large enough to hurtle to earth. Formerly, it was difficult to explain why a rain-cloud changed so rapidly from moisture-particles into rain-drops. It has now been discovered that when ascending currents carry parts of a cloud well into the freezing-level of the atmosphere, some of the super-cooled water-droplets are turned into ice-particles. These grow rapidly in size and begin to rush downward whereupon the formation of rain-drops spreads quickly through the base of the cloud. This sequence is common in towering cumulus clouds outside the tropics. Within the tropics, and during hot weather elsewhere, the freezing-levels of the atmosphere are relatively high and ascent sufficient to turn water-particles into snow or ice is less common. Under such conditions it appears that the ascensional and turbulent currents in a cloud are likely to lead to rain only when the cloud-droplets differ appreciably in size. The bigger droplets grow rapidly by coagulation or coalescence and, because of their different rate of movement, by impingement on the smaller particles. As these large droplets grow in size and number the rain-forming process spreads almost instantaneously throughout the cloud-base.

It will be seen that there are possibilities of stimulating artificially the formation of rain-drops by introducing into suitable clouds some kind of freezing or hygroscopic particles, such as ice or water-droplets of the requisite size. The particles usually used for 'seeding' clouds are silver iodide; 'dry ice' (solid carbon dioxide at a very low temperature); and water from a fine spray. One method is to 'seed' clouds from aeroplanes, especially with 'dry ice' or with water-droplets, and many of these operations have caused local showers. The most common methods, and by far the cheapest, are those worked from the ground which 'seed' the base of the cloud-layer by means of self-operating balloons or, as is more usual, by generators. These ground operations have been carried out on a large scale especially in the Middle West of the U.S.A. but it is difficult to determine how much of the ensuing rainfall was natural and how much, if any, was due to artificial stimulus. It seems certain, however, that large-scale 'seeding' holds hope of increasing rainfall and of shortening the dry season and breaking droughts in climates which are prone to cumulus clouds and thick cloud-layers.

LIST OF BOOKS

Authoritative summaries of the above topics will be found in Malone, T. F. (ed.), *Compendium of Meteorology*, 1951
Durbin, W. G. 'An introduction to cloud physics', *Weather*, 1961, pp. 71–82; 113–25
Brewer, A. W. 'Why does it rain?', *Weather*, 1952, pp. 195-8

Mason, B. J. *The Physics of Clouds*, 1957; *Clouds, Rain & Rainmaking*, 1962

For Evaporation
Penman, H. L. 'Evaporation over the British Isles,' *Quart. Journ. Roy. Met. Soc.*,
    1950, pp. 372–83; *Vegetation and Hydrology*, 1963
UNESCO. *Climatology*, 1958 (Arid Zone Research)
Oliver, J. 'Evaporation losses and rainfall regime in Central and North Sudan',
    Weather, 1965, pp. 58–64

For Snow and Hail
Péguy, C.-P. *La Neige*, 1952
Ludlam, F. H. 'The hailstorm', *Weather*, 1961, pp. 151–62

For Thunderstorms and Lightning
Simpson, G. C. 'Atmospheric electricity . . .,' *Weather*, 1949, pp. 104-8; 130-40;
    170-4
Byers, H. R. 'Structure and dynamics of the thunderstorm,' *Weather*, 1949, pp.
    220-2; 244-50; *The Thunderstorm*, 1949
Malan, D. J. *Physics of Lightning*, 1963
Schonland, B. *The Flight of Thunderbolts*, 1964

For Clouds
International Atlas of Clouds (2 vols.), 1956
Douglas, A. C. *Cloud Reading for Pilots*, 1943
Grant, H. D. *Cloud and Weather Atlas*, 1944

For Humidity and Settlement
Climate and Man, U.S.A. Dept. Agric. Yearbook, 1941
Price, A. G. *White Settlers in the Tropics*, 1939

For Artificial Rain-stimulation
Bannon, J. K. 'Rain making,' *Weather*, 1948, pp. 261-66
Ludlam, F. H. 'Natural and artificial shower formation,' *Weather*, 1952, pp. 199-
    204
Bowen, E. G. 'Australian experiments on artificial stimulation of rainfall,' *Weather*,
    1952, pp. 204-9
United Nations (Dept. of Public Information). *Rainmaking*, (pamphlet) 1954
Mason, B. J. 'Design and evaluation of large-scale rain-making experiments,'
    Nature, March, 1955, pp. 448-51; 'Recent developments in the physics of
    rain and rain-making', *Weather*, 1959, pp. 81-97

Note. The greatest recorded rainfall amounts are: in one year, 905 inches at
Cherrapunji; in one day (24 hours), 73·6 inches at Cilaos at 3,650 feet on the flanks
of a high volcano in the Ile de Réunion east of Madagascar, during a tropical
cyclone; highest mean annual rainfall, 471·7 inches at Mount Waialeale, Kauai,
Hawaii; greatest mean annual total of rain-days, 325 at Bahia Felix, near Strait of
Magellan, which had 348 rain-days in 1916.

CHAPTER FIVE

GENERAL CIRCULATION OF THE ATMOSPHERE

Causes of High and Low Pressure

ATMOSPHERIC pressure is of little direct importance to human life except at high altitudes where, until travellers have become acclimatized to the change, mountain sickness may be felt. The physiological effect of slowly changing, for example, from the pressure of 30 inches at sea-level to that of about 15 inches at 17,500 feet is insignificant, and only in mountain ranges such as the Himalayas and Andes are travellers forced to use passes of such an altitude that mountain sickness may be experienced.

The chief importance of pressure in relation to climate is its control of winds which are in essence air masses moving to compensate pressure differences. In Britain the rising barometer is hailed as a sign of finer, more settled weather, while a falling barometer signifies the possibility of rain and stronger winds. The causes of barometric changes in any locality are often exceedingly obscure, but generally speaking a decrease below the standard pressure at sea-level (29·9 inches) is usually associated with either or both—

(a) ascending air currents;
(b) an increase in the absolute humidity of the air.

The presence of ascending air currents means that the air is expanding and getting lighter; the presence of much moisture ensures that the air mixture will be less heavy since water-vapour is lighter than dry air. Conversely, an increase of pressure above normal is usually associated with

(a) descending air currents, whereby the air is compressed and made heavier;
(b) the presence of dry air.

The general world distribution of pressure at sea-level remains fairly true to these principles.

General Latitudinal Distribution of Pressure

Along the Equator the ascending air currents and high absolute humidity cause a marked low pressure belt (mean pressure 29·8 inches at sea-level). This pressure increases polewards to a maximum in about 35° N. and 30° S., where mean annual pressures of 30·1 inches are exceeded in some lowland regions. Thence the pressure begins to fall steadily to a minimum (29·7 inches) in 60° N. and 60° S. Beyond these belts it increases slightly in the polar circles.

The equatorial zone of low pressure forms the key to the major pressure and wind systems of the world. Here, where the world's circumference is greatest, the continuous heat ensures that the ascending air currents affect the whole of the lower troposphere. The continuous ascent and expansion of the lower air forces the higher layers of the troposphere to expand slowly outwards. Consequently, a gradual drift of the higher air towards the poles goes on perpetually from above the Equator. This poleward drift has no sooner left the Equator than it comes under the deflective influence of the rotation of the earth. The deflective force increases rapidly with progress towards the poles, and the outflowing air currents from above the Equator come more and more under its influence. Eventually the original poleward drift is deflected into an almost west to east drift, and a gigantic circumpolar whirl is set up in the upper troposphere[1] above latitudes 30°–35° N. and 30°–35° S. As soon as the air currents assume an almost west to east direction, they are affected by centrifugal force, and since their speed of rotation exceeds that of the earth below, they will tend to be forced back towards the Equator. In other words, near the Equator the upper winds move westwards and polewards, but in higher latitudes this direction changes to eastwards with a slight equatorward tendency. Consequently, the higher layers of the atmosphere in sub-tropical latitudes consist of air that is being piled up between the outflow from above the Equator and the centrifugal eastward movement. To summarize, the probable reasons for the sub-tropical high-pressure belts about 35° N. and 30° S. are:

(*a*) the accumulation of poleward upper-air currents on meeting the westerly whirl set up by the earth's rotation;

(*b*) the pressure exerted by this poleward drift of subsiding air in the upper troposphere and tropopause; coupled with

(*c*) the compression of a great mass of air that is moving from a wide zone into a much narrower one.

This sub-tropical high-pressure belt, or Horse Latitudes, is a zone

[1] This is the zone of commonest occurrence of jet streams, or strong upper-air westerlies, which blow most often over latitudes 30°–45° in summer and 20°–30° in winter, when velocities of well over 100 m.p.h. are reached locally.

of slowly descending air currents, of dry atmosphere and of relatively calm weather. It is the starting-point of the world's major wind systems.

Polewards of the Horse Latitudes the pressure lessens until the decrease is stopped by thermal conditions over Antarctica, and to a lesser degree over the Arctic regions. Here the intense cold all the year, especially on the ice-caps of Antarctica and Greenland, generates a weak high pressure.

Winds and Pressure Gradients

The 'engine-house' of the circulation of the troposphere is the great area of rising air currents above the Equator. Here the air movement is outwards in the upper troposphere and inwards at the surface, there being a fairly complete circulation between the low pressures at the Equator and the adjacent high pressures in Horse Latitudes. This circulation serves to illustrate that the most common cause of

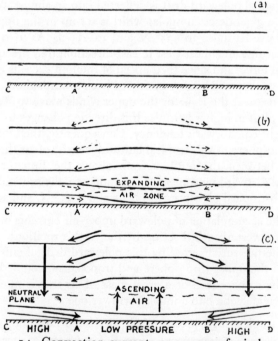

14. Convection currents as a cause of wind

The neutral plane is zone where horizontal movement is absent wind is the unequal distribution of heat at the earth's surface. The origin of air movements of this nature is explained in the accompanying diagram (Fig. 14). In (*a*) the isobars (or lines of equal atmospheric pressure) are horizontal, and there will be no air movement. If the surface above AB is heated the atmosphere in contact with it will expand and tend to rise. Continued heating will lead to further expansion, but the movement will be extremely slow and little or no change of pressure will occur at AB until the air expands sufficiently to move out over the neighbouring areas C and D (Fig. 14*b*). As soon as air begins to move down the barometric gradient towards C and D the pressure above AB is lessened. The outflowing air currents increase the pressure above C and D, and consequently the bottom layers of atmosphere are forced inwards towards the lower pressure at AB. In Figure *b* the dotted lines represent the position of the former layers of air, and the continuous lines the new isobaric surfaces. A continuation of the heating will heighten and intensify this system of *outflowing* upper air currents and *inflowing* surface winds until a regular atmospheric circulation is set up. The isobars for this stage are shown in Figure *c*, where ascending winds correspond with low pressure and descending winds correspond with high pressure at the earth's surface.

Normally the vertical subsidence of air currents is relatively slow, and in the Horse Latitudes, for example, the descent is thought to be about 300 feet a day. The speed of vertical currents of this nature compares with that of winds in much the same proportion as the movement of a glacier compares with that of a stream.

Although differential heating, as illustrated above, is the main cause of major pressure changes and of winds, it will be realised that some features of the atmospheric circulation cannot be explained on these grounds alone. The effect of the earth's rotation on the descent of upper-air currents is of great importance outside equatorial regions, and this dynamical influence prevents some areas of high temperature from experiencing marked convectional indrafts. It is most active and effective in the large high-pressure cells or anticyclones of sub-tropical latitudes.

From the diagram it will be noticed that where an isobaric surface has a gradient, air movement will occur. Normally, atmospheric pressure changes at sea-level are relatively small, pressure differences being usually measured in tenths of an inch. Consequently it is essential to view a major wind system as an air movement down a very gradual incline. The average height of the column of air that corresponds to one-tenth of an inch of pressure is about 90 feet; its variations, which depend mainly on pressure and temperature, are shown in the following table:

HEIGHT (IN FEET) OF COLUMN OF AIR CORRESPONDING TO $\frac{1}{10}$ INCH BAROMETRIC PRESSURE

Air pressure	Average Temperature °F.			
	20°	40°	60°	80°
Inches				
28	91 ft.	95 ft.	100 ft.	104 ft.
29	88 ft.	92 ft.	96 ft.	100 ft.
30	85 ft.	88 ft.	93 ft.	97 ft.

In studying the strength of the wind in relation to its barometric gradient the deflective force of the earth's rotation is of considerable importance. A slight pressure difference near the Equator will yield a strong wind, as, owing to the practical absence of deflection, the air currents will flow almost direct into the low pressure. In our latitudes, where pressure changes of one inch or more in a short distance oc ur, the rotation of the earth forces the winds to enter the low pressure on a very deflected course. Consequently the speed of the wind is greatly lessened, and it necessitates a comparatively great difference of pressure to cause a gale. The storminess of our climate is so proverbial that it is unnecessary to say that these changes do occur and that the barometer may rise to over 31 inches and fall to 27·4 inches in the same year. In temperate zones frequent temporary local changes of pressure have a great effect on the climate, whereas in sub-tropical and tropical lands seasonal changes of pressure are of paramount importance.

The Chief Wind Systems of the World

The essentials of the surface circulation of the atmosphere in each hemisphere are:
(a) the Trade Winds;
(b) the Westerlies;
(c) The Polar Winds (Fig. 15).

The Trade Winds. The descending air currents of the sub-tropical or Horse Latitude high-pressure zone give rise to basal winds that diverge towards the Equator and the poles. The surface

winds that converge upon the Equator are known as the Trades;
they blow from the north-east in the northern hemisphere and from
the south-east in the southern. Over the sea these winds are strong
and steady, and in the heart of the system, as at St. Helena, they are
almost constant in force and direction. In January, the Trades over
the ocean are strongest and most reliable between the Tropic of
Cancer and 5° N. But they have almost lost their character by the
time they reach the Equator, and on crossing into the southern
hemisphere they become light, variable north-westerly winds. In
July, the Trades blow strongest south of the Equator, and on crossing
it they are deflected to the right and carry a considerable volume of
air into the northern hemisphere.[1] The transference of air into the
opposing hemisphere is far greater in July owing to the great heat
and low pressures over the northern land masses at this season.

The characteristic steadiness of the Trade Winds is notably upset
by the differential heating of land and sea. The Trades depend for

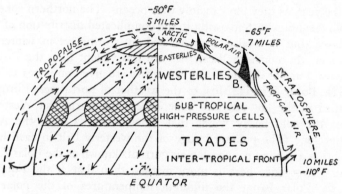

15. General atmospheric circulation

The diagram shows general surface winds (inside semi-circle) with main
frontal zones (dotted) and cross-sections of the probable circulation of the
troposphere. A. marks the general position of the Arctic Front and B. of
the Polar Front. The height and temperature of the base of the stratosphere
are also shown.

their continuance on the Horse Latitude high pressures. These high-
pressure zones are, however, broken by the heating effect of the sun
on land masses in summer. The low pressure formed over the

[1] Over the Atlantic the north-east Trades have a speed at the surface of about
13 m.p.h. in winter (January) and the south-east Trades have a velocity of 15
m.p.h. in July, the southern winter.

monsoon lands of south-east Asia completely obscures the northern
Trades in July. Yet, over the ocean, the high-pressure belt tends to
persist all the year. The two permanent areas of high pressure near
the Tropic of Cancer are centred upon the north Atlantic and western
Pacific. There are three similar permanent anticyclonic areas just
south of the Tropic of Capricorn over the Indian Ocean, South
Atlantic, and western Pacific. The Trades normally blow strongly
and steadily over only the eastern parts of these anticyclonic systems.

The Westerlies. The surface winds that move polewards from
the Horse Latitudes traverse a transitional zone of very variable
width. Roughly speaking, these westerly belts of low pressure
stretch from 40° N. and 35° S. almost to the Arctic and Antarctic
circles respectively. Both zones are characterized by endless variations
in the force and direction of the winds, caused by a constant procession of air eddies, mainly cyclonic, that move in an easterly
direction. The Roaring Forties, the winds of the westerly belt in the
southern hemisphere, are the more reliable, being stormy and strong
all the year, as they blow mainly over oceans. The northern zone has
a more complex system owing to the complicated distribution of land
and sea, but it is also characterized by storminess in winter. In
summer, the northern Westerlies are less violent than the Roaring
Forties. Further details of the Westerlies are given on pages 128–29.

The Polar Winds. Most of the weather of north-west Europe is
formed in the low-pressure zone, where the warm wet Westerlies
meet the cold heavy winds that blow outwards from the Arctic
anticyclone. The poleward edge of the westerly belt in each hemisphere is a zone of converging air currents known as the Polar Front.[1]
The convergence here is more marked than that at the Equator.[1]
At the Polar Front the different temperatures of the polar and
westerly winds bring about a zone of very indefinite width where
atmospheric eddies are formed. The higher temperatures of the air
above certain warm ocean currents carry the westerly influence far
northwards into the Arctic regions. Occasionally, however, the cold
polar air may overcome the westerly system and cause cold waves in
otherwise temperate lands. Intensely cold winds sometimes travel
down the Great Plains of America to the warm Gulf of Mexico;
similar winds occur east of the Urals, just as if the cold air was in-
vading the areas most shut off by mountains from the influence of

[1] At the Equator the converging air currents are less contrasted in temperature
and humidity; the Trades, too, are shallow and their influence seldom extends
much above 20,000 feet whereas the westerlies often extend up to the tropopause.

the Westerlies. The importance of the Polar Front is discussed in more detail on pages 122–24.

Factors Modifying the General Wind Systems: the Migration of Pressure Belts

The general pressure and wind systems of the world are greatly complicated by two main factors, firstly, by the seasonal migration of the pressure belts, or of air-masses and fronts, in accordance with the apparent movement of the overhead sun; and secondly, by the different heating qualities of land and water.

The seasonal migration of the pressure belts is of greatest climatic significance within tropical and subtropical latitudes. The pressure belts swing some 5° to 10° farther to the north in July and a similar distance farther south of the Equator in January. But just as the

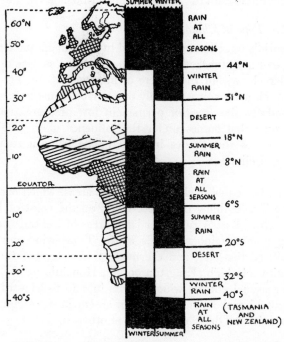

16. Climatic effect of the seasonal migrations of the main pressure belts on the western seaboards of continents

hottest time of a summer's day lags two hours behind noon so do the low-pressure and wind belts lag behind the overhead sun. Occasionally the limits of migration of the pressure belts are not reached till two months after the solstices. The climatic results of this seasonal migration are especially important in tropical and Mediterranean latitudes. The effect is roughly summarized in Figure 16, which applies only to the western sides of continents. This restriction is mainly due to the fact that the influence of the rapid heating of land masses on pressure and winds is so marked that the eastern sides of continents frequently have monsoon conditions. Thus, continents in sub-tropical latitudes in the northern hemisphere usually have two distinct coastal climates, namely, Mediterranean on the west coast and a monsoon or summer maximum rainfall type on the east coast.

The seasonal rhythm of climate caused by the migration of pressure and wind belts, or air-masses as they are often called to-day, is well seen in the Sudan and the Mediterranean lands.

The Sudan Type of Climate

The world's hot climates, as judged by a mean annual temperature of over 70° F., occupy rather more than half the surface of the globe and lie largely within the sphere of influence of the Trade Winds. These hot climates consist mainly of a narrow equatorial zone which corresponds to the areas of permanent low pressure, and a wider tropical zone that comes under the influence of the Doldrums in summer and of the Trades in winter. The tropical zone includes two types of climate:

(i) The hot equable climate of coastal areas, which occurs where the Trades blow on shore. Here, on the coast of Brazil and much of East Africa, over large areas of Central America and upon mountainous islands in the Trade-wind zone, there is rain all the year, and the mean annual temperatures remain uniformly high (Kingston 79° F., Honolulu 75° F.).

(ii) A seasonal climate which occurs in, and to leeward of, continental interiors, and consists of an equatorial régime for a few months and dry Trade Winds at other times.

This seasonal semi-continental climate covers large areas in both hemispheres, but it is developed most strongly in the Sudan over a belt of country stretching from 8° N. to about 15° N. (some 600 miles) and from the middle Niger almost to the Red Sea (Fig. 17 A3). Its characteristic vegetation of tall grassland interspersed with occasional trees and shrubs is known as savanna; northwards it grades into the Trade-wind deserts and southwards into the equatorial selvas. The vegetation reflects the main characteristic of the climate, which is the marked seasonal nature of the rainfall.

Dry Season. When the sun is overhead in the southern hemisphere, high-pressure systems develop over Asia and northern Africa, and the Sudan experiences Trade Winds that blow steadily from the north and north-east. After traversing arid land masses these winds have relative humidities of only 60 to 70 per cent., and so from November onwards drought, or the dry season, prevails in the Sudan. The luxuriant grass withers and dies; the soil soon crumbles into dust and the rivers dwindle. The first three months of the dry period, although called the 'cool season,' experience average monthly temperatures exceeding 70° F., and the midday sun is never lower than about 55° above the horizon. (*Cf.* our August.) The clear skies and dry air favour a considerable diurnal range, and at night the temperature may drop to 50° F.

Towards March the overhead sun begins to leave the southern hemisphere, and in the Sudan the weather becomes increasingly hotter. Soon the temperatures rise high above those experienced at the Equator, and when, in April, the sun approaches to an almost vertical position the heat becomes intense. Mean temperatures of over 90° F. for the hottest month are not uncommon, and in May daily maxima frequently exceed 110° F. At night the thermometer may drop to below 70° F., and the natives then shiver at the 'cold.' Under the terrific heat plants drop their leaves or remain almost dormant, only the larger rivers persist, and some animals hibernate because of the drought. When the Harmattan blows from the desert the relative humidity may fall as low as 10 per cent., and the dust and heat will dry and crack the skins of the already discomfited inhabitants. Fig. 158 explains this fierce sunbaking season.

Wet Season. The northward progress of the overhead sun is followed more slowly by the doldrum low-pressure belt, which in May has migrated to about 12° N. (Fig. 18). As the low-pressure trough gradually assumes control of the weather, violent storms or tornadoes usually herald the commencement of the rains. When the rainy season commences, the temperatures drop and all life springs up anew. The clumps of grass shoot upwards to heights of over six

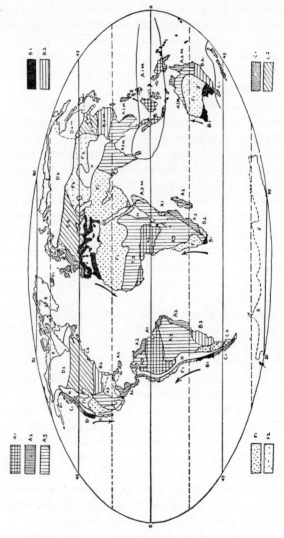

17. Generalized world distribution of chief climatic types

Arrows show position of cold ocean currents

A. Hot climates. With mean annual temperature above 70° F. 1. Equatorial (rain-forest), 2. Tropical marine. No marked dry season. 3. Tropical continental or Sudan. Marked dry season
B. Sub-Tropical Climates. No month below about 45° F. 1. Western Margin or Mediterranean. Winter maximum rainfall. 2. Eastern Margin. Rain all year
C. Cool-Temperate. One to five months below 45° F. 1. Marine. Rain all year, no marked maximum. 2. Continental. Summer maximum rainfall
D. Cold Climates. Six or more months below 45° F. Warmest month above 50° F. 1. Marine. Rain all year. No marked maximum. 2. Continental
E. Tundra and Polar Climates. All months below 50° F.
F. Desert Climates. Mean annual rainfall under 10 inches. 1. Hot Deserts. All months above 45° F. 2. Deserts with cold winters. One or more months below 45° F.
H. Highland Climates. Effect of relief predominates
M. in all cases denotes monsoon variety of the climatic type

feet, the baobab once more fills its spongy tissues with moisture, the stream-beds are filled with fast-flowing rivers, and the natives rejoice at the beginning of their farming year. The rains decrease in length and amount away from the Equator; in the south a rainy season from April to October yielding about 40 inches of rainfall is usual,

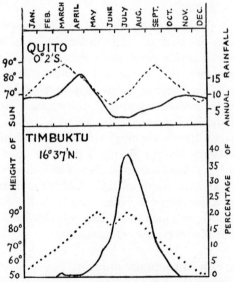

18. Relation between rainfall and zenithal sun at the Equator and in the Sudan, showing the typical climatic 'lag'

(*After Supan*)
Dotted line marks altitude of sun. Solid line shows percentage of annual rainfall falling in each month

while in the north (Timbuktu and Khartoum) the wet season diminishes to July and August and the amount to less than 10 inches (Fig. 18). The rainfall is of the convectional nature typical of the doldrums; the clear skies of the morning are covered towards noon by massive, billowy, cumulus clouds which heighten and darken until, to the accompaniment of lightning and thunder, a torrential downpour occurs. The rainfall differs, however, from that of the equatorial zone in being less in amount, more unreliable, and often interspersed by short spells of fine weather.

From late September onwards, when the overhead position of the sun begins to move southwards, the low-pressure belt gradually withdraws from over the Sudan and the dry Trades set in again. The departure of the rains is often marked by tornadoes and by a slight rise in temperature. The whole of the Sudan comes under the influence of the Trades by early November, and the seasonal rhythm of climate and vegetation begins anew.

Mediterranean Climates

The effect of the migration of pressure belts is well seen also between latitudes 30° and 40° in both hemispheres, where coastal areas on the west of continents come under the influence of the sub-tropical high pressures in summer and of the westerly low-pressure belt in winter. This Mediterranean type of climate is limited to western seaboards; on eastern seaboards the westerly winds in winter, having crossed the land mass, are dry and cold, while in summer a monsoon influence frequently predominates. The main areas that experience 'Western Margin Warm Temperate' climates are the coastal lands of the Mediterranean Sea, California, Central Chile (30° S.–37° S.), and the southern tips of Africa and Australia. The greater of these areas does not extend beyond a few hundred miles from north to south, or much more than 50 miles from the sea. The characteristics of Mediterranean climates are:

(a) winter rain and summer drought;
(b) hot summers and mild winters;
(c) high amounts of sunshine.

Summer. In summer, high pressures are situated over the Azores and south western Europe, while low pressures extend over the Sudan and southern Asia. At this season the Mediterranean is swept by northerly winds that, in the Aegean, assume almost the strength, steadiness, and regularity of Trade Winds. Mean temperatures rise to between 70° F. and 80° F. in the hottest month, and the intense heat persists throughout July and August in the western Mediterranean, and from June to September in the eastern. The excessive heat of the midday sun has led to the siesta habit. The houses are shuttered to keep out the dust and glare, and the vegetation, unless irrigated, is usually markedly xerophilous and adopts numerous devices to diminish transpiration. For week after week the sky is brazen and quite cloudless, and storms are so rare that some hotels offer to return visitors' fees for each rainy day.

During autumn the Azores high pressure decreases in intensity, and the low pressures over southern Asia are almost filled up by monsoons, hence the winds over the Mediterranean become less regular.

Winter. By October the westerly belt begins its southward migration, and the Mediterranean region gradually comes under the influence of winds from the Atlantic. Throughout the winter the warm Mediterranean sea attracts the passage of cyclonic depressions, many of which actually form over its own waters (Fig. 31). The weather now assumes much of the variability and changeability associated with the British climate. The temperature, however, is slightly higher than in Britain, and the sunshine hours considerably greater. Frosts and snow are rare near the littoral, although 15° of frost have been known in Rome, and there is a marked absence of the dull overcast skies and drizzle so well known to dwellers in Cornwall and Ireland. English winter resorts, such as Falmouth, have about 1,750 sunshine hours in an average year against the 2,570 hours at Athens. The difference is greatest in winter, when the Mediterranean is twice as sunny as southern England. The same fact can almost be deduced from the number of rain-days in the year. The Scilly Isles have less rain than Rome, but they experience twice as many days with rain. The annual rainfall of 31 inches at Nice falls on 67 days, while towns with an equal rainfall in southern England experience nearly 180 rain-days. In winter the Mediterranean sky is usually intensely blue and cloudless; the depressions bring heavy storms, but the rain is soon over and the sky blue again. The actual yearly rainfall amounts are relatively small, ranging from 15 to 35 inches, except where relief rains occur as at Catarro (180 inches) in the Dalmatian Mountains. The sharp intense nature of the rainfall offers the minimum of hindrance to human activities, but causes difficult problems of soil waste and flood control. Although the vegetation shows a rhythm based on the seasonal rainfall, the rhythmic feature is much less marked than in the savanna, where heat and moisture coincide. Nor have the farmers a corresponding rhythm in their labours, as the fruits employ them in the hot season and cereals and vegetables in the other months.

Several local winds of considerable interest occur in the Mediterranean Basin. During the west to east passage of a cyclonic depression especially in spring, the frontal winds from the south may carry with them the dust and heat of the desert. These hot southerly blasts, which are known as 'khamsin' in Lower Egypt, 'levêche' in Spain, and 'sirocco' in Sicily and Algeria, are greatly dreaded when the vines and olives are in blossom. In crossing the sea the winds become moisture-laden, but they are still no more welcome owing to their enervating, depressing nature.

Over the Mediterranean the winds in the rear of a cyclonic depression are northerly and will be expected to be relatively cold. Sometimes in winter it happens that a high pressure over Europe will assist these northerly winds to transfer a blast of cold air far to the south. The transference occurs most easily and most violently where a low gap in the mountain ranges that rim the Mediterranean Sea on the north allows the wind to sweep southwards. Thus the *mistral* of the Rhône valley and the *bora* at the head of the Adriatic are blasts of the cold air that was resting only a few score miles away on the plateau of central France or the mountain-girt Hungarian Plain. Although the air is descending, its temperature seldom rises much above, and sometimes remains below, freezing-point. The violent gusts appear all the more biting in a land noted for its winter warmth. The mistral has been known to blow railway carriages over, and the farmers plant tall cypress hedgerows on the northern side of their orchards to ward off the icy blast. When, as happens occasionally, the same cyclonic depression brings the mistral that chills Provence and the sirocco that wilts the Sicilian groves, the transitional nature of the Mediterranean climate is no longer in doubt.

The Effect of Differential Heating of Land and Sea

Land and Sea Breezes. The more rapid heating of land by day and its more rapid cooling by night tends to cause slight pressure differences between land surfaces and adjacent water surfaces. The difference may give rise to a wind that blows from the sea to land by day and from the land to sea by night. The sea breeze begins three or four hours after sunrise, and the land breeze some two or three hours after sundown; but usually on British coastlines these light air movements can only be felt during calm weather conditions, as at other times they are overcome by the prevailing wind. The influence of land and sea breezes seldom extends above 1,800 feet vertically and more than 25 miles inland, except in the subtropics and tropics.

Monsoons. The seasonal counterpart of this diurnal phenomenon is the monsoon, which is essentially a gigantic land and sea breeze with a half-yearly periodicity. The characteristic feature of a monsoon wind is its partial or almost complete reversal of direction from summer to winter. Monsoons owe their origin to the low pressures that develop over large continental masses in summer and to the high pressures in winter. The monsoon compensates for the difference in pressure between the continental interior and the nearest accessible ocean. Asia, which covers nearly half the land surface of the globe, is naturally the chief area of the monsoons (Fig. 17 M).

The Indian Monsoon

Well-marked monsoons are experienced throughout eastern and south-eastern Asia, in the East Indies, tropical Australia, and the extremities of southern Arabia, but the world's most famous monsoon is the summer wind of the Indian sub-continent.

This area, which covers one and three-quarter million square miles and extends through some 30° of latitude, is almost isolated climatically from central Asia by the Himalayas, and is situated between the world's greatest mountain barrier and one of the world's warmest seas (Fig. 11). The Indian weather year consists of—

1. *The season of the dry, land monsoon,* which comprises—
   (*a*) a cool weather season, January–February;
   (*b*) a hot weather season, March–mid-June.
2. *The season of the sea monsoon,* which also may be sub-divided into two seasons—
   (*a*) the rainy season, mid-June–mid-September;
   (*b*) the season of the retreating monsoon, mid-September–December.

The Cool Weather Season. From mid-December to the end of

February a very high-pressure system is situated over central Asia, but the strong out-flowing winter winds so characteristic of north China are not met with in India as the Himalayas isolate the Indian plains from the Asiatic atmospheric gradient. In consequence the winds in India are light and consist mainly of continental air that ensures dry, sunny weather. Rainfall is confined to the northern and southern extremities of the sub-continent. Upon the Indo-Gangetic plain shallow eastward-moving depressions, or 'lows', occasionally give rain or, on the hills, snow in Kashmir and the Punjab; in Ceylon and Madras, low-pressure waves moving westward from off the Bay of Bengal bring heavy rains especially to coastal tracts. At this season, mean monthly temperatures decrease from 80° F. in the warm south to about 55° F. in the cool north. Here, eastward-moving depressions are occasionally followed by southward intrusions of cold air that may cause frosts in the Punjab and temperatures as low as 40° F. in central India.

Hot Weather Season. In March the warmth increases, and during April and May the sun appears farther overhead in peninsular India. By May, the heat of much of central India exceeds 95° F., and shade temperatures of 105° F. become usual. Some parts of the Indus Plain, where the sky is cloudless and the air dry owing to lack of water surfaces, experience shade temperatures of 125° F. The sun shines like a yellow orb through the fine dust which rises high into the air. On the plains, the heat is at times made more oppressive by dust-storms and, in the north-west, by furnace-like winds from the desert which will raise night *minima* to over 100° F.

With such heat even small amounts of moisture will cause atmospheric instability. Upon the coastlands of the peninsula cool sea-breezes blow by day and in Malabar and Bengal shallow tongues of maritime air penetrate inland during April and May and encourage thunderstorms with heavy rain. In Bengal descending deluges may chill the air sharply by 20° F. and create fierce out-rushing gusts of wind. These early rains allow the farmers here and in Malabar to begin ploughing but the rest of India must endure the stifling aridity until June or even early July. Thus the inland progress of the monsoon seems held back until certain conditions allow it to advance suddenly in a dramatic 'burst'. The reasons for the abnormal time-lag and unusually abrupt change are not fully understood but certain interesting sequences or coincidences occur.

First, the intense insolation on the northern plains induces a low pressure which by May has deepened to 29·4 inches over Sind. Yet this heat-low is shallow and is surmounted by a dry continental air-flow from the north-west which precludes strong convection and rainfall. Thus the Sind low pressure is not potent enough to draw in the monsoon during April and early May as happens in Burma and is not now regarded as the primary factor in the advance of the Indian monsoon.

Second; throughout May and June the overhead sun stands near the northern Tropic and under its influence the Sind low expands

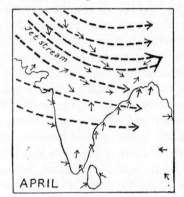 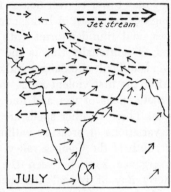

19. A possible cause of the 'burst' of the summer monsoon in India and Pakistan

Pecked arrows show winds at above 20,000 feet

and at the same time there develops a low-pressure trough across Bengal. Yet this trough, which some consider the equatorial trough at its northernmost migration, seems unable to exert its normal influence as long as it is surmounted by prevailing, dry upper-air westerlies.

Third; about the time of the 'burst' of the monsoon, these upper westerlies are replaced by easterly winds. It should be noticed that the upper-air westerlies, including the strong sub-tropical 'jet stream', normally migrate poleward in summer. The delay in this migration seems to be due mainly to the effect of the great, crescentic barrier of the Himalayas in forcing the upper westerlies, and especially the jet stream, to flow along its southern flanks longer than they would otherwise do.

Fourth; an equatorial-like trough, with easterly winds aloft, now extends along or just south of the Ganges valley and because of the

Himalayas must be filled mainly from the south. Moist maritime air-masses now surge forward from the Arabian Sea and Bay of Bengal and in the course of three or four weeks bring frequent downpours to all India and Pakistan except the far west and north-west where upper-air westerlies still prevail. If, as happens occasionally, high-level westerlies re-assert themselves for a short while over the Ganges plain, a temporary break in the monsoon may occur.

The Rainy Season. The burst of the monsoon is a remarkable event. The sky fills with majestic clouds, which above the Ganges valley may tower to over 60,000 feet, and the air is suddenly cooled. Along the west coast of India a doubling in the strength of winds and waves, and thunderstorms with torrential rain usually herald the burst. Traditionally this is the time for staying at home because travel is difficult and danger and damage from heavy rain common.

After the first violent rains, the monsoon usually settles down to alternate spells of wet and fine weather except on the wetter hill-slopes. These weather spells are associated with the movements of, and variations of pressure within, the trough of low pressure lying just south of the Ganges valley. Particularly during July this is the convergence zone between the south-westerly Arabian Sea and easterly Bay of Bengal branches of the monsoon and here many thunderstorms develop, especially along its more maritime eastern stretch. In addition, from mid-July to late September, rain-bringing depressions or low-pressure waves travel westward along the trough and very rarely penetrate as far as the deserts of Sind, yielding the equivalent of a year's rainfall in one day. The trough sometimes migrates northward on to the outer Himalayan flanks and then the eastern Himalayas receive up to 24 inches of rain in a day and rivers, such as the Brahmaputra, flood violently. Yet then, in the absence of the trough, no rain falls in the Plains and if such conditions persist, prolonged drought causes great damage to growing crops.

With the coming of the monsoon rains the temperature falls 6° or 7° F., and the furnace heat gives place to oppressive, sultry weather. The intensity of the rains is, however, the main feature of this season. The rainfall is cyclonic, convectional, and orographic, the last-named being important locally. The Arabian Sea branch of the monsoon gives enormous rainfalls where it strikes the Western Ghats. Bombay has $\frac{1}{2}$-inch of rain in May, 20 inches in June, and 24 inches in July. Thirty to 40 inches a month are common on the coastal plains to the south, while Mahabaleshwar (4,500 feet) has over 100 inches in July. Here, on the slopes of the Ghats, only six days from June to September are fine, and the average daily rainfall nearly equals 2½ inches.

The Bay of Bengal branch, which travels up the Ganges valley, impinges on the Burmese coastal mountains (Akyab over 200 inches of rain a year), and sends minor currents to the mountains of Assam (Cherrapunji 350 inches from June to September). In its journey up the Ganges plains its moisture content diminishes rapidly, so that precipitation in the wet season decreases from 48 inches near Calcutta to 36 at Allahabad, 22 at Delhi, and under 10 inches on the lowlands of north-west India. The Himalayan foothills extract much moisture from this monsoon and Darjeeling has 102 inches, Naini Tal 81 inches, and Simla 48 inches from June to September.

The Retreat of the Monsoon. The rapid advance of the monsoon in three weeks contrasts with its slow retreat of three months. As the overhead sun moves toward the southern tropic the maritime air-mass retreats southward and a high-pressure, with dry continental air from the north-west, slowly develops over India and Pakistan. The wet monsoon leaves the north-west plains about the end of September and by mid-December its influence has passed south of Ceylon. During its retreat, fresh surges of moist maritime air, associated with the westward passage of waves of low pressure from off the Bay of Bengal, bring rain to south-east India and Ceylon. More violent storms, or true tropical cyclones, also develop occasionally in the Bay of Bengal and cause great damage if they strike the coast. Thus the northern margin of the retreating maritime air-mass is associated with an autumn maximum of rainfall near Madras and in eastern Ceylon.

In brief the Indian sea-monsoon appears to be a complicated combination of the effects of the differential heating of land and sea, of high relief, of the migration of the thermal equator and of the interactions of surface and upper-air conditions. Its effect on human affairs is equally interesting; the combination of heavy rainfall and great heat leads to riotous plant growth. The typical monsoon products, such as rice and jute, yield prolific crops, which, coupled with the possibility of more than one crop in a year from the same patch of land, encourage a dense population. The physical conditions that govern the burst and retreat of the life-giving monsoon are, however, so complicated and so widespread that slight irregularities inevitably occur. In a farming land, where a crowded people lives near the

CLIMATIC

MEAN MONTHLY TEMPERATURES ° F.

TROPICAL CLIMATES	Lat.	Alt. (feet)	J.	F.	M.	A.	M.	J.	J.	A.	S.	O.	N.	D.	Yr.
Kayes (Sudan Type)	14° N.	197	77	81	89	94	96	91	84	82	82	85	83	77	85
Lindi (Sudan Type)	10° S.	268	82	81	81	81	80	77	78	77	78	79	81	82	80
Cuyaba (Brazilian Campos) (Sudan Type)	16° S.	541	81	81	81	80	78	75	76	78	82	82	82	81	80
Port au Prince[1] (Marine Type)	19° N.	120	76	77	78	79	80	81	82	81	81	80	78	77	79

MEDITERRANEAN (Western Margin Warm Temperate)

Gibraltar	36° N.	53	55	56	57	61	65	70	73	75	72	67	60	56	64
Athens	38° N.	351	48	49	52	59	66	74	80	80	73	66	57	52	63
Valparaiso (Chile)	33° S.	135	67	66	65	61	59	56	55	56	58	59	62	64	61
Sydney[2] (Eastern Margin Warm Temperate)	34° S.	138	72	71	69	65	59	55	53	55	59	64	67	70	63

MONSOON

Bombay	19° N.	37	76	76	80	83	86	84	81	81	81	82	81	77	81
Akyab	20° N.	20	70	73	79	83	84	82	81	81	82	82	78	72	79
Benares	25° N.	267	60	65	77	87	91	89	84	83	83	78	68	60	77
Peiping (North China)	37° N.	131	24	29	41	57	68	76	79	77	68	55	39	27	53

[1] Notice contrast in rainfall distribution between this marine station (West Indies) and Savanna and Campos areas.
[2] Notice contrast between rainfall of eastern and western margins in these latitudes.

margin of subsistence, even slight variations are liable to cause disastrous famines.

The Air-Mass concept applied to the Indian Monsoon. These variations in the time of the onset and close of the rainy season are typical of the erratic arrival and departure of air-masses (*see* pp. 112–14). The 'burst' of the monsoon is the arrival of the damp maritime air-mass and along its advancing edge typical variable weather-spells, with the frequent passage of low-pressure waves or shallow cyclonic depressions, may develop, especially near the ocean. Once the air-mass has fully settled in it gives weather characteristic of the area from which it came.

STATISTICS

MEAN MONTHLY RAINFALL (INCHES)

J.	F.	M.	A.	M.	J.	J.	A.	S.	O.	N.	D.	Total
—	—	—	—	0·6	3·9	8·3	8·3	5·6	1·9	3·3	0·2	29·1
6·1	4·1	7·5	4·7	0·1	—	0·3	0·5	0·6	0·7	1·8	4·8	32·0
9·8	8·3	8·3	4·0	2·1	0·3	0·2	1·1	2·0	4·5	5·9	8·1	54·6
1·2	2·5	3·7	6·5	9·4	4·1	2·7	5·4	7·3	6·6	3·4	1·3	54·1
5·1	4·2	4·8	2·7	1·7	0·5	—	0·1	1·4	3·3	6·4	5·5	35·7
2·0	1·7	1·2	0·9	0·8	0·7	0·3	0·5	0·6	1·6	2·6	2·6	15·5
—	—	0·6	0·2	3·5	5·8	4·8	3·2	0·8	0·4	0·1	0·3	19·7
3·7	4·3	4·8	5·6	5·1	4·8	4·8	3·0	2·9	3·2	2·8	2·9	47·9
0·1	0·1	—	—	0·7	19·9	24·0	14·5	10·6	1·9	0·4	—	72·4
0·1	0·2	0·5	2·0	13·7	49·4	53·7	42·5	24·6	11·6	5·0	0·6	203·8
0·7	0·6	0·4	0·2	0·6	4·8	12·1	11·6	7·1	2·1	0·2	0·2	40·6
0·1	0·2	0·2	0·6	1·4	3·0	9·4	6·3	2·6	0·6	0·3	0·1	24·9

LIST OF BOOKS

Haurwitz, B., and Austin, J. M. *Climatology*, 1944
Kendrew, W. G. *Climates of the Continents*, 1961
Weather in the Mediterranean, 1962 (H.M.S.O.)
Trewartha, G. T. *An Introduction to Climate*, 1954; *The Earth's Problem Climates*, 1961
Miller, A. A. *Climatology*, 1961
Sheppard, P. A. 'General Circulation of the Atmosphere', *Weather*, 1958, pp. 323-36
Tucker, G. B. 'General circulation of the atmosphere', *Weather*, 1962, pp. 320–40
Crowe, P. R. 'The Trade Wind Circulation of the World,' *Inst. Brit. Geog.*, 1949, pp. 37–56
Miller, R. 'The Climate of Nigeria,' *Geog.*, 1952, pp. 198-213
Monsoons of the World, Indian Met. Dept., 1958
Lockwood, J. G. 'The Indian monsoon', *Weather*, 1965, pp. 2–8
Walker, H. O. *The Monsoon in West Africa*, Ghana Met. Dept., 1958
Pédelaborde, P. *The Monsoon*, 1963
Garbell, M. A. *Tropical and Equatorial Meteorology*, 1947
Riehl, H. *Tropical Meteorology*, 1954
De la Rüe, E. A. *Man and the Winds*, 1954

F

CHAPTER SIX

THE MAJOR CONTROLS OF CLIMATE

TEMPERATURE, wind, and atmospheric moisture are the essential elements of climate; their combination forms weather conditions, the average of which, if computed over a considerable period of time, is known as climate. In other words, climate may be defined as 'the average state of the atmosphere at a given point on the earth's surface.'

It has been shown in the previous chapters that:

(a) local temperatures depend largely on latitude, altitude, and whether the surface is land or water;

(b) atmospheric moisture is largely a question of temperature and available water surfaces;

(c) winds are especially important as transporters of 'air masses,' which impose their own moisture and temperature conditions.

Therefore, generally speaking, it may be said that the climate of any land area depends on the following controls:

(i) Latitude;

(ii) Altitude;

(iii) Direction and character of prevailing winds;

(iv) Distance from sea and other bodies of water;

(v) Exceptional topographical, soil and vegetal features of the locality and its neighbourhood.

The climate of the Sudan, Mediterranean, and India, which have already been described on these lines, could each be distinguished by certain characteristics. These climatic types serve to illustrate the fact that the several controls of climate vary in potency from place to place. The characteristics of some localities depend mainly on latitude, of others on altitude, of others on distance from the sea or continentality, and of others on the character of the prevailing winds. It is not that any one control can be divorced from the whole group, but that one factor happens to be of outstanding importance in a certain area. In the following pages the typical effects of the main controls on the climate of various regions will be discussed. The general extent of the climatic regions is shown in Figure 17 (p. 70).

The Effect of Latitude: Equatorial and Polar Climates

The effect of latitude is least obscured by other factors in lands along the Equator, where days and nights are almost of equal length and seasonal variations in insolation are so small that, for our purposes, they may be disregarded. The persistently high altitude of the noon-day sun and the almost constant length of daylight lead to uniform temperatures through the year. Para has a mean monthly range of 3° F., Akassa of 4° F., while Jaluit, at sea-level, has a mean monthly range of less than 1° F. Even on the highlands the monthly temperature range is insignificant, that at Quito being less than 1° F., and at Bogota less than 2° F. On the lowlands, month after month and year after year bring mean temperatures of 79° F. to 82° F., and a passing cloud will make more difference to the heat than does the progress of the 'seasons.' The daily changes of temperature appear large compared with the yearly range. The thermometer rarely exceeds 100° F., and equally rarely falls to 60° F.; Para, for instance, has absolute extremes of 95° F. and 63° F. The nights are usually 15° F. to 25° F. cooler than midday, and the natives light fires to ward off the 'cold' and to dispel the unhealthy vapours rising from the ground.

The uniform temperature distribution over large areas results in calm or very light winds, and gives an exaggerated importance to the land and sea breezes that only affect a narrow coastal strip. Where there are no cooling breezes, the equatorial climate is the most oppressive and enervating in the world as the combination of constant heat, large water surfaces, and great transpiration from the dense foliage, leads to an excessive humidity, both absolute and relative. At Singapore the relative humidity probably never falls below 75 per cent., and at Para even the 'drier' season of the Amazon delta has humidities of 85 per cent.

The heat and high moisture content of the air cause abundant precipitation which is essentially convectional in nature. Bates[1] describes the typical afternoon rains of the Amazon delta:

'On most days in June and July a heavy shower would fall some time in the afternoon, producing a most welcome coolness. The approach of the rain-clouds was after a uniform fashion very

[1] *The Naturalist on the Amazons*, Chap. II.

interesting to observe. First, the cool sea-breeze, which commenced to blow about 10 o'clock, and which had increased in force with the increasing power of the sun, would flag and finally die away. The heat and electric tension of the atmosphere would then become almost insupportable. White clouds would appear in the east and gather into cumuli, with an increasing blackness along their lower portions. The whole eastern horizon would become almost suddenly black, and this would spread upwards, the sun at length becoming obscured. Then the rush of a mighty wind is heard in the forest, swaying the treetops; a vivid flash of lightning bursts forth, then a crash of thunder, and down streams the deluging rain. Such storms soon cease, leaving bluish-black motionless clouds until night. Meantime all nature is refreshed; but heaps of flower petals and leaves are seen under the trees. Towards evening life revives again. The following morning the sun again rises in a cloudless sky, and so the cycle is completed; spring, summer, and autumn, as it were, in one tropical day. With the day and night always of equal length, the atmospheric disturbances of each day neutralizing themselves before each succeeding morn; with the sun in its course proceeding midway across the sky, and the daily temperature the same within two or three degrees throughout the year—how grand in its perfect equilibrium and simplicity is the march of Nature under the equator!' In some areas the convectional thunderstorms are so regular that appointments are made 'before' or 'after the rains.'

Generally speaking, the rainfall is heavy and reliable. The Amazon basin contains the largest area in the world with over 80 inches annual rainfall, while a considerable portion of the Congo lowlands experiences from 50 to 70 inches. Wherever the normal convection rains are augmented by a relief or monsoon influence the yearly amounts usually exceed 100 inches. Iquitos has 100 inches a year, the windward hill-slopes of Java nearly 270 inches and the seaward side of Kamerun Peak practically 412 inches at sea-level.

The precipitation, being independent of latitude, shows considerable local variations. Athwart the Equator there is often a maximum about April and November after the overhead passage of the sun, but spring is usually the wetter period, as the doldrum belt has then moved from the ocean hemisphere (Fig. 18). Inland stations such as Manaus experience a well-marked drier season, but the driest months at Singapore and Iquitos have over 5 inches of rainfall.

The constant heat and moisture is reflected in the remarkable rapidity of plant growth, banana leaves growing 4 inches and bamboo shoots 9 inches in a day. The vegetation shows no seasonal rhythm, and in the same tangle of forest, different trees will be budding, flowering, fruiting, and dropping their leaves. These dense jungles, known as selvas, form a good index to the distribution of lowland equatorial climates. Selvas cover the vast expanse of the Amazon Basin and the wetter tracts of the Congo system.[1] Similar forests occur on the wetter parts of the coast of Guinea and of Zanzibar, and in Malaya and much of the East Indies. The poleward limits are roughly 9° north and south, but the width of the selvas in any one area seldom exceeds 500 miles.

From a human point of view, these 'glass-house' climates are not encouraging. Their unhealthiness and the profusion of plant-growth have hindered economic development and have had a retarding effect on the culture of the native peoples. White settlers need periods of recuperation in cooler or less humid regions if they are to maintain their standard of efficiency. The monotonous diurnal regularity of the climatic phenomena offers very little mental or physical stimulus, and it is not surprising that the modern development of these lands is being undertaken by people reared in the changeable westerly belt.

Polar Climates

No greater contrast of climate can be found than that between equatorial and polar regions. Everywhere within the polar circles at least one whole day has continuous light or continuous darkness; at the poles there are three months of darkness, three of twilight and six of light. Hence, the essential factor controlling the climate is latitude, which governs the length of time when the sun is above or below the horizon and the lowness of its altitude in the summer sky. The appearance of the sun after a long period of semi-twilight and darkness is the greatest climatic event in the year for the inhabitants. At first a narrow rim of gold shows for a brief moment on the horizon and then sinks out of sight; on following days the visible sun grows in size like a waxing moon until its whole circumference appears all day low in the heavens.

The actual climate experienced depends much on the local geography. The five million square miles of Antarctica is made homogeneous by the ice and snow-cover, and here the climates of perpetual ice extend beyond the continent roughly to the limits

[1] Large areas of the Congo basin are savanna.

of the pack-ice (Fig. 17E.). The vast plateau has a mean altitude of about 6,500 feet, and summits double that height. Hence latitude and altitude make this the coldest area on earth. The lowest recordings are $-127°$ F. (minimum) at 11,200 feet and $-97°$ F. (monthly mean) at 12,100 feet. Even at McMurdo Sound (78° S.) on the Antarctic coast, the mean monthly temperatures never reach 32° F., and remain below zero for seven months every year.

The Arctic region, on the other hand, consists of an irregular grouping of land and sea, which under the influence of ocean currents gives a complicated temperature distribution. At sea-level the length of the winter cold varies according to position with regard to warm currents; Spitzbergen (78° N.) has two months, Upernivik (73° N.) three months, and Barrow Point (71° N.) five months below zero.

During the long daylight of the summer half-year, where the snow cover is thick, the heat from the sun's rays is absorbed in the effort to melt the snow. Hence, on the perpetual ice-caps of Antarctica and of Greenland the air tends to assume the temperature of the ice-surface. Where the snow cover is thin enough to be melted, the rise of temperature in spring is retarded in the melting process, but immediately the ground is clear the mean monthly temperatures attain 40° F. to 44° F. In these areas, the tundra lands of the northern hemisphere, the soil is only thawed to a depth of a few inches.

Wind movement assumes a great human importance in lands of extreme temperature due to its effect of increasing evaporation from the body. Fortunately, ice-cap climates, owing to the anti-cyclonic conditions that originate from the chilling of the lower layers of the air, are usually associated with clear summer skies and light winds or calms. The conditions are especially stable towards the centres of the high pressures, and even the relatively small anticyclone over Greenland tends to fend off cyclonic depressions. Yet it not infrequently happens that when a deep 'low' passes near the periphery of the ice-cap the normal outward creep of cold air may be accelerated into a furious blast. These blizzards occur most frequently where relief features encourage the downward gravitation of cold air from the ice-cap. In Antarctica, Adelie Land is peculiarly liable to bitterly cold blizzards which are thick with whirling snow, partly new and partly picked up from the surface.

The low temperature and anticyclonic systems ensure a small precipitation that usually takes the form of fine powdery snow. Barrow Point experiences the equivalent of 6 inches of rain a year and Upernivik of 9 inches. Much remains to be discovered about ice-cap climates, but it seems that precipitation, although small, is not less than ablation. The ice-caps illustrate a common feature of meteorological conditions; namely, the tendency of a state of weather to feed itself and so to persist. The ice-caps are due largely to the cold of a six months' winter, but they, in their turn, greatly decrease the summer temperatures, and some outside factor would have to be called in to break up the partnership. What that factor might be is one of the main problems connected with the Ice Age.

Cold climates in their extremest, or ice-cap form, render plant life practically impossible and animal life is almost confined to the sea. But on the warmer seaboard strips of Arctic regions the long summer daylight encourages a growth of mosses, lichens, and, in most favoured localities, of stunted willows and even of carpets of flowers or 'bloom-mats.' The tundra forms the habitat of a scanty population of fishers, hunters, and reindeer herders, some of whom depend mainly on the sea, while others migrate southwards in winter to the shelter of the coniferous forests. The Eskimo of the New World and the various tribes of northern Eurasia wage a continuous struggle against cold and starvation. The following account describes a winter in northern Lapland, where the climate is less severe than in the tundra lands of Siberia and Canada.

'A part of the body exposed to a cold of 50° below freezing-point becomes frozen in a few seconds. When asleep, therefore, people must be carefully protected by sheepskins, fur gloves, a fur cap drawn over the ears, and a woollen cloth over the face. As a kind of hot-water bottle they like to take a dog into the sleeping-bag. It is much easier to protect people of mature age than small children, who often cry ceaselessly because of the intense cold. The whole country was turned into a region of stiffness and rigidity. Everything was frozen hard: the meat was sharp as iron, the milk had to be put in lumps into the coffee, the blood saved for dog-food had to be carefully cut into pieces with an axe (carefully, so that the axe itself should not break), the wood lost all its resilience and became as brittle as glass. Only a small circle around the tent fire formed a zone of plasticity. There a frozen salt fish, more like a steel weapon than human food, could be transformed into a delicacy. But on the coldest day I experienced in a Lapp tent, not even this was easy of accomplishment, as the fish, fully cooked on one side, was still frozen on the other.'[1]

[1] E. R. O. Koebel, *Geographical Magazine*, June 1936, p. 105.

Yet at the Equator the civilization of the native peoples was being equally oppressed by humid heat and by the profusion of plant growth. The curvature of the earth's surface, which is the same as saying altitude of the sun or latitude, has brought about this remarkable change in a distance of barely 6,000 miles.

The Influence of Distance from the Sea: Continental Climates

The diminution of temperatures away from the Equator is made irregular by the uneven distribution of land and sea, since the equable nature of marine temperatures contrasts vividly with the rapid heating and quick cooling of continental surfaces. Within tropical areas, such as the Sudan, the variations in the length of daylight are too small to cause marked seasonal changes of temperature even in the interiors of the land masses. But in the temperate latitudes (50°–70° N.) of the northern hemisphere, where the winter nights are long and the continents at their widest, the land masses have the greatest effect upon a climate (Fig. 20). A large area of Eurasia lies more than 1,000 miles from any sea-coast and over double that distance from the Atlantic, which is the main source of warmth in winter. In North America the distances involved are much shorter, but the great Rocky Mountain barrier almost shuts out the westerly oceanic influence within 200 miles of the Pacific coast. Consequently, the climatic results of a situation far from marine influence are well seen in the interior lands of Asia and of Canada.

During winter a vast anticyclone develops over the cold Eurasian land mass, and the pressures, when reduced to sea-level, become the highest known on the earth's surface. This continental 'high' fends off sea-winds on its western side and reinforces the winds on its eastern borders, thereby intensifying the land monsoons of north China and Manchuria. The rapidity with which the warmth of the westerly airflow is replaced by the chill of the land mass may be judged from the following mean January temperature of localities situated in much the same latitude, but extending over 4,000 miles of longitude.

	°F.	
Bergen	34°	Sea influence declining
Oslo	24°	
Helsinki	20°	
Leningrad	15°	Transitional
Tobolsk	−3°	Land influence supreme
Olekminsk	−31°	
Yakutsk	−46°	

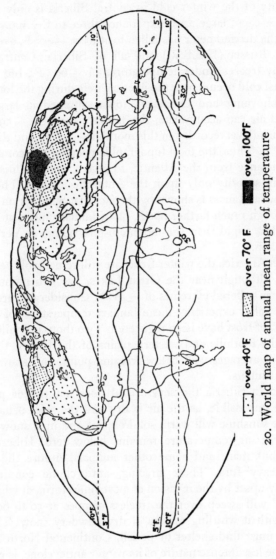

20. World map of annual mean range of temperature

over 40°F over 70°F over 100°F

The severity of the winter cold in central Siberia is only paralleled on polar ice-caps; large areas experience three to five months below zero, and the thermometer has fallen to at least — 40° F. over most of north-east Russia (U.S.S.R.) and all of Siberia. Central Siberia occasionally freezes under temperatures of —60° F., but the most phenomenal cold occurs near the Arctic Circle during the long winter. Near Verkhoyansk and Oymekon, towns situated in large valleys where cold air can accumulate, the January mean is — 59° F. The greatest warmth yet recorded in this month is — 13° F., and the greatest cold —90° F. Thus, the local topographic peculiarities combine with extreme distance from the Atlantic, to give a winter climate that is exceeded in severity only upon the Antarctic ice-cap. The chilling effect of the land mass is shown by the fact that Sagastyr on the Lena delta, although much farther north, has a January mean of —34° F., or 25° F. less frigid than the mountain-girt valleys at Verkhoyansk and Oymekon.

In North America the winter frosts are still remarkably severe, and Winnipeg, although near the Great Lakes, has two months below zero, and has suffered extremes of —46° F. Considerable areas in the Mackenzie valley experience mean January temperatures of — 30° F., and 102° F. of frost have been registered. Two thousand miles farther south on the foothills and western plains of the U.S.A. the winter weather still averages well below freezing-point, and snow cover is almost continuous.

Yet even in Siberia the severe cold does not render plant and animal life impossible, as the air is dry and the skies usually clear. The bright sunshine will warm solid objects and melt snow on roofs although the air temperature remains below zero. Hibernation is common, but man, and some other animals, endure the cold by wearing heavy furs. These bracing, anticyclonic conditions are occasionally upset by the invasion of a cyclonic storm, at which times the 'buran' will sweep over the treeless steppes at 50 to 60 m.p.h., bringing with it whirling masses of dry powdery snow. Then men and beasts must find shelter or perish. Continental North America, owing to the less intense nature of its winter anticyclone, is especially liable to these winds, here known as blizzards.

In spring, as the sun mounts higher in the sky, the snow and ice melt quickly, and, once the soil is exposed, temperatures rise rapidly and spring plunges into summer. Where the rivers flow northwards, as so many of them do, their upper courses melt several weeks before their mouths, and widespread floods result. Transport, which was so easy by sledge over the frost-bound surface of winter, halts for a while until the heat of late spring dries up the floods and morasses.

The long hours of daylight and absence of cooling sea-winds cause the summers to be appreciably hotter than at corresponding coastal areas. July means of 64° F. to 66° F. are common, and Verkhoyansk has the same July temperatures (60° F.) as south-western Ireland, which lies 1,000 miles nearer the Equator. Or, to take a Canadian example, Calgary, at a height of 3,500 feet, is as warm as the shores of Cornwall. The absolute maximum temperatures in these continental areas differ little from those of eastern England; Verkhoyansk has experienced 94° F., Moscow 99° F., and Yakutsk 102° F. It will now be seen that the Verkhoyansk neighbourhood can be distinguished from the polar climates of tundra lands because its July temperatures are well above 50° F., which is the lower climatic limit of tree growth.

The relatively warm summers and extremely cold winters give continental climates an enormous yearly range of temperature. The seasonal range of 15° F. at Valentia in south-west Ireland increases to 36° F. at Berlin, to 40° F. at Warsaw, and to 54° F. at Moscow. Then the seasonal change becomes excessive, and almost all Siberia, except the Pacific littoral, has a range of over 70° F., which increases to nearly 120° F. towards the 'Siberian Pole' of cold. Nothing is more characteristic of continental climates than this phenomenal range, but it is necessary to notice that most of the range is due to the excessive winter cold, as only at that season can sea-winds greatly modify the weather of adjacent coastal lands. A comparison of the statistics for Bergen and Yakutsk will demonstrate this fact (Tables p. 96).

The precipitation of coastal and interior lands in the westerly wind belt is easily distinguishable both in amount and time of fall. Over the more continental parts of Asia the extreme cold and anticyclonic conditions of winter favour a very low moisture content of the air and discourage the invasion of the westerly winds. Consequently, precipitation is small, seldom exceeding 7 inches, and often falling below 5 inches for the winter half-year. At Irkutsk snow falls on 65 days a year, but the average depth does not exceed $\frac{1}{2}$ inch a day. In areas such as this, where the flat plains offer no shelter from winds, snow cover is not continuous, but farther north, in the coniferous forest belt, snow may accumulate to a depth of 3 feet or more.

Much the same conditions prevail in the prairie provinces of Canada, where Winnipeg, for instance, has 4 feet of snow a year.

In summer the low-pressure system that develops over continents tends to draw the influence of the westerly winds far inland, and the decrease of rainfall away from the western ocean is much less pronounced than in winter. From July to August, Winnipeg has more rainfall ($8\frac{1}{2}$ inches) than East Anglia; Yakutsk, although 4,000 miles from the Atlantic seaboards, has little less than eastern Scotland. The following table illustrates the gradation to the marked summer rainfall maximum that is characteristic of the prairies, steppes, and taïga.

Climatic Type	Station	Total yearly Rf.	Percentage of total in	
			Summer	Winter
			June-Aug.	Dec.-Feb.
Western European	Bergen	84 in.	21	29
Transitional	Leningrad	19 in.	39	15
Steppe .	Tobolsk	19 in.	51	11
Taïga .	Verkhoyansk	5 in.	58	8

The characteristic vegetation associated with this climatic type consists of short grassland which merges northwards into the coniferous forest belt. The rolling expanses of grassland are notably devoid of woodland except near the water courses. In spring the steppes are bright with flowers, but the summer heat soon withers the vegetation except for the thistles and stronger plants. The vast taïga or coniferous forests of Siberia covers the lands with the coldest and most extreme climates. To-day the warmer and better-watered areas of these continental interiors form the world's great granaries, as their summer incidence of rainfall is especially suited to the cultivation of spring wheat (Chapter XXV).

The Importance of the Character of Prevailing Winds: Hot Deserts

The moderate precipitation of the Siberian taïga and steppes is not associated with aridity, as the evaporation is not excessive and the soils are snow-covered or frost-bound for several months a year. To the south, however, the lower rainfall and much higher temperatures of the mountain-girt basins of central Asia give rise to arid and semi-arid conditions. This vast area cannot partake of the rainfall of the summer monsoon primarily because of the high mountain barriers that surround it and partly because of its distance from the sea. The extent to which mountains deprive winds of their moisture is also illustrated in Patagonia, where a considerable desert lies in the lee of the Andes.

Yet an interior or mountain-girt position will not explain all arid regions, since many deserts stretch to the seashore (Fig. 17F). In these cases it will often be found that the prevailing winds blow from over the land mass and from cooler to warmer latitudes. Thus, the great hot deserts of the world, such as the Sahara, Arabia, and Kalahari, are situated on the lee side of continents in Trade-wind latitudes.

The presence of coastal deserts cannot, however, be entirely explained on the grounds of dry land-winds, as the seaboard strips of Peru and of the Kalahari have frequent sea-winds. At Iquique nearly 80 per cent., and at Walfish Bay nearly 50 per cent., of all winds come from a seaward direction. Yet much of these coastal areas experience less than 1 inch of rainfall a year, and at Iquique earthquakes are more common than rainstorms. The reason for the aridity is partly to be found in the cold ocean-currents that flow equatorwards along the western shores of continents in Trade-wind latitudes. The cold sea chills the winds traversing it and causes much condensation, but when the winds reach the hot land they are warmed and tend to absorb rather than to yield moisture. In Peru sea-mists commonly roll inland for a few miles, but they give no precipitation on the lowlands. This marine type of desert climate differs only in its cooler temperatures and cloudier skies from that of inland districts.

Away from the sea the aridity of hot deserts produces a number of secondary characteristics, the chief of which are:

(a) excessive midday heat;
(b) considerable diurnal range of temperature;
(c) abundance of sunshine.

Over the Sahara, winds from a northerly quarter blow with a remarkable persistency throughout the year. The arid nature of the land surfaces crossed by these winds and their southward direction into warmer latitudes make them extremely dry. At Aswan, where the mean relative humidity varies from 46 per cent. in January to 30 per cent. in July, the yearly evaporation is twenty times the rainfall. The ensuing aridity could scarcely be more complete. Occasionally rain may form in the upper atmosphere, but it is usually re-evaporated before it reaches the ground. When a storm does break, the convectional overturning that causes it is necessarily so strong that the rain is intense, but after a torrential downpour of a few hours the desert will lapse into as many years of complete drought. In some areas, especially near the coast, copious dew may form at night only to be quickly evaporated in the morning. At first sight it is puzzling why the terrific heat should not cause convection currents that would yield heavy rain. But it must be remembered that the air is so hot and so relatively dry that it would have to rise to a considerable height before being cooled to the dew-point, and in these latitudes the descending air currents of the upper troposphere tend to prohibit ascent to this extent.

Owing to the almost tropical latitudes of Trade-wind deserts the annual range of temperature is only about 30° F. in most localities. On the other hand, the diurnal range is always great because of the remarkable absence of cloud. In large areas of the Sahara and Arabia, well over 80 per cent. of the possible sunshine is experienced, and a summer day with less than twelve hours bright sun is a rarity. In the heart of the deserts, as at Asyut, the sky may be practically clear from June to October, and August is often absolutely cloudless. The intensely blue, clear skies allow the sun's rays to beat down with unmitigated heat upon the earth, and there is no vegetation or water surfaces to check by evaporation the rise of temperature. The barren stretches of sand and rock reflect the sun's rays with great efficiency, and so the arid conditions are augmented by the scenery they have created. The sun rises rapidly, and as soon as it appears in the sky the air temperatures jump upwards, until under the intense reflection from the bare landscape the heat becomes excessive. In the hottest season, shade temperatures soar to 120° F. and to 130° F. while 134° F. have been recorded at Death Valley, California, and 136·4° F. at Azizia in Tripolitania. Nor is this furnace heat of short duration; Insalah is scorched by a July mean of 99° F., while Death Valley, with a July mean of about 102° F., experienced 100° F. on one hundred and thirteen consecutive days.

After sundown the clear starlit skies allow the air to cool with great rapidity. Diurnal ranges up to 35° F. frequently occur, but ranges up to 60° F. are not unusual, and 74° F. has been recorded. In the coolest months slight frosts may occur at night, although the noon temperatures are as high as those of a very hot July in England. The absolute temperatures of the coolest months are 102° F. and

28° F. at Wadi Halfa, and 85° F. and 15° F. at Death Valley. Because of these extremes the Arabian and Saharan tribes wear heavy clothing to protect them from the chilly nights.

Except near the water holes and wadis, arid regions are practically without vegetation. Large areas are quite barren, but usually a few highly xerophilous plants survive, waiting maybe years for a fall of rain. The night dew is probably the most important source of moisture for plants, as the rare storms bring sudden floods that do more harm than good to the narrow area they affect. Semi-deserts and 'tame' deserts have a sparse vegetation, which is grazed in the cooler months by herds of nomad tribes, such as the Arabian Bedouin, whose life is linked to the water holes and oases. The desert climate is not unhealthy, nor is it unbearable, as the air is dry, and the cool invigorating night brings a welcome relief from the furnace heat of day.

However, one characteristic of desert climates—their windiness—is disliked by all forms of life. The absence of wind-breaks, the intensity of convection, a general position athwart persistent wind-belts, and, in some areas, proximity to frontal zones, combine to cause frequent winds. Dust-devils, or vortices of hot air, may arise in the daytime but more unpleasant are local and periodic winds, such as the shamal of Iraq which blows from the north-west for many days at intervals in summer, and the hot, dry khamsin of the Nile Valley, which rushes northward in spring when cyclonic depressions move eastward along the Mediterranean seaboard. These and other gusty air-movements tend to make desert travel an incessant struggle against the wind charged with dust and sand.

LIST OF BOOKS

See list at end of Chapter V: Miller, Kendrew, Riehl, and Trewartha
Crowe, P. R. 'Wind and Weather in the Equatorial Zone,' *Inst. Brit. Geog.*, 1951, pp. 24–76
Watts, I. E. M. *Equatorial Weather*, 1955
Beckinsale, R. P. 'The nature of tropical rainfall', *Tropical Agriculture*, 1957, pp. 76–98
Hare, F. K. 'Some climatological problems of the Arctic and sub-Arctic,' *Compendium of Meteorology*, 1951, pp. 952-64
van Rooy, M. P. (ed.). *Meteorology of the Antarctic*, 1957
Sutton, L. J. *The Climate of Helwan*, 1926
Gautier, E. F. *Sahara, The Great Desert*, 1935
For influence of climate: Missenard, A. *L'Homme et le Climat*, 1937
Buxton, P. A. *Animal Life in Deserts*, 1923
Adolph, E. F. and others. *Physiology of Man in the Desert*, 1947
Millington, R. A. 'Physiological responses to cold', *Weather*, 1964, pp. 334-37

CLIMATIC STATISTICS

MEAN MONTHLY TEMPERATURES °F.

	Lat.	Alt. Feet	J.	F.	M.	A.	M.	J.	J.	A.	S.	O.	N.	D.	Year	Range
EQUATORIAL Ocean Island	1° S.	177	81	81	81	81	81	81	81	81	81	81	81	81	81	0·3
Singapore	1° N.	10	80	80	81	82	82	81	81	81	81	81	81	80	81	2·3
Para	1° S.	42	78	77	78	78	79	79	78	79	79	79	80	79	78	2·7
Manaus	3° S.	144	80	80	80	80	80	80	81	82	83	83	82	81	81	3·0
POLAR McMurdo Sound	78° S.	Coast	24	16	4	-9	-11	-12	-15	-15	-12	-2	14	25	1	40
Spitzbergen (Oceanic)	78° N.	37	4	-2	-2	8	23	35	42	40	32	22	11	6	18	44
Upernivik	73° N.	65	-8	-9	-6	6	25	35	41	41	34	25	14	1	17	49
Sagastyr (Extreme tundra)	73° N.	11	-34	-36	-30	-7	15	32	41	38	33	6	-16	-28	1	77
CONTINENTAL Yakutsk (Taïga)	62° N.	330	-46	-35	-10	16	41	59	66	60	42	16	-21	-41	12	112
Tobolsk (Steppe)	58° N.	340	-2	-5	15	33	48	60	66	60	48	33	14	1	32	68
Winnipeg (Prairie)	50° N.	760	-4	0	15	38	52	62	66	64	54	41	21	6	35	70
Warsaw (Warmer continental)	52° N.	436	26	29	35	46	57	63	66	64	56	46	36	30	46	40
Berlin	53° N.	196	30	33	38	48	57	63	66	65	58	49	39	33	48	36
Bergen (Western oceanic)	60° N.	72	34	34	36	42	49	55	58	57	52	45	39	36	45	24
DESERTS Walfish Bay (Marine)	23° S.	10	65	66	66	65	62	60	59	57	58	60	61	64	62	9
Insalah (Trade-wind continental)	27° N.	920	55	59	68	76	86	94	99	97	92	80	68	58	78	44
Jacobabad (Trade-wind continental)	28° N.	186	57	62	75	86	92	98	95	92	89	79	68	59	79	41
Astrakhan (Desert with cold winter)	46° N.	-46	19	21	32	48	64	73	77	74	63	50	37	26	49	58

Notice: (a) Contrast between summer temperatures of Yakutsk (Taïga) and Sagastyr (Tundra).
(b) Warsaw is on western edge of continental climates in Europe; Berlin is typical of Central Europe; Bergen of colder seaboards of Western Europe.
(c) Astrakhan is typical of deserts of central Asia: Notice latitude and coldness of winters.

CLIMATIC STATISTICS
MEAN MONTHLY RAINFALL (Inches)

J.	F.	M.	A.	M.	J.	J.	A.	S.	O.	N.	D.	Total
11·5	8·9	8·6	8·1	5·6	5·1	6·8	3·9	5·2	5·6	5·7	8·9	83·9
9·9	6·6	7·4	7·6	6·7	6·8	6·8	7·9	6·8	8·1	9·9	10·6	95·1
12·5	14·1	14·1	12·6	10·2	6·7	5·9	4·5	3·5	3·4	2·6	6·1	96·2
8·3	8·0	8·1	8·4	6·6	3·9	1·8	1·3	1·4	4·6	4·5	8·2	65·1
Not available												
1·4	1·3	1·1	0·9	0·5	0·4	0·6	0·9	1·0	1·2	1·0	1·5	11·8
0·4	0·5	0·7	0·6	0·6	0·5	0·9	1·1	1·1	1·1	1·1	0·5	9·1
0·1	0·1	—	—	0·2	0·4	0·3	1·4	0·4	0·1	0·1	0·2	3·3
0·9	0·2	0·4	0·6	1·1	2·1	1·7	2·6	1·2	1·4	0·6	0·9	13·7
0·7	0·6	0·7	0·8	1·3	2·7	3·5	3·2	1·5	1·4	1·3	0·9	19
0·9	0·7	1·2	1·4	2·0	3·1	3·1	2·2	2·2	1·4	1·1	0·9	20·2
1·2	1·1	1·3	1·5	1·9	2·6	3·0	2·9	1·9	1·6	1·5	1·5	22·1
1·7	1·4	1·6	1·5	1·9	2·3	3·0	2·3	1·7	1·7	1·7	1·9	22·7
9·0	6·6	6·2	4·3	4·7	4·1	5·7	7·8	9·2	9·3	8·5	8·9	84·3
Practically nil												
Practically nil												
0·3	0·3	0·3	0·2	0·1	0·2	1·0	1·1	0·3	—	0·1	0·1	4·0
0·5	0·3	0·4	0·5	0·7	0·7	0·5	0·5	0·5	0·4	0·4	0·5	5·0

G

CHAPTER SEVEN

THE INFLUENCE OF TOPOGRAPHY ON CLIMATE

OF the major controls of climate, altitude has the most potent influence, but, like all other climatic factors, it cannot be considered alone. Rather the study must be of relief, which includes the nature of the local topography, whether plateau or enclosed basin, simple hill-range or complex mountain system, and of the wider setting in relation to the prevailing winds. Relatively minor topographic features, such as the Cotswold escarpment and the wide deep valleys of its dip-slope, exert a climatic influence that is important locally. Such influence increases with the magnitude of the relief features, until eventually the topography, instead of merely bringing about slight modifications in the climate, becomes the deciding factor. Consequently, in areas with considerable topographical variations, certain climatic peculiarities are developed, or, as is usually said, the 'mountains make their own climate.' The influence of relief upon climate is best seen in the complex topographies of the Alpine type of mountain systems. Here the abrupt altitude-changes within a small horizontal distance bring about complicated alternations of wet and dry slopes, of sunshine and shadow, and of shelter and exposure that form the chief climatic peculiarity of a mountain region as a whole.

Influence of Topography on Insolation and Temperature

Insolation. The peculiarities of mountain climates become strongly marked above 6,000 feet, a height that surmounts the lower layers of air containing half of the water-vapour and most of the solid impurities in the atmosphere. Solar energy reaches a high intensity in elevated regions as the clarity and purity of the air allow some 75 per cent. of the total insolation to penetrate to 6,000 feet. The total radiant energy from the sun received in a year at Davos (5,000 feet) is 50 per cent. more than at Potsdam (near sea-level). The intensity of the sun's rays may be judged from the fact that at Leh (11,500 feet in the Upper Indus Valley) water was boiled (b.p. 192° F.) by simply exposing it to the sun in a small bottle blackened on the outside and sheltered from the air by a larger transparent vessel. The bright sunlight quickly tans the bare skin, and in summer reflection from snow-surfaces is trying to the eyes. The high percentage of ultra-violet rays, which are twice as abundant as at sea-level, makes the sunlight especially stimulating to life and has given some elevated settlements a great reputation as sanatoria.

The thinness and purity of the atmosphere, although the main cause of high insolation, ensures a low atmospheric temperature, as the heat radiated from the earth's surface cannot be retained by the air. For the same reason there is a remarkable difference between rock and air temperatures, especially at great altitudes. Similar differences occur between the temperatures of shadow and sunlight and of night and day. At night, except on the permanently cold wind-swept summits, the air temperature drops abruptly. This is especially evident on a high plateau, such as Transvaal, where the noonday sun may be uncomfortably hot, but within a short time of sundown coats are necessary and frosts may occur in winter even in the tropics. The mean diurnal range at Pretoria in August is 34° F. On the Bolivian plateau above 10,000 feet the main characteristic of the climate is said to be discomfort. The seasons really occur every twenty-four hours, the sunshine being too hot to be comfortable and the nights cold enough to make the day welcome. The 'eternal spring' of places such as Quito is in fact a concoction of statisticians.

Effects of Shade and Exposure. The temperature differences between shadow and sunlight are important geographically. On Kilima Njaro at 16,000 feet objects in the sun may be warmed to 70° F., while those in the shade remain at 8° F. In the Alps the chill of the shade seems to range from about 20° F. at 6,000 feet to over 50° F. at 9,000 feet. On the *puna* of the High Andes the natives avoid the shade of their hovels and do their cooking out of doors in the sun-light. Parts of the body in the sun are hot and parts in the shade cold, while sunlit slopes are markedly warmer than those in the shadow.

The difference between the insolation of opposing slopes is negligible in Equatorial regions where the sun's rays, being almost overhead all the year, fall equally on both sides of a ridge. But in, for example, the Alps, where the relief system has an east-west trend and the altitude of the sun is below 45° for half the year, slope and exposure become of great importance. North-facing slopes remain largely in shadow while south-facing declivities receive almost vertical rays. So distinctive is the sunshine and warmth of these op-posing slopes that special names are given to them locally; in the Alps, *sonnenseite* and *schattenseite* or *l'adret* and *l'ubac*; in the Pyrenees, *solanas* and *ubagas.* This phenomenon is especially pronounced in some of the deeper east-west Alpine valleys such as the Engadine, where the high, abrupt valley-slopes keep out so much of the sun that settlements and crops seek the *sonnenseite.* Many Swiss villages are perched at considerable altitudes above the valley-floor, the two or three hours' increase of sunshine thereby gained being deemed a recompense for difficulties of transport. (Plates 1 & 2.)

Influence of the Shape of the Land-forms. The local variations in insolation depend on the steepness and orientation of the slopes, but the shape of the complete land-form also exerts a considerable influence on the climate. There is a marked difference between the temperatures of areas of convex and concave relief.

A ridge or isolated peak (convex relief) offers a relatively small area to the sun's rays, and usually suffers also from so strong a wind movement that the daytime heat from the rock surface is dissipated into the wider atmosphere and has little effect on local air temperatures. Above such an area the effect of surface cooling at night is equally small, and the wind again tends to control the temperatures. Consequently convex summits often experience small diurnal ranges, that on Mont Blanc, for instance, being only 6° F. in July compared with 20° F. at Geneva.

On the other hand, the air in valleys, cirques, and other concave relief features tends to be warmed on three sides during the day and to be sheltered from the wind. For these reasons the air is heated to a considerable extent, the warmth rising locally rather than being dissipated in the general circulation. At night, however, when the extensive ground surface of a concave land-form begins to chill the warm air resting upon it, the cooled air sinks into the basin. Normally, owing to its much higher initial temperature, the valley remains warmer than the higher slopes, but frequently a well-marked hollow may become a collecting centre for the cold air that gravitates from the neighbouring mountain slopes. As a result, a valley bottom or the floor of a depression may become extremely cold at night. The difference between the climates of convex and concave land-forms has been likened to that between marine and continental climates. The peaks (islands) normally have a smaller diurnal and annual range than the depressions (continental interiors), which are much warmer in summer but are often appreciably colder in winter, especially if they are almost or quite enclosed.

These differences are illustrated in the following statistics for

[A. Bertschinger, Klosters

1. A Swiss alp upon the Silvretta massif, on the sunny slope and above
the topographical shade in the nearby valley

2. The Rieder Alp, a sunny slope above the upper Rhône, Valais,
Switzerland
(By courtesy of Swiss National Tourist Office)

[Dorien Leigh

3. A volcanic island. The Nuuanu–Pali road near Honolulu, the Hawaiian Islands

[Dorien Leigh

4. Luxuriant growth of coral exposed at low tide on the outer edge of the Great Barrier Reef, Australia

Rigikulm, an isolated rocky summit near Lucerne, and Bevers, a village in the deep Upper Engadine Valley.

Station	Altitude	Mean Temps.—July °F.			Mean Temps.—Jan. °F.			Range °F.
		Daily	7 hr.	13 hr.	Daily	7 hr.	13 hr.	July–Jan.
Rigikulm	5,863 ft.	50	49	52	24	22	25	26
Bevers	5,610 ft.	53	48	62	14	8	22	39

The flat plateau landscape is intermediary in form between the convex ridge and concave hollow. In this case the climate also is intermediary, the diurnal and annual range being greater than on mountain slopes and less than in the valley bottoms.

Inversion of Temperature. It will now be seen that the mean temperature decrease of 1° F. in 330 feet of ascent is only of the most general application. Locally the rate of decrease varies with the land-form, with the season and with the time of day. In the eastern Alps the average decrease is more rapid in summer (1° F. in 290 feet) than in winter (1° F. in 360 feet) and on southern (1° F. in 300 feet) than on northern slopes (1° F. in 360 feet). Locally, in deep and enclosed valleys, the valley sides may actually be warmer than the adjacent valley floor, and this reversal of the normal thermal gradient is known as 'inversion of temperature.' The phenomenon originates from the differential cooling of convex and concave land-forms. During clear calm nights, especially in winter, the cold air upon the higher slopes may drain downwards into the neighbouring valley. The warmer air of the valley is displaced by the colder air and is forced upwards on to the higher valley slopes, which then become warmer than the valley floor. The large basin at Klagenfurt in the eastern Alps shows the effect of inversion even in its mean January temperatures. The following statistics for stations on one side of the high enclosing slopes show that an ascent of 2,000 feet is accompanied with an increase in warmth of 9° F. instead of a decrease of about 6° F.

Station	Height	Mean Jan. Temperature
Klagenfurt . . .	1,500 ft.	20° F.
Hüttenberg . . .	2,500 ft.	26° F.
Lolling . . .	3,600 ft.	29° F.
Stelzing . . .	4,600 ft.	25° F.

The extreme winter cold that accompanies inversion of temperature in the deep valley bottoms is not equalled on the high peaks; in
Switzerland, Bevers has experienced $-31°$ F., but the Säntis peak
(8,200 feet) has yet to record below $-15°$ F.; in the U.S.A., the
lowest known temperatures, $-65°$ and $-66°$ F., occurred in the deep
Yellowstone valley at and near Miles City, while Pike's Peak, some
11,000 feet higher, has not recorded below $-40°$ F.

The extraordinary cold of Verkhoyansk reflects the influence of
inversion upon extreme continentality.

The phenomenon, in a lesser degree, is extremely common in the
scarplands and undulating topographies of England. In the deep
Windrush Valley near Burford, western Oxfordshire, inversion was
observed on sixty-three nights in twenty-one months. Here the damp
alluvial soils of the wide valley floor radiate heat comparatively
slowly after sundown. At night, cold air creeps downhill from the
adjacent limestone uplands and displaces this warm valley air upwards until the uplands are $5°$ F. or so warmer than the vale (Fig.
21). Here, as elsewhere, the phenomenon only occurred on calm clear
nights that allowed the downward gravitation of cold air. In windy
weather the higher land is probably always colder than the valleys.

The effect of inversion on cultivated crops is of great importance
locally. The coffee plantations of São Paulo keep to the hillsides, and
the vineyards of Valais, in France, occupy the rocky slopes up to
3,000 feet in preference to the valley bottoms. In Britain, horti-
culturalists fully realize how less hardy plants may be killed by frost
in the valley bottom, only slightly injured on the nearby slopes, and
untouched on the plateau. This especially applies to delicate crops in
late spring and autumn when frosts are rare but disastrous. Striking
examples of the disastrous effect of 'frost drainage' on vegetation
occur in the Forest of Dean, where the topography is deeply dissected.
In occasional years severe frosts may occur in late spring, and then
in the deep valley bottoms all young foliage is completely blackened,
many young plants killed, and the growth of the larger trees severely
checked by the killing of the terminal buds. The upper limit of the
frost drainage is often marked by 'perfectly straight contour lines
along the valley sides, varying from 300 to 500 feet altitude, below
which all is literally black, and above the normal spring green.'

Influence of Topography on Winds

Formation of Local Winds. The downward creep of cold air on still nights that brings about an inversion of temperature is known as a katabatic wind in Britain. In more mountainous lands this air movement grows into a mountain breeze, and may be given special names such as the *Nevados* of high Ecuador. The mountain breeze,

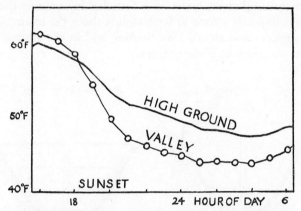

21. Mean temperatures during 'inversions' in summer in a tributary valley of the Windrush, Oxfordshire

(After Heywood)

however, consists of more than a mere katabatic creep down the valley-sides. It is a definite wind that blows down the valley-floor and originates from a combination of local and regional factors. The local factor is the rapid chilling of air on the valley-sides with the consequent formation of katabatic creeps, already discussed; the regional factor is the greater cooling of the whole mountain area compared with that on the adjacent lowlands where the atmosphere is less thin. On calm clear nights this regional difference in cooling causes a marked air-flow out from the mountains, the movement being canalized into the valleys. In Fig. 22 the composite mountain—or down-valley wind (represented by coarse stipple) is shown as moving down the valley-floor at right angles to the cold creep that gravitates from the shallow layer of chilled subsiding air (marked in fine stipple) in contact with the adjacent valley-sides.

The down-valley or mountain-breeze has its counterpart in the up-valley wind that blows occasionally during the daytime. During sunny spells, because of its thinner atmosphere, a mountain region has more insolation than adjacent plains and there is a general air-movement towards it from the lowlands. This regional inflow is reinforced by the local anabatic ascent of the layer of air greatly heated by contact with and radiation from the sunny valley-sides (fine stipple in diagram). The combined result is an appreciable wind (shown by coarse stipple in Fig. 22) up the floor of the valley. Sometimes in the Alps during the hottest hours of a fine day in summer the valley breeze is sufficiently strong to form clouds about the mountain tops. Towards noon these clouds may thicken, and in the afternoon may culminate in a violent thunderstorm.

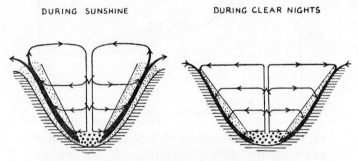

22. Explanation of up-valley and down-valley winds
(*modified from Wagner*)
Arrows show direction of air movement

The mountain and valley breezes partly explain the weather differences between valleys and highlands. Summits tend to be less cloudy in winter than in summer, and at night and in the early morning than in the afternoon. The lower valleys have less cloud in summer than in winter and at noon than at night. In winter, Davos enjoys the clear intense blue skies of the highland, but in summer partakes of the cloudy skies of the mountain tops. The contrasting weather at Zurich, a valley city, is shown in the following table:

	Altitude	Cloud Cover (tenths)		Sunshine (Hours per day)		Relative Humidity (Per cent.)	
		Jan.	July	Jan.	July	Jan.	July
Davos	5,000 ft.	4	5	3·2	6·7	84	80
Zurich	1,500 ft.	8	5	1·4	7·7	88	74

The relative clearness of uplands from cloud or fog holds good even in the English scarplands; in the Windrush valley thick fog envelops the valley floor on one night in every four on which inversion of temperature occurs, and during these times the uplands are clear.

Föhn Winds. The effect of topography upon the nature of air currents is admirably seen in the case of the föhn, a characteristic local wind of mountainous districts. The character of föhn-like winds depends upon the loss of water-vapour by precipitation during ascent on the windward slope and the consequent more rapid heating of the drier air on descending the leeward slope. On the windward side of a ridge the ascending currents are cooled at the rate for unsaturated air until the temperature reaches the dew-point. Further ascent results in precipitation and the liberation of latent heat, which means that the temperature now falls much more slowly. In descending, the drier air is warmed dynamically at the rate of unsaturated air, that is, at the adiabatic rate of about 16° F. in 3,000 feet, and so the wind on the leeward slope is considerably warmer than at corresponding heights on the windward side. Figure 23 shows a specific case where the air arrived at the south side of the range (at 1,000 feet) at a temperature of 65° F. The air is forced to rise 8,000 feet, during which, assuming that condensation occurred at 3,000 feet, its temperature fell to 36° F. The wind then descended 7,000 feet, and after allowing a few degrees for contact with the cold ground, its warmth increased by 34° F. In other words, its final temperature would be 70° F., although the leeward side of the range is 1,000 feet higher than the windward. The air would be relatively dry, and in any

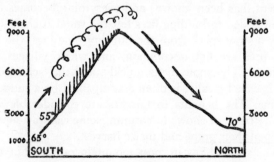

23. Atmospheric conditions during a föhn

future ascent would need to be cooled to almost 36° F. before reaching its dew-point.

Föhn tendencies are most felt in winter when, by their turbulence and gustiness, they may drive out the cold air stagnating in deep valleys and replace it with a warm wind. These winds have been closely investigated in Switzerland, where the name föhn originates. The strongest föhn here occurs when a 'low' north or south of the Alps draws air currents across the mountain system. The most typical wind, which comes from the warm south, causes rainy and cloudy weather on the Italian side but gives clear dry weather in the Swiss valleys. The wind tends to follow relief lines, being strongest in north to south valleys, such as those of the Rhine from Chur to Lake Constance, the upper Aar, Linth, and Reuss and the south bank tributaries of the Upper Rhône. Föhn winds also occur in the valleys of the Italian Alps, notably in the Ticino and Toce, but here the winds come from the north, and being cool in origin make less difference to the already warm lowlands.

The south föhn, the prime phenomenon, is experienced in Switzerland on about 100 days a year, being most common in spring and least in summer. The wind affects different areas at different times, and the Reuss valley, for example, has about 50 days with a noticeable föhn. The wind is usually announced by a roll of thick cloud, the 'föhn-wall,' along the mountain crests; then a few fierce gusts of cold air are felt in the valley, and after a short calm the warm blast of the föhn asserts itself. The clouds have now gone, the sky is clear, and the wind blows steadily, perhaps for several days. The warmth may increase 20° F. in a few hours and 40° F. in the day, but increases of over 60° F. have been known. The relative humidity rapidly decreases (14 per cent. has been known) and everything becomes tinder dry. Orders are issued forbidding fires lest the wooden chalets should be caught by a stray spark from the burning pine logs, the usual fuel. Even so, fires are not uncommon, and whole villages have been burned down. The snow melts quickly, and the 'schnee-fresser,' as the föhn is called locally, is often accompanied by a sudden spate of the streams. The benefits to farmers are considerable: in spring, pastures are freed of snow; in autumn, some cantons depend on the föhn to ripen their grape and maize harvest.

Föhn-like winds blow in most mountain ranges. The *Chinook* of the Rockies melts the snow and sets pastures free for grazing. Here, temperature increases of 40° F. in fifteen minutes have been known in winter, and although the thermometer seldom rises above 40° F. the contrast between the former frost makes the chinook appear very warm. The mean January temperature at Calgary (11° F.; 3,000 feet above sea-level) is high compared with that of Winnipeg ($-4°$ F.; 760 feet) partly because of this föhn effect.

Elsewhere in the world relatively dry warm winds of this kind are given distinctive names by the local inhabitants; the *Samun* of Persia, *Nor'-Wester* of the Canterbury Plains, and *Berg* winds of the shores of Cape Province are examples. It must be admitted, however, that some of these occasionally owe their föhn nature largely to the subsidence of air originally situated only just above the level of the local mountain summits. In this connection, it should be pointed out that chinooks and similar winds, even when they do not undergo condensation, give remarkably high temperatures on leeward slopes. Often these temperatures are too high to be explained solely by the normal process of adiabatic heating as outlined above. In these instances it appears that relatively warm, high-level air has been forced down to the ground by turbulent waves or eddies in the atmosphere to the lee of the mountains (p. 349). The comparatively tropical air becomes exceptionally warm by the time it has reached the foot of the leeward slope. This lee-wave subsidence is an important factor in all strong föhns but, needless to say, is itself auxiliary to the trans-montane air-movement which remains the chief immediate cause of föhn-effects in mountain ranges.

The difference between mistral and föhn winds may now be apparent; as the former do not cross highlands, they do not condense on ascent and re-acquire extra heat on descent as dry air; they are relatively cold at origin and do not subside far enough to be warmed appreciably.

Shelter from Winds. It is apparent that relief features which lie across prevailing winds may have an appreciable effect on the climate of an area by sheltering it from strong and cold winds. The Carse of Gowrie, near Dundee, is sheltered on the north by the Sidlaws; Eastbourne is sheltered from strong west winds by the South Downs. A gap in a mountain range may affect the climate of the districts on its leeward side. The fogs and sea influence from the Pacific pour inland through the gap in the Coastal Range at San Francisco. Stockton, opposite the gap, has a July mean of $73°$ F., while places north and south are $10°$ F. to $13°$ F. hotter.

Influence of Topography on Precipitation

Mountains not only make their own climate but, owing to their influence on winds, are also climatic barriers. The precipitation of a mountain range depends mainly on the absolute moisture-content of the prevailing winds and the magnitude of the relief feature. Generally speaking, however, a sudden abrupt elevation near a seaboard, such as in Snowdonia or the Coastal Range of California, will experience a disproportionately heavy rainfall owing to the rapid ascent of the air currents. These general points may be noticed:

(a) High relief increases the rainfall, or at least the possibility of rainfall. Notice, for example, the 'altitude oases' caused by the Ahaggar and Tibesti Hills in the Sahara.

(b) On high mountains there is a zone of maximum rainfall above and below which the precipitation decreases. This zone is usually 1,000 feet to 2,000 feet higher in summer than in winter; its average height is about 6,500 feet in the Alps, 4,500 feet in the Western Ghats, and 5,500 feet in the Sierra Nevada of California (Fig. 13).

(c) Relief features, especially those which run at right angles to the prevailing wind, give rise to marked rain shadows. Many of the deep valleys of the interior Alps have small rainfalls. Most of the Upper Engadine experiences less than 30 inches a year, and the Upper Rhône valley, with about 24 inches, is the driest region in Switzerland. Here the Sion district in summer supplements its small rainfall by an ancient irrigation system of wooden conduits that lead from the melting snows down the mountain precipices to the vineyards and fields on the warm valley slopes below.

(d) A large proportion of the precipitation on the higher areas is in the form of snow, and the amount in intermontane valleys is probably increased by irregularities of the relief. The length of snow-cover increases rapidly with altitude; in the Alps, snow covers the ground for about six weeks a year at Altdorf (1,500 feet in the Reuss valley), for over six months in the Upper Engadine valley (5,500 feet), and for ten months on the Säntis (8,200 feet). The permanent snow-line lies at 9,000 feet in the Western Alps, and at 10,500 feet on the drier interior ranges. The extension of the snow-cover down the mountainside in autumn and its retreat in late spring is a notable control of life in mountainous districts. In many areas the seasonal migration of the snow-line is reflected in the seasonal movement of flocks and herds. During the retreat of the snow-cover in late spring the advantages of the *sonnenseite* become most obvious, slopes in the shade retaining their snow-cover while sunny areas several thousand feet higher will be clear. Near Innsbruck, the limit of snow-cover in March is nearly 2,400 feet on the *schattenseite* as against 3,200 feet on the *sonnenseite*, while in September the difference (9,200 feet against 10,700 feet) is even greater.

Influence of Topography on Vegetation and Human Life

The influence of mountainous climates upon vegetation and cultivation is mainly connected with

(*a*) the great variety of plant associations caused by the rapid decrease of temperature in a short vertical distance;

(*b*) the migrations of the snow-cover.

The temperature and vegetation zones of the Andes in the tropics are known as *tierra caliente* (up to 3,000 feet), *tierra templada* (3,000–6,000 feet), and *tierra fria* (6,000–10,000 feet), above which are the *punas* and the tundra-like *paramos* (10,000–16,000 feet) (Fig. 161).

In the Alps the limit of crops and pastures varies widely in the same area, and the vegetation of the *schattenseite* may be a fortnight behind that of the *sonnenseite*. Altitude has a marked retarding effect on the time of flowering and ripening. Cherries ripen in late July near Davos, or nearly two months later than in southern England, while the outdoor vine in Swiss valleys seldom matures before late September. The belt of cultivation and of predominantly broad-leaved trees grades at 4,000 to 5,000 feet into the coniferous forest belt, which at 6,000 to 7,000 feet is replaced by Alpine pastures. These in their turn are limited by the bare rock and ice-fields above the permanent snow-line.

Transhumance, or the seasonal migration of flocks and herds, is commonly practised in the mountains of southern Europe. In April when the snow leaves the lower areas, the Swiss mountain peasants prepare the ground for the crops; by May the cattle can be led to the Mayen or pastures in clearings of the forest on the higher slopes above the village. This, the real commencement of the farming year, is made the occasion of an annual festival. As the snows retreat, the cattle are driven higher up the slopes, until in July the Alpine pastures are reached. The end of August brings the return migration through the forest clearings, and by the end of October the herds are back again in the village where they may be confined to the chalets till the following spring.

In a few localities the direction of migration is reversed. In the Val d'Anniviers, the two villages of Chandolin and St. Luc are perched high above the valley floor, at about 5,000 feet altitude, where the sunshine is of long duration and of high intensity. From these villages the herds migrate in spring *down* to the May pastures of the lower slopes and ascend in late autumn to their winter quarters.

LIST OF BOOKS

Standard works are given at end of Chapter V. See also:

Hann, J.　*Handbook of Climatology*, Vol. I (trans. by R. de C. Ward), 1903
Früh, J.　*Geographie de la Suisse*, I, 1937
Peattie, R.　*Mountain Geography*, 1936
Garnett, A.　*Insolation and Relief*. Institute of British Geographers, 1937
Wallington, C. E. 'An introduction to lee waves', *Weather*, 1960, pp. 269–76; *Meteorology for Glider Pilots*, 1961

For inversion:

Geiger, R. *The Climate near the Ground*, 1950
Heywood, G. S. P.　'Katabatic Winds in a Valley,' *Quart. Journ. Roy. Met. Soc.*, LIX, No. 248, Jan. 1933
Cornford, C. E. 'Katabatic Winds and the Prevention of Frost Damage,' *Quart. Journ. Roy. Met. Soc.*, 1938, pp. 553–85
A detailed account of Swiss föhns exists in a monograph by Emil Walter, No. 140, *Neujahrsblatt* for 1938 of *Naturforschenden Gesellschaft* in Zurich

For rainfall:

M. de Carle S. Salter. 'The Relation of Rainfall to Configuration,' *British Rainfall*, 1918, pp. 40–56
Poulter, R. M. 'Configuration, Air mass and Rainfall,' *Quart. Journ. Roy. Met. Soc.*, 1936, pp. 49–79
Sawyer, J. S.　'The physical and dynamic problems of orographic rain', *Weather*, 1956, pp. 375–81

For full details of pastoral migrations:

Carrier, E. H.　*Water and Grass*, 1932
Blache, J.　*L'Homme et la Montagne*, 1933

Temperature Statistics

	Lat.	Alt. (ft.)	J.	F.	M.	A.	M.	J.	J.	A.	S.	O.	N.	D.	Year
A															
Bevers	47° N.	5,610	14	20	24	33	42	49	53	51	46	36	26	16	34
Säntis	47° N.	8,202	16	16	17	24	31	37	41	41	37	30	23	17	27
B															
Quito	0°	9,350	55	55	55	55	55	55	55	55	55	55	54	55	55
Darjeeling	27° N.	7,376	40	42	50	56	58	60	62	61	59	55	48	42	53

A illustrates contrast between concave and convex topography in mountainous districts
B illustrates normal decrease of temperature with increase of altitude.

CHAPTER EIGHT

THE CYCLONIC INFLUENCE

Cyclonic Belts of the World

THE formation of air eddies or of surface-winds that tend to circulate about a local-pressure centre is, for reasons stated on pages 25–26, almost prohibited in equatorial latitudes. Yet, at the tropics, the remarkable steadiness of the Trades over the sea may be occasionally upset by terrific storms of wind and rain, called hurricanes or typhoons. These tropical cyclones originate over the warmest parts of oceans where the Trades are displaced seasonally by the doldrum air-mass. They are commonest where that air-mass migrates 10° or even 25° poleward and are absent, for example, in the South Atlantic where the doldrum trough does not move south of the Equator. In such a deep, moist air-mass many factors may create a weak local low-pressure. When, as happens rarely, this coincides with upper-air conditions that favour divergence aloft, the low pressure deepens, accelerates the inflow of surface winds and develops into a large vortex, energised by its own intense rainfall.

On an average about fifty typhoons occur in a year throughout the tropics, and of these, twenty-two are to be expected in the China Seas, seven in the Bay of Bengal, and most of the remainder off the West Indies, Madagascar, Queensland, and western India. The storms revolve about a centre or 'eye,' some 20 miles or so wide, where the pressure is low and the air calm. The weather is much the same before and after the 'eye,' and the tempestuous winds around its edge decrease in violence outwards until within about 100 miles in any direction light winds are prevailing.

Polewards of the tropics, cyclonic activity becomes increasingly common until it reaches a maximum frequency in seaboard areas at about 60° N. and 60° S. In Mediterranean latitudes the winter weather is largely controlled by the westerly wind system which forms the world's chief carrier of cyclonic disturbances. Polewards of this zone, or of 40° N. and S., the westerly wind system prevails most of the year except where the continental mass of Asia introduces a monsoonal effect on its eastern seaboards in summer. The air eddies in these 'temperate' latitudes are usually accompanied by less intense weather phenomena than in tropical cyclones. They are a grouping of winds about a local low pressure rather than a circular motion about a definite centre. Consequently, they are better called 'depressions' than cyclones.

These depressions are extremely frequent; the westerly belts may be carrying in a single winter's day more cyclonic depressions or 'lows' than the tropics experience in a whole month. The climates of lands such as the British Isles, Norway, British Columbia, and New Zealand, are distinguished from all others by rapid changes of barometric pressure and of wind direction that accompany the passage of depressions, solitary or in families.

Depressions or 'Lows'

Air-masses. The position of the zones where depressions commonly form is determined largely by the same general climatological factors that control the main wind and pressure belts. To-day it is becoming increasingly important to realize that wind and pressure belts are not continuous zones encircling the globe but are broken into separate sections or cells. Over the land these cells are seasonal whereas over the oceans they are fairly permanent although they move or expand in accordance with the seasons.

As a rule the cells coincide with vast uniform areas of the earth's surface upon which the general surface uniformity allows a uniformity of air conditions to develop. Thus, areas such as the southern North Atlantic or the great Siberian plain will, when meteorological conditions are favourable, encourage the creation upon themselves of an air-mass, throughout which the horizontal distribution of temperature and humidity is almost uniform. When such an air-mass moves or expands it will tend to give to the areas newly covered by it much the same weather as was experienced at its source. As we have seen, the Indian sea-monsoon brings weather characteristic of a moist equatorial region.

Air-masses may be classified in many ways but the main distinctions are always between marine and continental; and between polar and tropical sources. Their general distribution in the northern hemisphere is shown roughly in Fig. 24 where the classification is:

A	Arctic		
Pm	Polar maritime	Pc	Polar continental
Tm	Tropical maritime	Tc	Tropical continental
E	Equatorial		
M	Monsoon (m in summer; c in winter)		

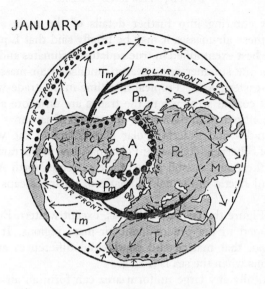

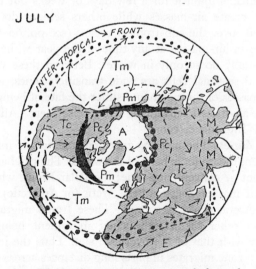

24. Main air-mass sources and frontal
zones of the northern hemisphere.

It should be noticed that some authors do not distinguish E and M from other
tropical air-masses. Obviously, it is hard to differentiate between M, Tm and E in
summer and between M and Pc in winter.

Without entering into further details of the main sources and nature of these air-masses it may be briefly said that Equatorial air-masses are best exemplified in true equatorial climates and in savanna climates in the summer-rainy season: monsoon air-masses in India and south-eastern Asia; Tropical maritime in the trade-wind oceans and in wet coastal areas where the trades are on-shore all the year; Tropical continental in the winter dry season of savannas and, all the year, in trade-wind deserts; Polar continental in the winter anti-cyclones of Siberia and Canada; Polar maritime in the raw air-masses of the northern and coastal parts of the extreme North Atlantic and Pacific; and Arctic in the cold climates of polar ice-caps and frozen polar seas.

In this Figure the transitional areas, except western Europe, have been included in the predominant air-mass regions. It should be noticed, too, that marine and continental differences are small in polar regions when the sea is frozen.

Theoretically any large uniform area can form an air-mass if the same air remains upon it for a few days or weeks but in fact some areas often create air-masses while others seldom do. The high-pressure cells over the sub-tropical oceans (see pp. 60–66 and Fig. 15) often form air-masses at any time of the year while continental interiors usually do so only in winter. Between these various 'permanent' and seasonal cells are wide transition zones, which rarely create air-masses of their own and suffer the fate of being the battle-ground of air-masses advancing from elsewhere. All peninsular Europe falls into this category.

Frontal Zones. The meeting zones of different air-masses are, as we have seen, the cyclonic belts of the world. The chief are the inter-tropical front or convergence zone; the Polar Front, where tropical and polar air-masses meet; and the Arctic (or Antarctic) front where polar and Arctic air-masses meet. These fronts migrate with the seasons and other factors, the chief movement being markedly sympathetic with that of the overhead sun. Thus the inter-tropical convergence zone migrates in parts long distances across the Equator into the opposing hemisphere. This zone is feeble as regards cyclo-genesis and usually gives rise to shallow cyclonic disturbances without well-marked 'fronts' in contrast to the many deep depressions that arise on the Polar Front (pp. 122–24).

Formation of Depressions. The origin of cyclonic depressions is generally explained by the 'frontal' theory of V. and J. Bjerknes

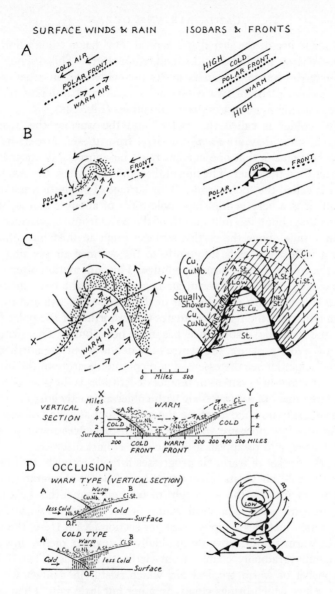

25. Life-cycle of a depression

A, B, C and D (right) are horizontal plans at ground-level, in which the warm front lies east and cold front west of the warm air (broken arrows). Rain-area is shown by dots and, in C (right), cloud-cover by initials and shading. In the vertical sections, rain is shown falling but winds in advance of the warm front could not be shown as they blow to northerly quarters. In D the two common types of occluded front are depicted.

who made use of the fact that a typical 'low' has a trough line with a rapid change in wind direction and temperature.[1] This discontinuity is where the warmer air-mass meets the colder, or, in other words, the junction plane of two winds or air-masses of different temperatures moving in almost opposite directions (Fig. 25A).

The colder is called the 'polar,' and the warmer the 'tropical' current, since they are *usually* derived from these sources; but the essential point is that one current must be appreciably warmer and moister than the other. The line of discontinuity where the two winds meet is called the 'polar front.' This surface of separation is not like a wall; it is a very gentle slope polewards of about 1 in 100, which means that about 100 miles north of the polar front the warmer air is about 1 mile above the earth's surface. Such an unstable condition over a wide zone leads frequently to the warmer air pushing itself into the colder air and forming a bulge. This indentation often grows in intensity, and as the warm air begins to rise the cold air on the north side is drawn into the rear of the bulge (Fig. 25B and C). The indraught from the rear accentuates the distortion of the polar front, and the cyclonic disturbance has now been born. The rising air ensures a low pressure, the centre of which is about the northern tip of the bulge. Hence the essential and common feature of the birth of local low-pressure systems in temperate latitudes is the gradual ascent of a large mass of saturated air with colder air partly surrounding it. The sinuosity now moves along the polar front as a kind of wave and carries the low pressure with it. Some disturbances never pass this initial stage, and disappear after moving a short distance.

As the pocket of warm air progresses it tends to push itself farther over the cold air on its advancing edge while, at the same time, its rear is cut into by the indraught of cold air that forms behind the low-pressure wave. The ideal young depression is shown in Figure 25C, which includes a typical weather section (XY). Here, the warm air, or warm front, slopes up gradually in advance of the 'low', and the clouds gradually change from cirrus to nimbus. Then follows the pocket of warm tropical air, which is still in contact with the surface and which brings cloudy weather but little rain. Then comes the zone, called the cold front, where the polar air is being drawn

---

[1] It should, however, be noted that although most depressions are frontal in nature and origin, some 'lows' are non-frontal. These arise mainly from strong thermal convection over continents in summer and from eddies in the lee of great mountain ranges lying athwart the paths of moving air-masses, as on the Mediterranean side of the Alps and on the eastern foothills of the Rockies.

wedgewise under the tropical air. In this narrow but fairly steep zone of separation the upward movement of the warm air is sometimes quite violent, and may be accompanied with sudden, heavy 'clearing showers' as distinct from the steady rain at the forward edge of the depression. The cold front may also be marked by squalls and by a line of discontinuity, but in any case the weather rapidly clears after its passage. As long as the warm air remains in contact with the earth's surface the depression retains the vigour and rapidity of movement that is characteristic of young cyclonic disturbances.

Older Depressions. With the progress of the low-pressure wave along the polar front the currents of cold air on each side of the warm

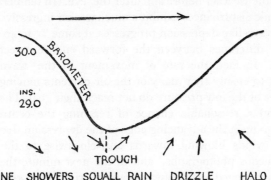

FINE   SHOWERS SQUALL RAIN   DRIZZLE   HALO
26. Weather sequence of an occluded depression
(*Modified from Abercromby and Goldie*)

sector gradually work closer together. Eventually the warm air is surrounded, and, helped by its own tendency to rise, is completely shut off from the earth's surface. The depression is now said to be 'occluded.' It still brings the same weather effects but with diminishing intensity. There is still the increasing cloud and rain of the advancing edge, and the sharp shower when the cold front passes, but as the tropical air is off the ground the intervening period of warm, damp, cloudy weather is not experienced. This is the weather scheme shown in the plan and the weather sections of Figures 25D and 26. It applies particularly to 'lows' that pass near the British Isles, as these disturbances are formed far out over the Atlantic and have matured before they reach the land. In these cases the isobars are often oval-shaped, because the line of discontinuity has almost or quite disappeared.

It will now be realized that a cyclonic disturbance has two major types of movement:

  (i) Ingressive (or convergent), which seems to give a grouping of wind, cloud, and rain around a low pressure;

 (ii) Progressive, which seems particularly associated with the more sudden weather changes that occur at the passage of the fronts.

In a tropical cyclone the ingressive rotational factors are well marked, the winds being terrific, but the rate of progress is only about 2 to 10 m.p.h., and consequently there is little difference between the weather before and after the 'eye.' In temperate latitudes the cyclonic disturbances exhibit a much weaker ingressive movement, but the wave-like depression progresses at some 15 to 40 m.p.h., and here the difference between the forward and rear sectors is considerable. In fact, the rate of movement of the 'wave' is usually sufficient to ensure that many of the air currents moving in towards the centre of the low pressure do not reach their goal. The only winds that stand a reasonable chance of reaching the centre are those blowing to meet the advancing edge of the depression. Isobaric charts showing winds blowing inwards anti-clockwise to the centre are 'instantaneous photographs,' and in the next minute the depression would have moved farther away from the winds in its rear. The mean rate of movement of depressions over the British Isles varies from 18 m.p.h. in May to about 27 m.p.h. in December, and consequently the progress of a 'low' is usually faster than the speed of the winds blowing towards its centre.

Anticyclones or 'Highs'

Characteristic Weather. In many ways anticyclones or 'highs' are the reverse of 'lows,' as in them the isobars are widely spaced and the system generally extends over a large area. The high-pressure system is sluggish in movement; it may remain stationary for long periods or merely expand or contract slowly. The clockwise movement of the surface winds is always light, and it has been proved mathematically that strong winds could not form in a clockwise circulation. The weather associated with anticyclonic conditions may be of any type, excepting violent atmospheric changes (Fig. 27). In summer most 'highs' bring brilliant sunny weather, and in winter, days that are clear, crisp, and invigorating. Yet at other times in winter a uniform sheet of cloud may cover the whole sky, and drizzling overcast weather may persist for long periods. The difference between summer and winter conditions is shown in the diagram. Anticyclones, owing to their calm nature, form 'radiation weather,' when phenomena such as land and sea breezes, mountain and valley winds, fog, and inversion of temperature can be most easily developed locally.

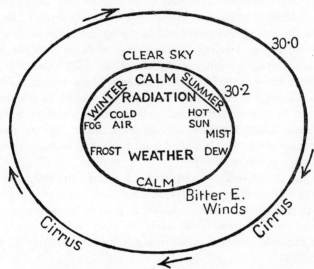

27. Weather associated with an anticyclone or 'high'
(*After Abercromby and Goldie*)
On left, winter characteristics; on right, summer characteristics

Formation. The origin of anticyclones poses different problems from that of cyclonic depressions. In the former the air-mass as a whole is subsiding; it dissipates on the surface and not in the upper air; it is expansive rather than progressive; dispersive rather than convergent; inert rather than energetic. In brief, the 'low' is an upward-moving pocket of air while the 'high' is an air-mass slowly subsiding to ground-level.

The world's great anticyclonic air-masses arise from two main factors: the subsidence of air from high levels in the atmosphere; and the chilling of air resting upon cold land-surfaces. The second main cause is illustrated by the great winter anticyclone of Siberia where snow- and frost-bound soils chill the lower air which gradually subsides upon it. The air-mass may not much exceed 7,000 feet in thickness yet it usually ensures intense cold, scanty precipitation, clear skies and calm weather.

Anticyclonic air-masses due to subsidence of air from high levels in the troposphere above an area form mainly in sub-tropical latitudes, as near the Azores. These air-masses on expanding polewards often bring warm, dry sunny weather.

It will be obvious, however, that all air-masses will begin to change once they leave their source. The change is usually slow and depends much on the rate of movement and on the nature of the surface being traversed. Hence, locally, although the weather brought by an anticyclone depends mainly on its source not a little depends on its subsequent movements and modifications. As a rule a cold anti-cyclonic air-mass moving into warmer regions tends to become less stable and to lose its strong anticyclonic characteristics; whereas a warm anticyclonic air-mass moving into colder regions tends to become more stable as it is chilled from below and so does not readily lose its high-pressure qualities.

Over the British Isles an anticyclone is usually the extension of, or a projection from a larger anticyclonic air-mass. Such a high-pressure system persists longest when it is fed, as it were, by its major source. Thus during the great frosts early in 1947 a frigid 'high' from Eurasia expanded and remained over much of the country for nearly two months.

Intermediate Types of Pressure Systems

The cyclonic eddies and calm anticyclonic backwaters of the atmosphere are the two main air disturbances that can be distinguished by peculiar isobaric shapes. But during the continual passage of these systems certain intermediary isobaric shapes have become associated with special weather conditions. The most important of these intermediate types are the secondary depression, the V-shaped depression, the wedge of high pressure, and the col.

Secondary Depression. Within a large 'low' and around its edges there frequently develop one or several small eddies which are known as secondary depressions. These may be shown by a slight curve of the isobars or they may be completely enclosed, so forming a miniature depression (Fig. 28). The weather they bring is typically cyclonic, except that it tends to be more consistently unsettled and rainy than in a bigger depression. Secondaries form the most frequent of all pressure types in the British Isles, and it is seldom that any primary depression exists long without a smaller eddy beginning to move counter-clockwise round it.

V-shaped Depression. A V-shaped depression is usually associated with the warm front or cold front of a 'low' (Fig. 29a). If associated with the former, the weather will be wet followed by mild

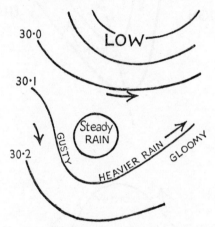

28. Weather associated with a secondary
depression

cloudy conditions; if associated with the cold front, as is most usual in our latitudes, the rain ends with a clearing shower and brighter, cooler weather ensues. The front or the trough, where the weather conditions change, is often associated with squalls or sudden alterations in wind direction and storms, hence it was often known as the squall-line. It is shown in Figure 29a.

Wedge of High-Pressure. A wedge, which projects between two depressions, is generally recognized as being one of the most consistently fine of all pressure types. It is a transient, rapidly moving formation, that at its worst can only give showers in its rear. The characteristic weather is explained in the diagram (Fig. 29b).

Col. It occasionally happens that 'lows' and anticyclones are separated from each other by a rather indeterminate belt of pressure. This isobaric col is a variable, and as would be expected, a transitory feature. Its winds are usually light and favourable to 'radiation weather' in winter and to thunderstorms in summer.

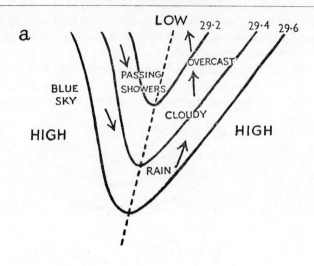

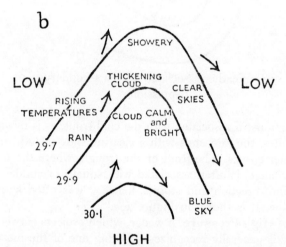

29. Weather associated with (*a*) a V-shaped depression,
(*b*) a wedge

(*After Abercromby and Goldie*)

The Polar Front

The polar front, or zone where the westerlies and polar winds converge, is very unstable especially in the northern hemisphere. The two Arctic areas from which outbursts of polar air are most common, are:

(i) the north of Alaska, whence cold waves spread south and south-eastwards across North America;

(ii) a less important area between Spitzbergen and Novaya Zemlya.

The temperature differences between the poles and tropics are greatest in winter, and then the polar front is farthest south. The Norwegian school of meteorologists place the mean position of the polar front in winter along a line running roughly from Brittany to Florida (Fig. 30). Hence, according to this view, the British Isles lie

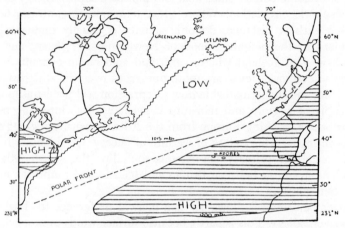

30. Mean position of the polar front in winter

Serrated line shows approximate boundary of relatively warm sea (*see* Fig. 37).
Greenland is usually covered with a local high-pressure

well within the polar front in winter. At this season, too, the continental high pressures may almost be regarded as part of the polar system. The polar front is a vast zone where the air from the polar circles and westerly belt intermingles. Here, the warm westerly winds interrupt the surface current of cold air that normally would flow from the poles towards the low pressure at the Equator. The air eddies that ensue allow vast quantities of tropical air to invade the atmosphere of polar regions and of probably greater amounts of polar air to escape into tropical latitudes. Thus the cyclonic disturbances of the polar front are an integral part of the world wind system, and provide for the interchange of surface temperature and air between the poles and tropics on a scale necessary to carry on the general

atmospheric circulation. The westerly wind belt much resembles a
vast ocean drift that as a whole progresses eastwards, but in which, at
any point, the turbulent water may be moving horizontally or
vertically and escaping into the ocean through which it flows.

Cyclonic Paths

The general movement of depressions is largely determined by the
main air stream, but the path locally is greatly influenced by the
nature of the earth's surface.

Warm sea surfaces especially favour the development and deepening of cyclonic eddies, since heat and moisture supply energy to any
atmospheric circulation. Where great temperature contrasts occur
within a short distance, cyclonic activity is usually intense. This
happens in the neighbourhood of Newfoundland, Japan, and the
south-west of Iceland where warm and cold ocean currents meet.
Even in summer, 'lows' are more numerous over the sea than over
the land, and it is remarkable how a depression will persist as long as
a supply of warmth and moisture is available. In winter the many
depressions reaching Europe from the Atlantic usually die out over
the continent owing to lack of warmth or energy. On the other hand,
it is usual for primary depressions to follow the North Atlantic
Drift in its course between Britain and Iceland and round the north
of Scandinavia.

Over land masses, 'lows' show a marked preference for lowlands
and river valleys. This is partly because surface winds cannot achieve
a circulation in mountainous regions. The cyclonic disturbances of
North America show a pronounced tendency to leave the continent
by way of the St. Lawrence lowland. The tendency of depressions
to be fended off by the Greenland ice-cap has already been mentioned. Many of the main cyclonic tracks of western Europe show a
marked avoidance of the highland areas (Fig. 31).

Weather in Cyclonic Belts

The weather conditions of cyclonic belts consists of a continuous
succession of local pressure systems which give a bewildering variety
of weather according to their size, intensity, and path. The character
of the winds experienced in any locality depends mainly on the source
of the air-mass which they are bringing, and is usually almost
independent of their direction locally.

In the British Isles, winds that originated over the continent are cold in winter, hot in summer, and relatively dry at all seasons, no matter whether locally they blow from the east or have been deflected into southerly winds. On the western seaboards of Britain the climate is remarkably equable and wet as the percentage of maritime winds is high, but the oceanic influence diminishes rapidly inland as the winds lose their winter warmth and moisture. Thus even a constant procession of cyclonic disturbances cannot prevent signs of continentality in the rainfall of Norfolk or in the temperature of London.

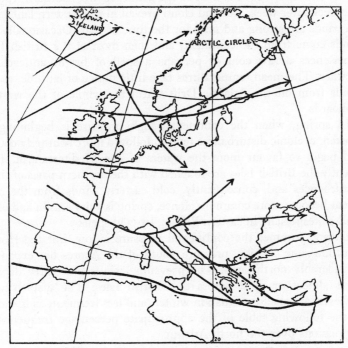

31. Main cyclonic tracks over western Europe

The climate of the British Isles illustrates the importance of the nature of variable air-streams and pressures. Three major pressure systems affect the weather of these islands:

(i) the almost permanent sub-tropical high south of the Azores;
(ii) the Eurasian continental pressure which changes with the seasons;

(iii) The 'low' just south of Iceland from which a constant procession of cyclonic disturbances passes eastward.

The interactions of these three variables give a highly erratic weather distribution, but certain useful generalizations may be made.

In winter, when the continental pressure over Eurasia is high, the most frequented track of depressions is south of Iceland. These major cyclonic storms tend to follow a course just off the western coast of Britain, which is then swept with their wet southern fringes. For week after week, wet and cloud prevail in the western lands and the winds are strong and mild for the time of year. Occasionally the winds come from the continent, and then most of the British Isles experiences a raw cold or perhaps a spell of bright anticyclonic weather. The mean temperatures take little notice of latitude, as the winds from off the Atlantic Drift warm the whole of the western seaboards.

In spring, when the continental high pressure is beginning to weaken, cyclonic disturbances tend to follow a more southerly course and begin to favour more the waters of the Mediterranean. Frequently the British Isles are touched with the northern parts of these disturbances and, consequently, cold easterly winds from the continent become more common. Hence, spring is a dry season and often marked, especially in eastern Britain, by cold spells.

During summer the continental pressure is low, but the Horse-Latitude zone of high pressure south of the Azores has expanded considerably northwards. The prevailing winds now carry depressions almost or quite across the British Isles. The disturbances, however, are shallower than in winter, and less frequent, as is shown in the following table of the approximate percentage frequency of depressions over the north Atlantic.

Winter (Dec.–Feb.)	Spring	Summer	Autumn
54 per cent.	18 per cent.	7 per cent.	21 per cent.

Summer gradually fades into autumn, which is usually wet, as winds from a warm sea are blowing on to a relatively cold land. About November the cyclonic control again becomes strong.

The mean meteorological elements that comprise our climate are shown in any atlas, but the real climate is depicted in the Daily Weather Charts. Only in the other seaboard areas of the westerly belt does the weather show such a tendency to revert suddenly to that of previous seasons, and only in the equatorial belt may so much cloudiness and so many rainy days be expected.[1] The 'western European' type of climate, as it is often called, is indeed the perfect combination of oceanic and cyclonic influence.

The Human Response

The regions of greatest cyclonic activity are usually covered in their warmer, wetter parts by broad-leaved deciduous forests, while the colder, drier, and more infertile areas bear a natural growth of coniferous trees. The more exposed situations may support only moorland and heather associations. The region is exceptionally favourable to the cultivation of grass, but the variations in vegetation brought about by climatic and human factors are so infinite that generalizations become merely misleading. The human importance of the zone to the world lies in manufacturing and not in the products of the soil. The rapid climatic changes are said to be stimulating to the mind and body. Yet the inhabitants' liability to lung complaints and germ-borne diseases may be traceable to the strain placed on the human system by too sudden changes, especially in winter and early spring. It is, however, probably more than a coincidence that the world's greatest manufacturing belts—north-eastern U.S.A., Europe, and central Japan—lie in the regions of greatest cyclonic activity. Some writers carry the coincidence further, and point out that during the rise of the early civilizations in Africa and Asia the polar front was probably considerably farther south than at present.

LIST OF BOOKS

Hare, F. K. *The Restless Atmosphere*, 1956
Willett, H. C. and Sanders, F. *Descriptive Meteorology*, 1959
Abercromby, R., and Goldie, A. H. R. *Weather*, 1934
Handbook of Aviation Meteorology, 1960 (H.M.S.O.)
Brunt, D. *Weather Study*, 1941
Petterssen, S. *Introduction to Meteorology*, 1958. *Weather Analysis and Forecasting*, 2 vols., 1956
Wexler, H. 'Anticyclones,' in *Compendium of Meteorology*, 1951, pp. 621-9
Miller, A. A. and Parry, M. *Everyday Meteorology*, 1958
Sawyer, J. S. *The Ways of the Weather*, 1957

British Isles
Climatic data. *The Books of Normals*. Met. Off. of Air Ministry. (H.M.S.O.)

[1] Average number of rain-days in the British Isles just exceeds 200. The number ranges from about 150 in the drier east to over 250 in the extreme west. Equatorial stations are much the same.

Best accounts of climate are
Bilham, E. G.   *The Climate of the British Isles*, 1938
Manley, G. M.   *Climate and the British Scene*, 1952
Climatological Atlas of British Isles, 1952. (H.M.S.O.)
Lamb, H. H. 'Types and spells of weather . . . in the British Isles,' *Quart. Journ. Roy. Met. Soc.*, 1950, pp. 393-438; 'British weather around the year,' *Weather*, 1953, pp. 131-6; 176-82; *The English Climate*, 1964
Taylor, J. A. and Yates, R. A.   *British Weather in Maps*, 1958
Thomas, T. M. 'Precipitation within the British Isles in relation to depression tracks', *Weather*, 1960, pp. 361-73

For photographic illustrations:
Gayle Pickwell. *Weather*, 1938
Grant, H. D. *Cloud and Weather Atlas*, 1944
Ludlam, F. H. and Scorer, R. S.   *Cloud Study: A Pictorial Guide*, 1957

For upper Westerlies and Jet Stream see NOTE below and:
Hare, F. K. 'The Westerlies', *Geog. Rev.*, 1960, pp. 345-67
Lamb, H. H. 'The Southern Westerlies', *Q.J.R. Met. Soc.*, 1959, pp. 1-23
Riehl, H. *Jet Streams of the Atmosphere*, 1962
Sawyer, J. S. 'The Jet Stream', *New Scientist*, 1959, pp. 947-49
Reiter, E. R. *Jet-stream Meteorology*, 1963 (full bibliography)

Note. A circumpolar zone of westerly airflow prevails in each hemisphere in middle latitudes (p. 66). These westerlies, with their variable surface direction locally and their numerous frontal depressions differ markedly from the shallow, steady Trades. The latter are often under 10,000 feet thick and contain a layer of subsiding air (upper inversion layer) which largely prevents convectional ascent through it except where it is pierced by high mountains and where the airflow thickens to 18,000 or 20,000 feet towards the equator. Hence weather in the Trades is usually dry and reliable. Observations made on mountains over a century ago showed that the Trades were overlain by winds blowing in an opposite direction, the anti-Trades.

Unlike the Trades, the middle latitude westerlies occupy all the troposphere (5 to 7 miles thick) and usually increase in velocity and westerly direction with height. Thus at 16,000 to 18,000 feet and beyond, the surface westerlies may be said to merge into the *upper westerlies* and these extend in the upper troposphere (partly as anti-Trades) often to within 10° or 15° of the equator.

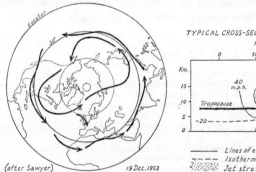

31a. Jet streams of the northern hemisphere

Thick solid line is tropopause. Thick dotted lines mark zone of relatively rapid air temperature change, or frontal surface, in the troposphere. Cross-section extends from north to south across about 20° of latitude. The vertical scale is greatly exaggerated and jet stream is ribbon-shaped air-flow, moving west-east at right-angles to the cross-section.

Jet Streams. The upper westerlies in parts acquire great velocity and become jet streams, or sinuous air currents usually thousands of miles long, a few hundred miles wide and a few miles thick, with speeds of over 30 metres per second or 67 miles per hour. Such jet streams occur most persistently over the subtropics (about 25°–35°) at 30,000 to 40,000 feet where they form part of the general circulation (pp. 61–66) and affect subtropical weather (p. 77). Another zone of frequent occurrence is at 16,000 to 30,000 feet above surface westerlies especially near the polar front. A typical horizontal distribution of jet streams at over 16,000 feet altitude in the northern hemisphere is exemplified by conditions for 19 December 1953 shown in the left-hand diagram above (Fig. 31a).

The common position of the subtropical jet stream migrates latitudinally mainly in conjunction with the seasonal migrations of the main pressure belts or thermal equator. The usual zonal location of the less persistent circumpolar jet streams undergoes smaller seasonal migrations but greater irregular movements, both horizontal and vertical. The vertical oscillations have a wavelength of several thousand miles and where the jet stream locally shows a downward tendency there seems a tendency for anticyclones to form at the surface. In such areas 'blocking anticyclones' sometimes persist for longer than expected and fend off the normal rapid succession of depressions.

The horizontal oscillations, coupled with changes in velocity, of the circumpolar jet streams may also affect cyclogenesis at the surface, especially as jet streams often occur above or near major fronts (Fig. 31a). A jet stream condition favouring divergence aloft would explain the removal of the vast masses of rising air that allows the depression to continue to deepen. Similarly a jet stream causing convergence aloft would tend to depress ascent and to discourage cyclogenesis.

The most direct human effect of jet streams is on air travel. As they often exceed 100 m.p.h. and sometimes reach 200 to 300 m.p.h., when adverse they seriously increase fuel consumption and decrease the speeds of aircraft. Hence the desirability of flying in the stratosphere.

I

Section Three

THE OCEANS

CHAPTER NINE

GENERAL CHARACTERISTICS AND CIRCULATION OF OCEAN WATERS

General Characteristics

The Salinity of Sea Water. The origin of the saltness of the sea remains a cause for speculation. The quantities of mineral salts present are enormous, and if they were withdrawn, sea-level would fall by about 100 feet. Their composition shows relatively little variation, and, on an average, about 35 grams of mineral matter in solution are to be found in every 1,000 grams of sea water. This is composed mainly as follows:

(Dittmar in *Challenger Reports*)

	Gms. per 1000	Percentage of total mineral matter
Sodium chloride	27·21	77·8
Magnesium chloride . . .	3·81	10·9
Magnesium sulphate	1·66	4·7
Calcium sulphate	1·26	3·6
Potassium sulphate	·86	2·5
Calcium carbonate	·12	·3
Magnesium bromide	·08	·2
	35·00	100

In addition, minute traces of numerous other chemical elements have been detected in sea water.

The vast amount of common salt (sodium chloride) may have been derived by solution from rocks in contact with the sea, but the major portion of it was probably derived from river water. Analyses, however, reveal that the matter in solution in river water normally consists largely of lime (calcium carbonate), and that the proportion of common salt is small. Most rocks do not contain common salt, although some, such as granite, contain small amounts of sodium, which is an element of salt. It happens that when chlorine, which not infrequently occurs in trivial quantities in ground water, unites with sodium, common salt is produced. This process, however, calls for the destruction of tremendous bodies of rock to yield a little salt.

The answer to the problem may lie in the time involved, which must necessarily be enormous. Perhaps, however, in the present incomplete state of the knowledge of the world's rivers, the average salt content of river water may be higher than is supposed.

The large amount of calcium carbonate found dissolved in river water contrasts markedly with the low proportion in sea water. A possible explanation may be that many marine organisms extract lime to form shells or skeletons for themselves. It seems, too, that some marine animals can transform calcium sulphate into calcium carbonate, in which form they use it for their shells. The large quantities of lime extracted are, on the death of the organisms, deposited on the sea-bed at all depths less than about 18,000 feet.

In addition to the extraction by living things, large quantities of lime are precipitated by chemical action. These factors keep the amount of calcium carbonate in sea water at a minute proportion,

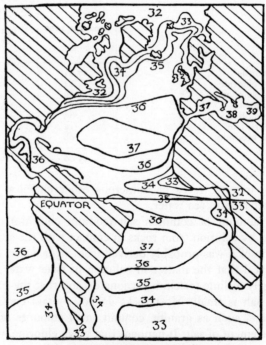

32. General distribution of salinity of the
surface of the Atlantic Ocean
(*After Schott*)

whereas salt, which is not used or extracted by any marine life, has gone on accumulating slowly throughout the existence of the oceans.

Variations in Salinity at Surface of Oceans. Generally speaking, the salinity of the surface waters of the oceans varies from about 37 per 1,000 parts at the tropics to 32 per 1,000 parts in the polar areas (Fig. 32). Salinity at the surface of any locality mainly depends on:

the amount of evaporation,
the amount of rainfall,
the influx of fresh river water, and
the presence, if any, of ocean currents.

The intense evaporation and small precipitation of Trade Wind areas bring about the highest degree of saltness known in the open oceans. Salinity is also especially high in partly or completely enclosed seas that lie within hot latitudes and have little rainfall or river water coming into them. The salinity of the Mediterranean increases from 36·5 per 1,000 parts near the Strait of Gibraltar to over 39 near the coast of Asia Minor. That of the Red Sea rises to over 41 per 1,000 parts in the Gulf of Suez, while that of the Dead Sea, the most salt of all the larger lakes, is about 237·5.

In contrast, low evaporation and a great influx of river water and ice give the Arctic Ocean a low salinity. Similarly the surface waters of the Gulf of Guinea are diluted by the great volume of fresh water from the Congo and Niger. Salinity reaches its minimum in almost enclosed seas in cold latitudes where the rainfall is considerable and the influx of river water great. In the Baltic, the saltness diminishes inland from about 8 per 1,000 parts near the island of Rügen to 2 at the heads of the gulfs of Bothnia and Finland, where the almost fresh water very appreciably encourages the early formation of ice.

The saltness of the surface water of oceans and seas is subject to continual alteration largely based on climatic factors. These inequalities of salinity usually cause the more saline, and consequently the denser, water to displace the lighter, and so there is formed a constant motion of sea water, which is partly vertical and partly horizontal. Movements of this nature are so slow as to be imperceptible.

Temperature of Oceans and Seas

Surface Temperatures. The surface temperatures of sea water range from about 80° F. to 82° F. in the tropics to 28° F. in the polar seas. The decrease, however, is irregular owing to the effect of ocean currents, of river water, and of the extent to which any part of an ocean is cut off from the general circulation. The tempering effect of the cold current off Peru and of the warm North Atlantic Drift off western Europe may be judged from the course of the isotherms over these oceans.

Temperatures beneath the Surface. Normally, some 90 per cent. of the heat that penetrates the surface of the sea is absorbed in traversing the upper 60 feet of water. The sun's rays on an average probably have no direct effect below 600 feet, and so, in spite of movement of the water, the vast bulk of the sea is relatively cold. Probably some 80 per cent. of water in the oceans has a temperature permanently below 40° F. The following table is a rough approximation to the mean conditions.

Depth.	Temperature.
600 ft.	60·7° F.
1,200 ft.	50·0° F.
3,000 ft.	40·1° F.
6,000 ft.	36·5° F.
13,200 ft.	35·2° F.

Even in the tropics the temperature rarely exceeds 40° F. at 5,000 feet depth, and here, as elsewhere, the temperature below 14,000 feet decreases slowly downwards from 35° F. to about 32° F. The dense icy water that creeps along the ocean floor from polar regions towards the Equator helps to keep the bottom layers of the oceans at this low temperature. Wherever a sill or submarine ridge tends to hold back the cold creep from the poles, the deepest waters of the area thus protected remain noticeably warmer than those on the

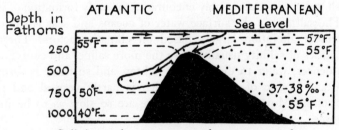

33. Salinity and temperature of sea water at the
Strait of Gibraltar

Dotted area has salinity exceeding 36·5 per thousand

poleward side of the ridge. An example occurs in the submarine ridge that connects the Shetlands with Iceland; here the level of the sill holds back the icy bottom waters on the north, but at the same time allows the passage northwards of the warm surface waters from the south. Similar cases of the isolating effect of submarine ridges occur at the mouth of the Red Sea, where the temperatures inside the sill remain at 70° F. or more, while those of the Indian Ocean decrease with depth to under 35° F., and at the Strait of Gibraltar where the waters of the Mediterranean do not cool below 55° F. (Fig. 33).

The effect of currents and of submarine topography causes the decrease of temperature with depth to vary considerably, and especially in the upper 600 feet. The distribution of temperatures in the sea somewhat resembles that in the atmosphere; the decrease of warmth away from the surface, the irregularities due to relief, and the imported temperatures of currents are all to be found. The ocean, too, has its 'troposphere,' 'tropopause,' and 'stratosphere.'

Movements of Sea Water

The movements of sea water may be grouped into—

(a) waves, tides, and tidal currents, which are extremely superficial;

(b) ocean currents and drifts, which are practically limited to the upper 100 fathoms;

(c) creeps, which may flow in waters as deep as the ocean floor.

The following account deals only with the movements of currents, drifts, and creeps, all of which affect the temperature of oceans and seas.

In the first instance the movement of sea water is based on variations in density caused by differences in salinity and temperature. The two factors work together. Warm currents, in spite of a high salinity, may be kept for long distances at the surface by their warmth. Thus the waters of the North Atlantic Drift keep on the surface right into the Arctic Ocean; only after crossing the submarine ridge that stretches from Iceland to Scotland does the Drift cool sufficiently to sink beneath the cold fresher surface waters to form a warm under-current. Consequently, in these parts of the Arctic Ocean there is a warmer layer at about 130 to 550 fathoms, above and below which cold conditions prevail.

On the other hand, polar currents may, in spite of their cold, be retained for long distances at the surface by their low salinity. In moving equatorwards the current may increase slightly in warmth, but it will finally sink beneath the surface as a relatively cold under-current. In this way, the cold water of the Labrador Current moves on the surface far down the coast of the eastern U.S.A. before sinking beneath the warm waters of the Gulf Stream.

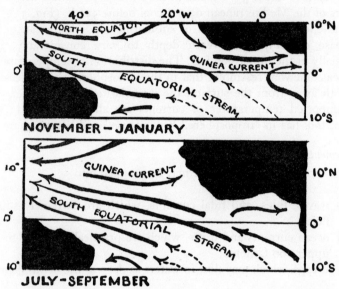

34. Seasonal swing of equatorial currents in the Atlantic

A marked transference of waters due to differences in density occurs at the Strait of Gibraltar, where water of relatively low salinity passes at the surface into the Mediterranean and a denser under-current creeps into the Atlantic (Fig. 33). The Baltic Sea offers a contrast; through the Skagerrak relatively fresh water flows outwards at the surface, while an undercurrent of denser water flows inwards from the North Sea.

In the open oceans, the constantly varying density of sea water ensures a continual but very slow circulation, the main feature of which should be a warm surface movement from the Equator to the poles, and a cold bottom creep from the poles to the Equator. This scheme is, however, completely upset on the surface by the action of winds which convert the poleward drift of waters into a circulation comparable with that of the atmosphere. Winds are the main directing agents of surface currents, and in most cases they hasten and distinguish the ocean movement, but they are by no means always the primary causal factor.

The connection between winds and ocean currents may be judged from the similarity of oceanic and atmospheric surface circulations, but the relationship is especially emphasized by the fact that—

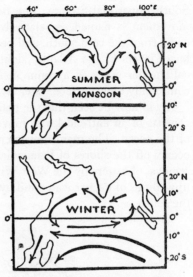

(a) the equatorial ocean currents swing north and south according to the seasonal position of the Trade Wind belts (Fig. 34),
(b) the currents of the Indian Ocean are reversed with the change over of the monsoon (Fig. 35).

35. Seasonal change in direction of currents in the Indian Ocean

The other chief factors that help to determine the general circulation of the oceans are the rotation of the earth and the shape of the ocean basins.[1] The typical features of ocean movements are to be found in the Atlantic Ocean.

Atlantic Ocean. Under the continuous impulse of the Trades a warm current flows westwards on each side of the Equator (Fig. 36). This westward flow of equatorial waters forms the basis of the Atlantic circulation.

South Atlantic. The South Equatorial Stream impinges on the north-east coast of Brazil, and is divided by Cape São Roque into two main branches, one of which passes north-westwards to join the North Equatorial Stream, and the other moves south-westwards. The latter, called the Brazil Stream, proceeds to between latitudes 20° S. and 40° S., and is gradually deflected eastwards by the earth's

[1] There is also the compensatory attraction to areas of continually outflowing currents, which gives rise to "counter-currents."

rotation. It then begins to mingle with the surface drift formed by the strong west winds or Roaring Forties of the southern ocean. The west wind drift is gradually deflected to the left, and a branch of it flows northwards along the west coast of South Africa to join the South Equatorial Stream. The Benguela Current, as the northward movement is named, is relatively cold for its latitudes since it has come from the cold south, and because cold water wells up from the lower layers of the Atlantic along this coastline. This upwelling of cold under-water is a common feature off coastlines where winds continually blow surface water away from the shore.

It will now be seen that the main circulation of the South Atlantic consists of an anti-clockwise movement of waters around the centre of the ocean. The main movement, contrary to the general scheme, occurs off the shores of Patagonia and Argentina. Here the eastward removal of waters by the West Wind Drift is compensated by coastal upwellings of deeper water and by a surface influx of cold water from the south.

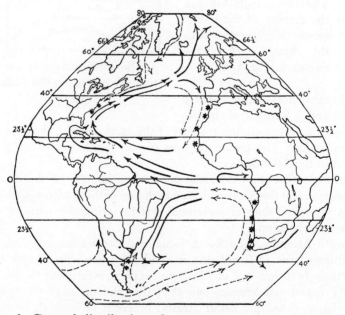

36. General distribution of ocean currents in the Atlantic

(After Schott)

Relatively cold currents are represented by broken arrows.
Asterisks denote upwelling of cold water

North Atlantic. The circulation of waters in the North Atlantic compares closely in general plan with that of the south. The waters move clockwise about a central area that coincides with a high-pressure or calm-weather cell of the atmosphere. But the different shape of the northern Basin introduces the following remarkable differences in detail:

(*a*) The much greater width of tropical ocean and the large Caribbean Sea lengthen the time during which the equatorial waters of the North Atlantic remain in hot latitudes. Moreover, the circular shape of the Gulf of Mexico, and the southward projection of Florida, which terminates in a strait that narrows to 50 miles and shallows to 60 fathoms, further conspire to retain and pile up the already warm water.

(*b*) The doldrum belt remains all the year north of the Equator in this part of the Atlantic. Hence there is a continual movement northwards of part of the South Equatorial Stream, and this makes the volume of warm water circulating in the North Atlantic exceptionally large.

(*c*) In contrast with the wide opening of the South Atlantic to the influence of the Southern Ocean, the northern entrance of the North Atlantic is relatively narrow and is partly blocked by islands. Consequently, the influence of warm currents moving polewards is not curtailed so abruptly as in the South Atlantic.

The circulation is roughly as follows. The North Equatorial Current, with its adjunct from the South, impinges on the Antilles, and part of the stream enters the Caribbean Sea. The remainder skirts the outer shores of the West Indies, and much of it gradually curves into the Sargasso Sea area. The branch that enters the Caribbean and Gulf of Mexico is subjected to high air temperatures, and its temperature increases to about 81° F. and its salinity to over 36·5 per 1,000 parts. The piling up and heating results in a warm current that issues[1] through the narrow strait of Florida and proceeds northwards as the Gulf Stream. About Cape Hatteras (or 35° N. to 40° N.) the Gulf Stream, now somewhat wider and deeper, passes out into the Atlantic (Fig. 37). The eastward tendency is due partly to the direction of the coastline, partly to deflection by the earth's

[1] At a maximum speed of about 5½ miles per hour. Beyond the strait this current is reinforced by warm water from east of the Antilles.

rotation, and partly to meeting cold water from the north. Soon after leaving the American coast the movement ceases to be a well-defined current and becomes a general eastward drift of relatively warm surface waters. The progress is now probably not more than 10 or 15 miles a day. Eventually, under rotational and coastline influences, one portion of the drift turns southwards, and assisted by the North-east Trades

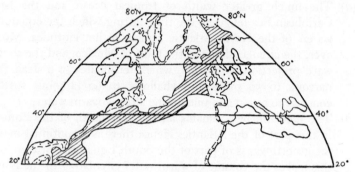

37. Temperature excess of Gulf Stream Drift at a depth of 600 feet
(*After Wüst*)
Shaded area is 4° F. warmer than average temperature of sea water in its latitude. Dots show continental shelf

completes as a relatively cold current the circulation of the North Atlantic.

The other branch of the eastward drift, under the influence of the prevailing south-westerly winds, continues in a north-easterly direction. The resultant North Atlantic Drift proceeds between the British Isles and Iceland, and sends minor branches into the shallow seas around Britain. About the latitude of North Cape the warm waters have cooled considerably, and they then sink and continue as a warm undercurrent which keeps the seas free of ice as far east as Murmansk.

In addition to the general clockwise scheme of the North Atlantic, there is the Labrador Current which brings cold water from near Baffin Bay and Davis Strait, and which receives additions from a cold drift down the eastern coast of Greenland. The icy water tends to hug the American coastline, and some of it proceeds as far as Cape Cod, beyond which it gradually disappears beneath the warmer Gulf Stream. This polar current carries much ice to the neighbourhood of

Newfoundland and beyond. Two other features are shown in the diagrams (Figs. 34 and 36).

(a) The centre of the North Atlantic circulation corresponds to an extensive area of ocean with little surface movement. Patches of the sea here are covered with floating weeds (Sargassum) which, however, are seldom dense, and cannot hinder the progress of ships.

(b) Between the North and South Equatorial Currents a counter-stream flows eastwards to fill up the lowering of surface level caused by the constant westward drift of waters.

Pacific Ocean. The circulation of the Pacific resembles that of the Atlantic, except that the shape of the coastline near the Equator does not bring about any marked transference of warm water from one hemisphere to the other. Consequently the Kuroshio, the warm current of the North Pacific, is considerably weaker than the Gulf Stream.

Influence of Ocean Currents on the Climate of Coastal Lands

Temperature. Generally speaking the circulation described above ensures that in low latitudes (0°–40°) there is warm water on the western side of an ocean and cold water along the eastern. In the former, prevailing winds tend to blow on-shore; in the latter, off-shore. The temperature contrast brought about by this difference is reflected in the following statistics for two coastal stations almost on the southern tropic:

	Jan. Mean (Monthly)	Yearly Mean
Rio de Janeiro	78° F.	73° F.
Walfish Bay (S.W. Africa)	66° F.	62° F.
Difference	12° F.	11° F.

Nor must it be imagined that only Rio has on-shore winds; at Walfish Bay winds blow frequently from off the cold Benguela Current, and travellers seldom fail to remark on the unpleasant chill of the sea breeze. Yet these bitter sea winds, like most others from off cold currents, seldom make themselves felt more than 50 miles inland, whereas the influence of on-shore winds from off warm drifts seems much more persistent.

Polewards of 40° latitude the east coast of an ocean becomes rapidly warmer than the west, largely because the former is washed by warm drifts and the latter by cold polar currents. Although the west coast of the Atlantic commonly experiences continental winds, it is obvious that the great cyclonic activity will ensure a fair proportion of winds from off the cold current. The following statistics illustrate in all cases the modifying influence of warm sea winds on the winter weather of western Europe:

		Lat.	Mean Jan.	Mean July	Annual
A	Washington . .	39° N.	33° F.	77° F.	55° F.
	Lisbon. . .	,,	49° F.	70° F.	59·5° F.
	New York . .	41° N.	30° F.	74° F.	52° F.
	Oporto . .	,,	47° F.	65° F.	58° F.
B	St. John's (N'f'dland)	47·5° N.	24° F.	59° F.	40° F.
	Brest . . .	,,	44° F.	64° F.	54° F.
	Nain (Labrador) .	56·5° N.	−7° F.	47° F.	23° F.
	Glasgow . .	,,	39° F.	58° F.	47° F.

In group A the winter temperature of the European coast is about 16° F. the greater. Yet, in summer, when the westerly oceanic drift is cooler than the land, sea winds lower the temperatures of Portugal, and the American seaboard is the warmer. In group B the warming effect in winter has increased to as much as 46° F., a remarkable difference not equalled elsewhere in the world. Yet, in this latitude, the summers of coastal north-western Europe appear to be warmer than those of the St. Lawrence mouth. The cooler summers of the American seaboard are due to the fact that the Labrador Current cools winds from the sea more efficiently than the warmer Atlantic Drift cools sea winds in north-western Europe.

Precipitation. Winds blowing from warm ocean currents usually give increased humidity and rainfall to the lands they cross, especially when the land is colder than the sea. The frequent rain of Ireland and the wet monsoon of India are examples.

Winds blowing from cold seas to warm lands are evaporators of moisture rather than rain-bringers. As already discussed, copious precipitation may form over the cold coastal current and upwellings off Peru, but the moisture is lost to the mainland.

The meeting of warm and cold currents greatly predisposes to fog as, for example, near Newfoundland. Warm winds from the Gulf Stream so readily form fog over the Labrador Current that when winds are southerly the International Ice Patrol cruises about in the Gulf Stream to avoid the fog. An instance of fog formation over a cold current occurs off the shores of central and northern California, where the magnificent coastal forests are nourished in summer by sea fog.

The Effects of Currents on Marine Life

All the elements needed for plant growth are present in sea water, but some of them occur in very minute quantities compared with the needs of the organic life. The problem of the density of life in the sea is largely bound up with the abundance and circulation of these 'limiting substances,' of which nitrates and phosphates are the chief. As a rule, nitrates and phosphates accumulate in winter and then are used up rapidly by sea life in the spring. With the practical exhaustion of this supply, plant life would grow less and much of the animal life dependent on plants would disappear. It happens, however, that when marine organisms die many of them sink and decompose on the floor of the ocean, thus giving back to the bottom layers some of the 'limiting substances.' Moreover, for chemical reasons, there tends to be more nitrate compounds in cold waters than in warm. Consequently, cold currents and upwellings of bottom water make extra supplies of nitrates available to life on the surface. In other words, they supply additional nutriment all the year and render possible a much greater density of organic life at all seasons. On the Pacific coast of North America the upwellings of cold water provide food for a continual growth of minute marine plants on which a wealth of small fishes and crustacea feed. These in their turn provide food for larger life, such as salmon. The waters of the cold current and upwellings off Peru are equally rich in plankton, and the birds that feed on the fish yield guano, to the great profit of the Peruvian treasury. It has long been recognized that polar seas are richer in organic life than inter-tropical waters, and that their abundance of bigger life, such as whales, seals, and penguins, reflects the abundance of minute organisms.

The great fisheries of the world—off the St. Lawrence mouth, north-western Europe, and Japan—are at the meeting and mixing place of major and minor currents on the continental shelf. Nutritive elements are brought by the currents, especially the cold, but in addition the constant overturning of the shallow waters by cooling, wave action, and tides makes the full resources of the whole mass available for plant growth. The physical conditions such as light and heat are also highly favourable to growth, and so there arises an exceptional wealth of organisms and the fisheries become of outstanding importance. Yet even in these fishing grounds the localities that yield most fish are usually those where the mixing and upwelling of waters are most pronounced.

LIST OF BOOKS

There are several excellent text-books on Oceanography:

King, C. A. M. *Oceanography for Geographers*, 1962
Ommaney, F. D.   *The Ocean* (Home University Library), 1961
Oceanography.   Bulletin of National Research Council of Washington D.C. No. 85, 1932. (By committee of experts. Rather difficult in parts.)
Vallaux, C.   *Géographie générale des Mers*, 1933. 776 pp. (With much on human geography.)
Dietrich, G. *General Oceanography*, 1963

The standard treatises are

Sverdrup, H. U., Johnson, M. W., and Fleming, R. H.   *The Oceans*, 1942
Defant, A. *Physical Oceanography*, 1961 (2 vols.)
Hill, M. N. (ed.) *The Sea*, 1962–3 (3 large vols.)

For reference purposes:

Challenger Reports, especially *Narrative*, Vol. I, and *Summary*, First Part. 1907
Discovery Reports.   1929, 20 vols., especially Vol. VIII, 1933 (for sections of South Atlantic) and Vol. XIII, 1936 (for Peru Coastal Current)
Sears, M. (ed.) *Progress in Oceanography*, 1963–4 (2 vols.)
Riley, J. P. & Skirrow, G. (eds.) *Chemical Oceanography*, 1965

Regional studies:

Schott, G.   *Geographie des Atlantischen Ozeans*, 1942
          Geographie des Indischen und Stillen Ozeans, 1935
Flint, J. M.   *Oceanography of the Pacific* (Smithsonian Institution, 1905)
Deacon, G. E. R.   'New Work on the Gulf Stream,' *G. J.*, August 1941. 'The Sargasso Sea,' *G. J.*, January 1942
Stommel, H.   *The Gulf Stream*, 1958

For influence of ocean currents on climate:

Sverdrup, H. U.   *Oceanography for Meteorologists*, 1942

For life in sea:

Russell, F. S. and Yonge, C. M.   *The Seas*, 1936
Chapter in *Oceanography*. Washington, 1932, and much in King, Ommaney, and Sverdrup and Johnson.
Carruthers, J. N.   'Some inter-relationships of Meteorology and Oceanography,' *Quart. Journ. Roy. Met. Soc.*, 1941, pp. 207–46
Armstrong, E. F. & Miall, L. M. *Raw Materials from the Sea*, 1946
Hardy, A. C.   *The Open Sea: The World of Plankton*, 1956; *Fish and Fisheries*, 1959
Le Danois, E.   *Marine Life of Coastal Waters: Western Europe*, 1957
Ovey, C. D.   'Some effects of weather and climate on surface life in the open oceans', *Weather*, 1959, pp. 339–44

CHAPTER TEN

Nature and Movement of Waves

Wave Propagation. The calming effect of pouring oil on troubled waters serves to remind us that ocean waves are caused by the friction of winds, the gustiness of which quickly deforms the water surface. The wave movement thus propagated travels along in the direction of the prevailing wind until eventually the wave is broken on a coastline or it dies out, probably as swell, in distant parts of the ocean. The restlessness of the sea is to be expected as its surface is highly favourable to air movement, and at the same time waves are transmitted from all the neighbouring waters just as ripples spread over the whole face of a pond.

Water Particles in Ocean Waves. An object floating in the sea moves no more horizontally under the action of the wave movement than does a cork bobbing up and down on the ripples of a pond. Ocean waves progress in a certain direction, but the particles of water in them oscillate rather than travel forward. In a true, or *free*, wave motion each particle of water performs a small circular or elliptical movement and then returns almost to its original position. Yet it is important to notice that, although the circular motion greatly predominates, there is a small mass transport in the direction of propagation, especially in high, steep-waves. That is to say, that each successive circular orbit completed by a particle takes it a little farther in the direction of the wave-advance. Figure 38 shows a free wave advancing eastwards. C^1 to C^3 mark the positions of the successive trough, crest and trough of one wave. The circles show the orbits in which the water particles move. In Figure 38*a* the solid line of the circles denotes the circular motion already completed by the particles, while the dotted line indicates the orbital motion yet to be undertaken by them.

If the whole wave be studied, $C^1C^2C^3$, the water particles appear to move in the direction shown by the arrows in Figure 38*a*. Thus the water particles in the upper part of a wave move in the direction of its advance while those in the lower part move in the opposite direction. Although the normal wave is largely a rise and fall in the surface waters, when the wind is strong the top of the wave may be slightly broken and then the 'white-horses' tell of a distinct forward displacement at the crest.

Except in shallow water the circular motion of the water particles prevails. The amount of movement, however, always diminishes rapidly downwards. In relatively shallow water the up-and-down motion decreases more quickly than the to-and-fro and the orbit of the particles assumes an elliptical shape (Fig. 38b). On an average the

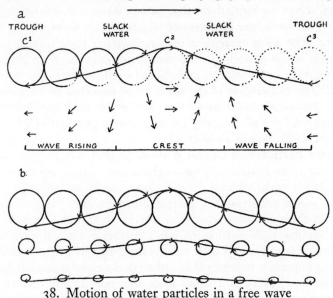

38. Motion of water particles in a free wave

b also shows decrease of wave motion with depth in relatively shallow water

movement decreases by one-half for each one-ninth the length of the wave that the depth increases. In simple terms this means that at a depth equal to the length of a wave, or distance from trough to trough, the movement is about $\frac{1}{500}$ that at the surface. According to these rules, a wave with a height of 20 feet and a length of 360 feet at the surface would produce a movement of about 15 inches at 160 feet and of half an inch at 60 fathoms. Some authorities place the extreme depth of wave motion at less than 200 feet, and all agree that waves are very shallow phenomena.

The height of waves from trough to crest may attain, but very rarely exceed 50 feet. The height is mainly determined by the strength of the wind; strong winds that have traversed wide seas, or have a long 'fetch,' produce the biggest waves. Winds of over 60 m.p.h. are needed to form waves exceeding 40 feet in height. The water that on sea coasts is hurled 100 feet and even 150 feet against the glass of lighthouses has long lost its true wave motion.

Waves on a Coastline. The speed at which a wave progresses depends partly on its length and partly on the depth of the water. Where the water is shallow relative to the length of the wave, its velocity tends to vary with the square root of the depth. In deep water, however, the velocity tends to depend on the square root of the length of the wave, which is usually less than 750 feet.

The way in which crests of waves appear to roll shorewards at right angles to a shelving beach is due to the speed of the landward end of the waves being checked on meeting the shallow water. This action gradually turns the waves towards the coastline.

In shallowing water the progress of the forward part of the wave is slowed down, and the water begins to pile up upon itself. The increase in height forces the water particles to travel in ever-growing orbits, while at the same time the supply of water in front of the wave is gradually lessening. The time arrives when there is not sufficient water to complete the orbit and the ever-steepening front finds itself unsupported. The approaching wave then collapses near the top of its circle and falls forward as a breaker. Consequently the 'waves' that are seen breaking on coastal beaches have been changed from vertical oscillations into a definite forward translation of waters. The observer at the seaside will notice that the water recedes as the true wave advances; the wave grows as it approaches closer to the beach, and eventually its waters collapse and rush forward up the shingle.

If the water is fairly deep close inshore, the waves may break actually against the coast and be hurled up the shore face. In this case the motion of the water particles that strike the shore was naturally forward and upward; the impact suddenly sets these particles free from their circular orbit, and being no longer bound by cohesion to the water in the front of the circle, they leap upwards to a considerable height. This is common on a rock-bound coast and frequently leads to the formation of cliffs.

Tides

Their Causes; the attraction of the moon. The main characteristic of tidal phenomena is their periodicity, which is associated with the orbital movements of the moon and the apparent movement of the sun.

The common statement that the moon revolves round the earth is not entirely true. What happens is that the earth and moon yield to each other's attraction and revolve round a common centre of gravity which, owing to the greater mass of the earth, lies about 1,000 miles beneath the earth's surface. During their revolutions round this common centre both planets are affected by two main forces:

(a) a gravitational force that keeps them about their common centre and is directed inwards;
(b) a centrifugal force that tends to make them fly away from each other and is directed outwards.

Owing to the solid nature of the earth the centrifugal force at all parts of it may be taken to be equal and parallel. The gravitational pull of the moon, however, varies with the distance to the various parts of the earth.[1] At the centre of the earth the centrifugal force is exactly equal to the pull of gravity (Fig. 39). On the side of the earth

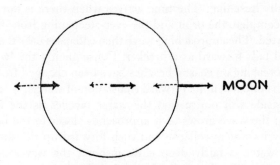

39. Explanation of tidal pull

Solid arrows represent pull of the moon and
dotted arrows the centrifugal force

nearest to the moon, the moon's attraction is greater than at the centre, and at the far side of the earth it is less than at the centre. Consequently, on the side nearest to the moon, the gravitational force predominates over the centrifugal and there is a surplus force directed towards the moon. On the side farthest from the moon the gravitational force is less than the centrifugal, and there is a tendency for this side to bulge away from the centre of the earth. The solid parts

[1] The gravitational pull of a body varies directly with its mass and inversely as the square of its distance. $\dfrac{M}{1} \times \dfrac{1}{Distance^2}$.

of the earth yield but little to these differential stresses, and the movement is too small to be felt. The mobile waters yield more easily and are drawn up into the bulge of waters called high tide.

It will be seen that the tide on the side of the earth nearest to the moon is caused by an extra pull on the waters, whereas on the opposite side the tide is produced indirectly by a decrease in the average pull of the moon. In other words, the tide-raising force is represented by the difference between the pull on the centre of the earth and on the parts of the earth nearer to and farther from the moon. The distribution of these tide-raising forces can be represented geometrically by the arrows shown in Figure 40.

The high tides that form on opposite sides of the earth draw the water away from the intervening parts of the oceans and give them a period of low water. The earth, however, rotates on its own axis, and so every meridian in its turn comes under the greatest and least influence of the moon. Consequently most sea surfaces experience two high and two low tides daily.

At the same time as the earth is rotating on its axis, the moon is proceeding along its orbital course. When any point of the earth's surface has completed one rotation, it has to rotate a little farther before coming again directly beneath the full influence of the moon. The moon traverses about one twenty-eighth of its orbit in a day,

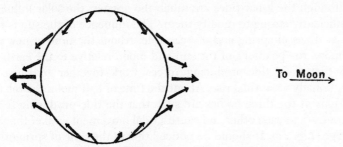

40. General distribution of tide-raising forces

and any meridian on the earth has to pass through this extra distance before regaining its original position with regard to the moon. The distance is accomplished in one twenty-eighth of 24 hours or in about 50 minutes. Consequently high tide is about 50 minutes later each day. In actual practice it is found that the intervals between the high tides, which should be 26 minutes where two tides occur daily, vary very considerably from the theoretical.

The Attraction of the Sun. The sun, although so large, has less effect on our tides than the moon, because the tide-raising force is determined by the difference between the attraction at the centre of the earth and at other points on the earth's surface. The centre of the moon is about 240,000 miles from the centre of the earth and about 236,000 miles from the near side. The relation between its gravitational pull on these points is therefore as $240,000^2$ is to $236,000^2$, or as 31 is to 30. In other words, the *difference*, or tide-raising force, is about one-thirtieth of the pull at the centre of the earth.

The sun is 93,000,000 miles away, or about 389 times as far as the moon, while its mass is 26,000,000 times that of the moon. Its attractive force at the centre of the earth is therefore $26,000,000 \div 389^2$, or about 170 times that of the moon. The sun, however, is so far away that the difference between its pull at the centre and at the near side of the earth is relatively small. The difference is in the proportion of $93,000,000^2$ to $92,996,000^2$, or 1,000,086 to 1,000,000, which is $\dfrac{86}{1,000,000}$ of the sun's pull at the centre of the earth. The tide-producing force of the sun is therefore $\dfrac{86 \times 170}{1,000,000}$ or $\dfrac{1}{68}$, of the pull of the moon at the centre of the earth. This is $\frac{30}{68}$ or $\frac{4}{9}$ of the tide-raising force of the moon.

Although the lunar tides are much the greater, the solar influence is sufficiently strong to modify them. The influence of the sun is best seen at times of spring and of neap tide. About the time of new and full moon the positions of the sun and moon relative to the earth are such that their tide-producing forces work together to cause an exceptionally great tidal rise. About the time of half moon the relative positions of the three bodies are such that the tide-producing forces act contrary to each other and cause a tidal movement weaker than the average (Fig. 41). It should be noticed that at the time of spring tides the moon, earth, and sun are not quite in the same straight line. The moon's orbit is inclined to the plane of the Equator, and this ensures that an eclipse of the sun or of the moon does not occur at each spring tide.

There are numerous factors that influence the height and time of high tide locally. A few examples must suffice. The moon, owing to the inclination of its orbit to the plane of the Equator, is overhead sometimes north and sometimes south of the Equator; at the same time its orbital distance from the earth varies slightly during the month (perigee and apogee) in much the same way as the distance of the earth from the sun varies slightly during the year (perihelion and aphelion). All told, about two dozen constituents of the tide-producing forces may cause slight variations in the nature and time of the tide locally.

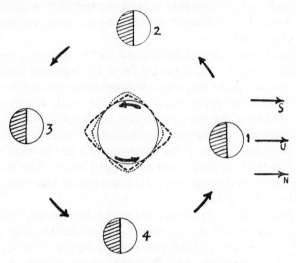

41. Cause of spring and neap tides

Dotted line shows tide due to moon's attraction. Broken line shows actual tide due to pull of sun and moon. 1 and 3 are spring tides at new and full-moon; 2 and 4 are neap tides at half-moon

Influence of Continental Masses. Tides are so profoundly modified by terrestrial features that the character of the rise and fall at any given point is largely determined by regional factors. Times of high and low water show a disregard for longitude, and the height of the tide has no apparent relation to latitude. At St. John's (Newfoundland) the high tide, of about $3\frac{1}{2}$ feet, occurs 7 hours after the moon's passage, whereas Brest experiences a rise of 16 feet some $3\frac{1}{2}$ hours after the moon's passage. Colon, on the Atlantic side of the Panama Canal, has a tidal range of 1 foot and only one high and one low tide daily, whereas Balboa, 30 miles away on the Pacific, has a rise of 14 feet and two high and two low tides every 24 hours 50 minutes. The main reasons for these differences lie in the varying depths of the oceans and in the obstruction of tides by continental masses.

The length of the tidal wave is so great that its rate of travel depends on the depth of the sea and its progress is frequently hindered by shallow water. Further complications are caused because the natural rate of travel of the free tidal wave is very much less than that forced upon it by the earth's rotation. Hence it is not surprising that the rise lags considerably behind the meridian passage of the moon and sun. In the Atlantic, spring tides are often one day behind the moon's passage.

The land masses prevent the normal east to west progress of the tidal wave, except in the ocean girdle of the southern hemisphere. The old idea was that a major tide moving westwards in the southern seas sent branch waves into the oceans to the north. For example, in the Atlantic Ocean the wave progressed northwards and its progress could be shown by progressive co-tidal lines. These lines, which join places having high water at the same time, tended to show that the wave moved faster in deep water and lagged behind near the coasts. The main wave sent subsidiary waves into coastal seas such as those around the British Isles.

This 'Progressive-wave' theory has to-day been largely superseded by a 'Stationary-wave' theory, which explains much more success-fully the peculiarities of tidal rise and fall locally. The stationary-wave idea has been likened by Johnstone to the motion of water in a shallow dish. If the dish is gently tilted the water rises at one end, lowers at the other, and remains almost level at the middle. The one tilt forms a series of waves of this nature, which gradually die away by dissipating their energy. If tilting from end to end is suddenly stopped and the dish is rocked from side to side, the waves immedi-ately change their direction and rise and fall from side to side. Experiments show that the direction of the wave changes with the rocking motion, but the central parts of the water remain at approximately the same level.

In the case of an ocean basin the water is made to oscillate by the tide-generating forces, and the oscillation changes with variations in the position of the moon and sun relative to the earth.

The theory calls for the existence in each ocean of a nodal point in which the water streams but does not rise or fall. In practice it is found that there may be several such 'amphidromic' points, and that smaller seas may develop their own oscillations. It is certain, too, that the oscillations of one ocean interfere with the tides of adjoining water surfaces. From the amphidromic points, co-tidal lines can be drawn radiating outwards and swinging round the points. The height of the tidal wave increases towards the extremity of these lines (Fig. 42).

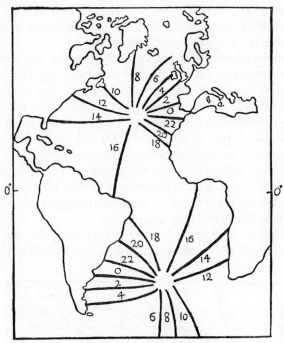

42. Co-tidal lines in the Atlantic Ocean
(After Sterneck and Johnstone)
Tidal wave is zero near centre and highest at other
extremity of line. Numbers show hour of day

Tides on the Continental Shelf

Tides are experienced in all large bodies of water. In the Mediterranean the tide does not exceed a few feet even at the heads of the gulfs. In the open ocean the tide is an imperceptible rise and fall of water probably not exceeding two or three feet. Yet around the coasts of Britain the difference between high and low water is commonly twenty and even forty feet.

The British Isles are open to the Atlantic, and the stationary-wave motions formed there send wave movements to the continental shelf. On reaching shallower water the progress of the tidal wave is hindered, and simultaneously, its length diminishes and its height increases. The front of the wave gradually steepens and the rear waters tend to gain on it. As the water becomes increasingly piled up, its true wave motion is eventually broken and changes into a definite forward transference of the water particles. This is similar to the breaking of wind waves, except that the tidal wave has a much greater length and takes several hours to ebb and flow.

This marked increase in height always occurs, but considerable variations in the character of the tides are to be expected on a continental shelf. In Long Island Sound and the Bay of Fundy the height of the tide increases towards the head of the estuaries, but there is little difference in the time of the tide. This seems to indicate a stationary-wave type of movement. Around the British Isles most of the ports and estuaries experience high tide at successive times, which points to the existence of a progressive wave on the continental shelf. The latest tidal charts, however, show that the tides of the east coast of Britain are related to small stationary-wave movements formed in the North Sea (Fig. 43).

The influence of regional features is seen along the mouth of the English Channel, where ports such as Southampton, Weymouth, and

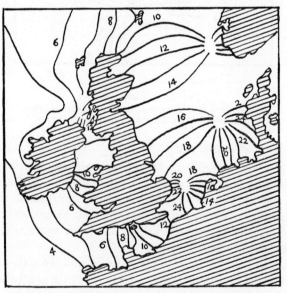

43. Co-tidal lines around the British Isles
Numbers show hour of day; 24 equals 0, or midnight

Honfleur have a remarkable prolongation of high tide, thought to be caused by peculiar stationary-wave oscillations natural to this part of the English Channel. The influence of local features may be exemplified by the action of funnel-shaped estuaries in piling up the tide. At spring tides the tidal difference at Avonmouth is 49 feet and at the head of the Bay of Fundy over 60 feet. Where rivers enter funnel-shaped estuaries the meeting of tidal and river water may lead to the formation of bores. For example, at spring tide when the wind is blowing upstream, the sea water may crowd up the narrowing bed of the Severn. At the meeting of river and tide the sea water hurls itself into a foaming wall that may rise to four feet in midstream. Similar phenomena rush up the Trent (*eagre*), the Solway, the Seine (*le mascaret*) and, above all, up the Tsientang Kiang in China, where the wall of water occasionally attains 12 feet.

Tidal Currents. In the open ocean the drawing up of the waters to form a tide does not give rise to perceptible horizontal movements of the water. On the continental shelf, however, where the rise and fall may be 30 feet and the places with high and low water only 100 miles apart, the horizontal currents often become strong. The vertical rise and fall of the water is called the tide, while the horizontal movement is called the tidal current. In a typical tidal wave on a continental shelf, the currents flow in the direction that the tide is travelling for about two hours before and after high water, and in the opposite direction for about two hours before and after low water. These flood and ebb currents have much the same general direction as is shown by arrows in Figure 38a.

The nature of tidal currents is modified and complicated by the nature of the coastline. A tidal wave that approaches the shore at right angles causes currents that ebb and flow to and from the shore, but if the tide approaches along the coast the currents flow parallel with the shore. In shallow seas the flood currents are usually the strongest, and the forward drift of shingle and débris is consequently not completely neutralized by the ebbing tide. Frequently the forward currents possess a little extra stimulus, such as of wind, that increases their shifting powers. Consequently it is usual for tidal waters to drift material in the direction in which the tide is moving. In narrow channels, such as the Menai Strait, the Pentland Firth, and Hell Gate near New York, the tidal currents may become a race of waters strong enough to hamper shipping.

The Effects of Tides on Estuaries and Ports. Tides assist in the disposal of sediment at the mouths of many of the world's rivers. The tidal scouring of estuaries usually needs some degree of artificial control before it becomes an unmixed blessing, but its general effect is beneficial. The ebb flow of a river estuary consists of the retreating tide combined with the dammed-back waters of the river. Consequently it is relatively powerful and is able to transport away with it the material brought in by the tide as well as that carried seawards by the river. It has long been recognized that tidal streams are usually most amenable to improvement for navigation purposes. The Clyde is a notable example.

The additional depth given once or twice a day to a shallow water channel is of great benefit to big shipping. With this aid, commerce is allowed to progress much farther inland. Most of the world's great ports—London, Hamburg, New York, Rotterdam—are on tidal rivers, and many a town owes some of its importance to a position at the upriver limit of high tides. The type of harbour often reflects the extent of tidal rise and fall; the closed docks of London and Avonmouth contrast with the open docks at Southampton and Marseilles. In addition, tides by bringing in salt water and creating a commotion of the waters tend to keep seaports relatively free from ice. The Bay of Fundy is never frozen, and New York is open all the winter although the river Hudson freezes.

The geological importance of tidal currents as agents of transport and erosion is discussed in Chapter Twenty; the biological effect of tides has already been dealt with. From a human point of view there is also the energy inherent in the tidal rise and fall, an enormous potential source of power as yet almost untouched by man.

LIST OF BOOKS

Kuenen, Ph. H. *Marine Geology*, 1950
Sverdrup, H. U. and others. *The Oceans*, 1942
Marmer, H. A.   *The Sea*, 1930
Ocean Waves
Cornish, Vaughan. *Ocean Waves and Kindred Geophysical Phenomena*, 1934
Russell, R. C. H. and MacMillan, D. H. *Waves and Tides*, 1954
King, C. A. M. *Beaches and Coasts*, 1959; *Oceanography for Geographers*, 1962
Tides and Tidal Currents
See also Tricker, R. A. R. *Bores, Breakers, Waves and Wakes*, 1964
Marmer, H. A., in *Oceanography*, Washington, 1932; *The Tide*, 1926
Bauer, H. A. 'A World Map of Tides,' *Geog. Rev.*, April, 1933
Admiralty Manual of Tides, 1941. (H.M.S.O.)

CHAPTER ELEVEN

SUBMARINE RELIEF AND ISLANDS

VIEWED in the widest sense the oceans cover large parts of the solid globe with a film of water, the depth of which is in the same proportion to the earth as a layer of perspiration is to one's hand. The topography of the Atlantic and Pacific floors cannot satisfactorily be represented by cross-sections, as the vast horizontal distances involved dwarf the vertical heights. When viewed, however, in a more local sense the relief differences of the ocean bed are comparable with, or even greater than, those of the lithosphere.

The Continental Shelf

For mankind the most important part of the ocean floor is the continental shelf, or the extension of the continent that is covered by water less than 100 fathoms deep. A rise of 300 feet would link Britain to the continent by dry land; while the dome of St. Paul's would not be covered by the depth of water in any part of the present English Channel.

The width of the continental shelf off the British Isles varies from about 30 miles west of Achill Head to about 200 miles west of Land's End. In the Atlantic a narrow continental shelf runs from the Bay of Biscay to the Cape of Good Hope, while on the American side the shelf is usually very much wider. In the Pacific the shelf is wide off Asia and narrow off the Americas, where, for instance, depths of 2,500 fathoms occur within 150 miles of the coastline of Chile.

The relief of the floor of the continental shelf differs little from that of the neighbouring land. The bed of the North Sea closely resembles an undulating plain gullied by river valleys (Fig. 44). The Dogger Bank falls rather steeply to 30 fathoms on its north side and slopes down gradually on the south in the manner of a scarpland. The Lincolnshire end of the Bank rises to within 8 fathoms of the surface, and it seems probable that the whole feature may be the submerged continuation of the Yorkshire Wolds. Immediately south of the highest part of the Dogger Bank is a trench 50 fathoms deep, known as the Outer Silver Pit. This trench, which appears to connect the northern and southern, or deeper and shallower parts of the North Sea, may be a channel of the predecessor of the river Rhine,

to which the rivers of northern England flowed as tributaries. A
similar elongated trough opposite the Wash may mark the course of
a more southerly tributary.

South-east of the Dogger a number of little hills rise about 30 feet
above a flat plain lying at 20 to 25 fathoms; it has been suggested that
these are morainic dumps planed off by wave action. A study of an
atlas will reveal many other features characteristic of terrestrial
landscapes such as the submarine valley furrowing the floor of the

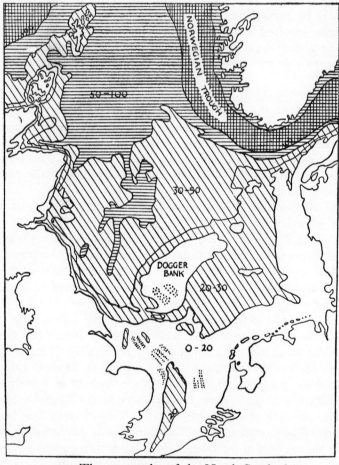

44. The orography of the North Sea bed
(*Mainly after Krümmel*)
Dotted areas mark larger ridges and banks of sand. Depths in fathoms

English Channel and trenches, probably deepened by glacial action, in the Irish Sea. The origin of the remarkable gully that lies near the Norwegian coast and descends to 480 fathoms (2,880 feet) in the Skagerrak cannot be satisfactorily explained.

Mention should also be made of somewhat similar submarine

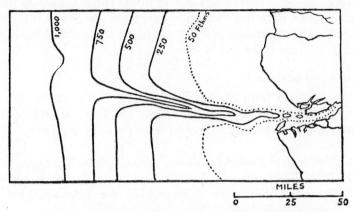

45. The submarine valley off the mouth of the Congo
(*Modified from Schott*)
Depths in fathoms

canyons at the edge of the continental shelf in other parts of the world. Off the coast of the U.S.A. submarine canyons of amazing proportions have been charted. Many have a sinuous trend, steep sides and narrow bases characteristic of river canyons on land. Most have walls rising 1,000 feet, and some to as much as 6,000 feet above their valley floors. Off California fourteen submerged canyons almost reach the coast, but elsewhere on the seaboard of the U.S.A. the canyons usually penetrate the continental shelf for a few miles only or terminate just beyond its outer edge. Yet the canyon off the mouth of the river Hudson continues across the continental shelf for 93 miles as a shallow gently-sloping valley that joins up with the present river bed. On the European coastline a submerged canyon prolongs the valley of the river Adour, and others somewhat similar occur off the mouths of the Gironde and Tagus. These and other canyons appear at first sight to be the seaward continuation of river valleys, but if this is so, the tremendous subsidence needed to bring about their present state of submergence cannot be explained. Around the U.S.A., canyons occur most frequently off the shores of

New England and off the western coast north of 32° N. latitude. These coastlines bordered on ice-sheets in the Ice Age, but the connection between heavy glaciation and submarine canyons breaks down in the case of the Congo, off the mouth of which a vast gully extends seawards (Fig. 45). Faulting along two parallel series of fissures and the abrasive effect of very turbid currents slipping down steep inclines have been tentatively suggested as causes. Students interested in the puzzling problem will find considerable detail in the references given at the end of the chapter.

Continental Islands

Islands rising from the continental shelf are usually structural continuations of the adjacent land mass. The structural relations of Britain with Europe may be judged from the following facts:

(*a*) South-west England is similar geologically to Brittany.

(*b*) The Weald terminates in the Boulonnais, a narrow wealden structure that faces westwards in the neighbourhood of Boulogne (Fig. 46).

(*c*) The asymmetrical fold-belt of the Isle of Wight re-appears at the surface in the Pays de Bray.

(*d*) The line of the South Wales and Kent coalfield is continued eastwards across northern France.

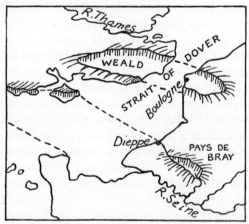

46. Main structural relations of south-eastern
England and the Continent

(e) The Highlands of Scotland are related geologically to those of Scandinavia.

Less complex examples could be taken from almost any continental shelf, Trinidad, for instance, being a particularly clear extension of the relief of the mainland.

Some difficulty may be experienced in classifying islands such as Madagascar and New Zealand as continental. Structurally they are related to the nearest mainland, but the intervening sea is much deeper than 100 fathoms. These isolated islands are usually termed 'old continental,' to give the idea that the separation is ancient and complete.

The Deep Sea

At its seaward edge the continental shelf drops rather abruptly to the deep-sea plain. The transition from 100 to 1,000 fathoms is usually a well-marked feature that is called the continental slope. Beyond it stretches a relief which differs from that of the continents only in the absence of sudden minor irregularities. Although gentle gradients prevail over vast areas, great mountain formations are not absent. Where the Brest–New York telegraph cable reaches the deep sea on the north side of the Bay of Biscay (48° 30′ N. lat.) there occur several isolated mountains with slopes of over 30°. Many islands rise steeply from the ocean bed; the slope of the upper 1,000 feet of the submerged base of St. Helena and Tristan da Cunha reaches nearly 35° in some parts, while the lower bases of these and many other oceanic islands slope down at an angle of about 15°. By way of comparison the volcano, Fuji Yama, has a maximum slope of 35° near its summit and of about 12° at its base. Sufficient, however, has been said to indicate the essentially continental aspect of submarine topography. The point is further emphasized by a study of the floors of the various oceans.

Atlantic Ocean. The main characteristic of the Atlantic Basin is the fact that 24 per cent. of its floor is less than 3,000 feet deep. The shallowness of many of the tributary seas and the width of the continental shelf mainly account for this high proportion.

The Atlantic is traversed lengthwise by a 'Central Rise' that, in a course of 8,000 miles, seldom falls below 2,000 fathoms. The Rise is generally some 9,000 feet above the bottom of the basins which it separates. Locally, it widens out into great plateaux, such as the

Telegraph Plateau (51° N.) and the platform on which the Azores stand. Ridges branch off from it, perhaps the most notable being the Walfish Ridge lying between the Rio Grande and Walfish Bay, which tends to divide each basin of the South Atlantic into two parts.[1]

According to Murray there are 19 'deeps,' or areas below 3,000 fathoms, in the Atlantic. The greatest depth, 4,800 fathoms, occurs just east of Santo Domingo on the seaward side of the insular arc of the West Indies.

Indian Ocean. In the Indian Ocean a definite central rise also occurs, and in the same way it also separates deeper basins and forms a basal platform for oceanic islands. The ocean, which is on the whole slightly deeper than the Atlantic, exceeds 3,000 fathoms only off Madagascar and the Australian–East Indian coastlines.

Pacific Ocean. The Pacific is the greatest and deepest of all oceans, and the contrasts of its floor are not equalled in any other ocean (see Table following). Only about 7 per cent. of its area is less than 3,000 feet, and the major portion of the ocean lies below 15,000 feet. Depths of well over 30,000 feet occur, the greatest as yet measured being 35,410 feet in the Mindanao Deep, about 40 miles east of the Philippines, and 36,204 feet off Guam in the Mariana Islands.

All of the deeps, well over thirty, recorded in the Pacific are situated comparatively near to land, such as those bordering on the Andes, the Aleutian, Kurile, and Philippine Islands and Japan. The depressions, which are typically trough-like in shape, usually stretch

Depths in Fathoms	Atlantic Ocean Basin		Pacific Ocean Basin	
	Approx. area square miles	Percentage	Approx. area square miles	Percentage
0–500 . .	9,948,000	23·8	4,861,000	7·0
500–1,000 . .	1,588,000	3·8	2,431,000	3·5
1,000–2,000 . .	7,607,000	18·2	12,362,000	17·8
2,000–3,000 . .	19,770,000	47·3	45,143,000	65·0
3,000–4,000 . .	2,884,000	6·9	4,445,000	6·4
Over 4,000 . .	17,000	0·04	208,000	0·3
	41,814,000		69,450,000	

[1] With this result on sea temperatures at 15,000 feet:

Argentine Basin	.	.	.	33° F.
Brazil Basin	.	.	.	34° F.
Cape Trough	.	.	.	34° F.
African Trough	.	.	.	36° F.

along the seaward side of either insular arcs or of high folded mountain ranges. Thus the Tuscarora Deep below the 3,000 fathom line
extends over 1,700 miles, and yet seldom diverges more than 100
miles from land. The Kermadec and Tonga Deeps are even more
trough-like (Fig. 47). It seems that deeps in general are the result
of marginal warpings of the ocean floor, the degree of warping
apparently being roughly proportional to the extent of the ocean bed.
The amount of crustal movement should be measured from the
summit of the exposed mountain range to the floor of the adjacent
deep, and the whole earth-fold is called a geosyncline. Geosynclines
often show relief differences of as much as 35,000 and 40,000 feet
within a short horizontal difference, and consequently they form the
greatest folds of the earth's crust.

Ocean Deposits

The floor of oceans and seas is covered by deposits that vary in
composition largely in accordance with distance from land surfaces
and the depth of the sea-water. Muds and sands may accumulate to
a considerable thickness upon the continental shelf, but beyond the
continental slope the normal rate of deposition must be extremely
slow. Only recently have methods been devised to measure the
thickness of deep-sea oozes.

Ocean deposits may be classified as follows:

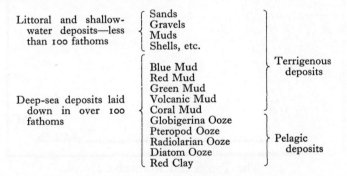

The littoral and shallow-water deposits are dealt with in Chapter
Twelve, but the more distinctive mud deposits, although terrigenous
in nature, are found well beyond the continental shelf, and are
described in broad outline below. It should be remembered, however,
that the deposits grade into each other, and that their distribution

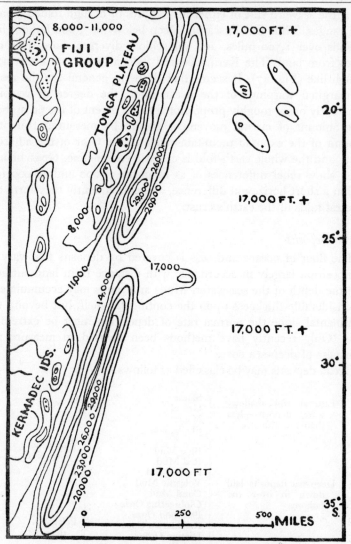

47. The Tonga and Kermadec Deeps
(*After Daly*)

is far too irregular to be shown in any but a highly generalized way
in atlases.

Green muds and sands owe their greenish tinge to the abundance
of fragments of glauconite, a silicate of iron that fills the cavities of

the shells of calcareous organisms known as foraminifera. The deposit occurs generally upon the continental slope at 100 to 900 fathoms. It seems to accumulate especially along coastlines that are relatively free from large rivers and where ocean currents vary in strength with the seasons. Considerable areas occur off Cape Hatteras and off the coasts of Portugal and California. Among the constituents of green mud is potassium (in the form of phosphatic nodules) which has been withdrawn from solution in sea-water.

Red muds and blue muds are very similar in composition to the green. Blue muds commonly cover the floor of enclosed seas and of the deep-sea plain near the edge of the continental slope. The deposit, which occurs usually at 100 to 2,900 fathoms (average 1,500 fathoms), consists mainly of organic matter and of the finer particles of land-derived material. The blue colour appears to be due largely to the presence of finely divided sulphide of iron. A considerable

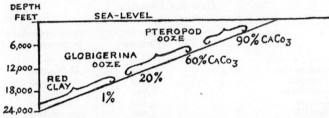

48. General vertical distribution of deep-sea deposits. Showing decrease of lime ($CaCO_3$) content with increase of depth[1]
(*After Murray*)

area of the floor of the Pacific between Acapulco and the Galapagos Islands is covered with this deposit.

Red mud owes its reddish tinge to iron oxides that are transported seawards by large rivers, and are then redistributed by ocean currents. The deposit predominates over much of the Yellow Sea and along the Brazilian coast.

Pelagic Deposits. True deep-sea deposits are distinguished from terrigenous deposits by the paucity of land-derived matter, except for the finest particles. Probably, too, the bottom material

[1] It should be noticed that this diagram does not indicate that the lime-content of individual oceans necessarily decreases regularly with depth, as much depends on surface temperatures and on the kind of bottom deposits. Thus in the Atlantic the vast expanses of Globigerina ooze at 1,500 to 2,000 fathoms cause the lime-content of the sea-floor to be greatest at that depth; below it the general lime-content here then decreases regularly.

derived from life in the sea is actually less in a deep ocean. On a continental shelf the great quantities of terrigenous matter hide the deposition arising from the relatively rich marine life, while upon the deep-sea plain the marine deposition, however small, greatly predominates. The pelagic deposits, wherever possible, are named after the predominant type of shell, and are either largely calcareous, siliceous, or argillaceous in nature.

The chief calcareous deposit is globigerina ooze, which consists mainly of the calcareous shells of pelagic foraminifera (Fig. 48). The tiny shells, which only average about the size of a pin's head, distinguish the deposits over a very large area of ocean floor, especially in the Atlantic, southern Pacific, and western Indian Ocean. The most favourable depth for its accumulation seems to be 1,500 to 2,000 fathoms, but the deposit occurs at least 1,000 fathoms above and below these limits.

DEEP SEA DEPOSITS

Kind	Average Depth (Fathoms)	Approximate Area (square miles)
Globigerina . . .	2,000	48,800,000
Pteropod . . .	1,000	600,000
Radiolarian . . .	2,900	2,700,000
Diatom	2,100	12,000,000
Red Clay	3,000	39,500,000

Pteropod ooze, the other calcareous pelagic deposit, is the least abundant and probably one of the least distinctive of deep-sea oozes. It consists largely of the conical shells of floating molluscs that seem to prefer warm seas. The shells range up to one-half inch in length, but they are so fragile that they are easily dissolved by sea-water. As a result large patches of pteropod ooze only form on submarine elevations that rise to within 1,500 fathoms of the surface, and are situated far from land masses. The patches of calcareous ooze on the Central Rise of the southern Atlantic have accumulated in such a position.

There are two chief siliceous oozes—Radiolarian and Diatom.

(a) Radiolarian ooze occurs chiefly between 2,000 and 5,000 fathoms in the tropical seas of the eastern Pacific. The siliceous content arising from the abundance of shells of radiolaria is increased by the presence of the needles of sponges.

(*b*) Diatom ooze is characterized by the abundance of skeletons of a kind of pelagic algae which seem to show a preference for waters of relatively low salinity. This siliceous deposit covers vast areas of the Antarctic Ocean and of the extreme North Pacific, where it lies at depths ranging from 600 down to 4,000 fathoms.

Red Clay. The second most extensive of pelagic deposits and the typical covering of the abyssal floor is Red Clay, which consists of the residual matter that has withstood dissolution. The clay is found below 2,000 fathoms, especially in the Pacific and Eastern Indian Ocean. It occurs in most other pelagic deposits, but at these great depths the solvent action of sea-water removes practically all of the calcareous organisms and the clay remains as the predominating constituent. The clay, once thought largely volcanic and aeolian in origin, is now considered a slow accumulation mainly of colloidal clay particles, too fine to be flocculated by salt-water, and carried in suspension vast distances from the lands.

Oceanic Islands

Islands that rise from out of the deep sea may be formed by one or more of the following agents:

Earth folding
Vulcanism
Coral growth.

These factors are, however, closely related; volcanic action, for example, is typically associated with earth movements, and corals usually grow around or upon a volcanic base.

The insular arcs or festoons of islands typical of the western Pacific correspond to the unsubmerged portions of folded mountain ranges, that in most cases are definitely linked to structural lines on the adjacent continent. The curved chain of the Aleutian group, the festoons or arcs in the East Indies, and the sweeping curve of the West Indies illustrate the nature of insular arcs. Not infrequently the arcs enclose a relatively shallow sea and border seawards on an oceanic deep. In practice it will be found that the individual islands forming the arc are in most cases largely and in many cases entirely volcanic and coral in composition.

Volcanic Islands. Vulcanism forms oceanic islands in two ways: firstly, by further elevating the crest of a geosyncline, and secondly, by building huge volcanic cones upon submarine plateaux or upon the floor of the deep-sea plain. The first type is essentially part of an insular arc, while the second gives rise to islands that are either quite isolated or are scattered haphazardly amid deep ocean. Together they form the great majority of oceanic islands. In the Atlantic, Ascension Island, Tristan da Cunha, and St. Helena are typical volcanic islands, being small but rising to considerable heights. St. Helena rises to a height of 2,800 feet above sea-level, yet the island covers barely 47 square miles. Ascension Island, which is considerably smaller, rises to about the same height. In the Pacific, the great Hawaiian chain of isolated volcanic cones stretches for over 2,000 miles and ocean depths of over 12,000 feet exist between individual members of the chain (Plate 3).

Coral Islands. In tropical seas coral commonly fringes the coastlines of volcanic islands and caps the summits of submarine peaks. Coral islands rise abruptly from the deep-sea plain, but the coral itself is presumed to be a covering of varying thickness of a submarine elevation, usually volcanic in origin.

The coral polyp is a fixed animal, or kind of fleshy anemone, that secretes a calcareous deposit at its base to form its resting-place. Some corals live alone in coral cups, but others live in colonies and build reefs. The polyp feeds largely on plankton and to a slight extent on food it extracts chemically from sea-water. Corals exist widely at medium depths in fairly warm seas, such as the Mediterranean, but they only form sizeable reefs where the salt-water is

(a) not more than a degree or two below 70° F.;
(b) not deeper than about 40 to 50 fathoms; and
(c) relatively free from, and especially from sudden influxes of, sediment.

The most massive reef builders live in the warmer oceans at less than 30 fathoms, especially where there is a rather brisk circulation of sea-waters. The most favourable position for growth is on the seaward side of reefs, and as a result there is a marked tendency for corals, once they have reached the surface, to grow outwards at sea-level. Indeed, when the reef is at the surface, outward growth alone is possible as the action of rain-water and exposure above sea-level for any length of time kill the coral polyps. The normal rate of growth of reefs only reaches 6 to 8 feet a century in the warmest parts of the Indo-Pacific Ocean, and is a mere fraction of this in the West Indies. Owing to the relative softness of coral deposits this rate would hardly keep pace with the erosive action of the waves were it not for the presence of calcareous algae that secrete a much harder form of lime than coral. These 'nullipores' incrust large areas of the seaward flat of many coral growths with a thin layer of durable lime that strengthens and cements parts of the reef face. Thus, in the Indo-Pacific seas, although the reefs are composed almost entirely of ordinary coral, the strengthening effect of the nullipores allows the more exposed portions of the reefs to withstand better the battering of the waves. Certain calcareous algae will live down to 150 fathoms, and as they are not restricted to such warm waters as the usual reef-builders they often form much of the 'coral' of temperate seas. In parts of the reefs round the Cape Verde Islands and off the Brazilian coast the nullipores are very important *builders* of reefs.

The main areas of coral growth lie between latitudes 30° North and South, the main exception being Bermuda (32° N.) where calcareous algae and other organisms occur in a higher proportion than usual.[1] Reefs especially abound in the warmer part of the Pacific and Indian Oceans; but perhaps the most remarkable feature of their distribution is the deficiency on the eastern coasts where upwellings of cold water strongly discourage growth.

The typical reef is only a few hundred yards wide, and consists of a low exposed portion that extends seawards as a flat a few inches above or below low-water mark. The seaward edge of the coral growth drops abruptly to the ocean floor and at this edge the breaking waves, even in the calmest weather, throw themselves over the flat, smoothing the surfaces of its deeply fissured mass. At high tide parts of the exposed reef may be awash, while at low tide the whole flat may be

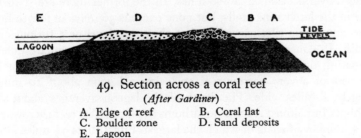

49. Section across a coral reef
(*After Gardiner*)

A. Edge of reef   B. Coral flat
C. Boulder zone   D. Sand deposits
E. Lagoon

[1] The influence of the Gulf Stream is apparent in the case of the Bermuda Island

exposed simultaneously. Locally the outer edge of the reef may be raised a foot or so above the flat owing to the protective capping of calcareous algae, but the whole coral growth is usually trenched by channels through which the tide rushes. At a short distance from the seaward edge of the more exposed reef the waves may pile up blocks of broken coral, so forming a zone of 'boulders' which lie loose on the solid coral (Fig. 49). Behind this, if the reef is wide enough, the higher waves may sweep up a thin covering of sand and pelagic deposits that soon becomes cemented by the lime in the water. This sandy strip may continue to grow, but the accumulation of the reef rarely exceeds 10 to 20 feet above high-tide level. The higher part of the reef slopes down gently or falls by a short step into a lagoon that is usually less than 30 fathoms and seldom more than 50 fathoms in depth. When the sea and wind bring seeds, vegetation springs up upon the sandy accumulation on the reef overlooking the lagoon and completes the scenery of the typical 'low' island. Perhaps the greatest contrast in the warm Indo-Pacific seas is that between the flat palm-clad reef of the 'low' island and the towering volcanic peaks of the 'high' island (Plates 3 and 4).

Four main types of coral reef can be distinguished. The *atoll reef*, which forms the majority of coral structures in the island groups of the Indo-Pacific, is like a horse-shoe or circle in shape. The reef is usually breached by channels, and appears at the surface as a number of small elongated islands arranged about a central lagoon. The lagoon at Suvadiva (in the Maldives), the biggest atoll in the world, measures some 42 by 32 miles. The enclosing reef stretches for 121 miles and carries no less than 102 separate islets. The lagoon varies in depth to 50 fathoms and communicates with the open ocean through forty channels, one of which is 38 fathoms deep.

The Gilbert and Ellice groups consist entirely of atolls that are spaced over a distance of 800 miles. In the former there are sixteen, and in the latter nine atolls, but none exceeds 50 miles in length nor does any single islet exceed 15 feet above sea-level or 5 furlongs in width. In consequence the total area is less than 180 square miles. Funafuti, one of the Ellice Islands, is a typical atoll (Fig. 50). It consists of a very narrow reef enclosing a lagoon about 10 miles long by 8 miles wide. The depth of the lagoon averages about 20 fathoms, but almost reaches 30 fathoms in places. The reef is crowned by a number of islets of which the largest, Funafuti, is $7\frac{1}{2}$ miles long and 600 yards wide at its greatest extent. The reef is broken in several

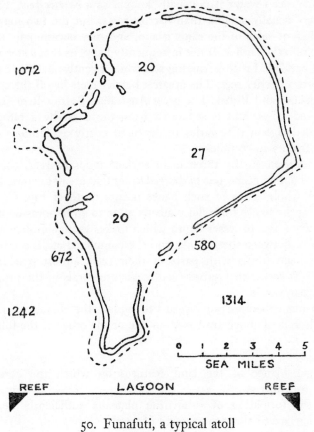

50. Funafuti, a typical atoll
(*After Daly*)
Depths in fathoms. In the section, water is shown in black

places and especially on the western side, where two four-fathom openings occur. The more continuous eastern reef may correspond to the higher rim of an underlying volcanic peak. On this side, however, the south-east Trades blow steadily from March to November, whereas the western side is exposed, from December to February, to monsoonal winds which at times cause highly destructive seas.

The other types of reefs are connected with the shore-lines of islands and continents. Those which lie close to the shore and are separated from it by only a shallow lagoon, are called *fringing reefs*. If the reef lies some distance from the shore and is separated from it by fairly deep water it is usually known as a *barrier reef*. Fringing reefs are usually smaller than barrier reefs, but the two may occur off different sides of the same island, and may occasionally actually merge into each other. It not infrequently happens that a steep coastline is paralleled with a fringing reef while a gently shelving coastline supports a barrier reef. The greatest barrier reefs lie off the coasts of Queensland and Brazil. The great Australian Barrier Reef stretches for 1,200 miles, and is separated from the coast by a rather flat-bottomed lagoon that varies in depth to nearly 40 fathoms and in width from 7 to 50 miles.

In addition to the three main surface types of reef, coral seas usually contain numerous *submerged banks* that are not exposed at low tide. A whole series of such banks occurs north of Fiji. Generally speaking the banks form flat plateaux at 20 to 30 fathoms depth, but may have a rim of lesser depth which makes them simulate an atoll in shape. Between these banks and the completed atolls are hosts of intermediate forms with parts of their reefs almost reaching the surface. It seems that submerged banks are corals in the process of growing upwards.

The main theories put forward to explain the characteristic coral reefs described above are based on one or the other of the following ideas:

(*a*) subsidence of the land features on which the coral was growing;

(*b*) the formation of submarine plateaux sufficiently near the surface to allow the growth of coral;

(*c*) changes in sea-level.

(*a*) Darwin propounded the theory that reefs were built up on landforms that were undergoing gradual depression. According to this idea the fringing reef, barrier reef, and atoll correspond to successive stages in the subsidence (Fig. 51). The first two types would occur around unsubmerged landforms, and the latter around submerged isolated peaks. The widespread depression of land demanded by this theory cannot be substantiated by clear evidence. Some coral reefs show tilting and some appear to show subsidence, but, on the other hand, in, for example, the Pelew and Fiji groups of the Pacific, elevated coral structures appear. Of late years this theory has been greatly strengthened by W. M. Davis, who lays emphasis

on the existence of drowned valleys, or embayments, around
the coasts of many islands in coral seas.

(b) Murray and others set out to explain the formation of atolls
on grounds other than subsidence. They suggest that the level
of submarine peaks and platforms is slowly raised by the
accumulation of pelagic deposits (Fig. 51). When the submarine structure is near enough to sea-level, coral growth will
begin and will continue until an atoll is formed. In addition,
some island tops would be eroded away by wave erosion, and
around other landforms submarine platforms would be cut
on which corals would grow upwards. It will be seen that to
postulate pelagic accumulation in the one instance and wave

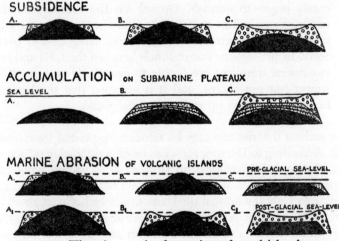

51. Theories on the formation of coral islands

In the sections showing marine abrasion, the solid line A B C represents the level
of the sea during the Ice Age

erosion in the other is rather contradictory. Murray, even after
allowing for the more rapid growth of coral at the seaward
parts of a reef, found it difficult to account for the considerable
depth of the lagoon. He suggested that much dead coral was
removed as lime in solution, whereas to-day it is generally
recognized that deposition goes on in the lagoon. The main
difficulty of the theory, however, was to explain convincingly
the incredible number of peaks that must have been elevated
sufficiently for coral growth. Gardiner assisted this theory by

his suggestion that calcareous algae, which will flourish down to 150 fathoms, could raise a submarine base of that depth.

(c) To overcome the main difficulties encountered in previous theories, Daly and others suggest changes in sea-level rather than widespread movements of the earth's crust and accumulation. It has long been observed that submarine platforms may result from the action of waves on volcanic peaks. Daly incorporated this fact into his 'Glacial Control' theory as follows: During the Ice Age, sea-level must have fallen by 40 or 50 fathoms, and then wave action cut away dead pre-glacial reefs, cut wide terraces round the large islands, and gradually planed off the summits of smaller islands (Fig. 51). At the retreat of the ice-caps, sea temperatures and sea-level rose and corals began to flourish. Growth on the submerged island summits resulted in atolls, while similar growth on the edges of wave-eroded terraces brought about fringing and barrier reefs. In the lagoons and channels between the reefs and land, conditions were less favourable to growth. Here growth was slower but not absent, and it is assumed that the lagoon is being slowly filled up rather than deepened.

The various theories can only be satisfactorily tested by extensive boring. According to Darwin, reefs can be of great thickness; according to Murray only the upper 30 fathoms or so should be of coral in position of growth, and the lagoon should not be deeper than this; according to Daly the normal thickness would be between 40 and 50 fathoms and the lagoons should be of approximately equal depth.

Borings down to 1,114 feet in Funafuti and 3,556 feet on Bikini atoll passed through reef limestones and calcareous sands without reaching the volcanic base. On Eniwetok atoll the basalt base was struck at 4,610 feet and the reef pierced consisted mainly of shallow-water varieties of coral and algae. The results undeniably favour Darwin's theory. Over a long period of time either the sea-bed had sunk or sea-level had risen slowly. These findings, however, relate to the central Pacific where numerous seamounts, or submarine mountains with summits at depths of 3,000 to 6,000 feet, seem to indicate a prolonged crustal subsidence. In eastern Indonesia and on island-arcs elsewhere in the Tropics, some reefs are merely thin cappings probably of recent age; some are slightly above sea-level and rest on bases that seem to be up-warped.

Thus no single theory will account for the wide range of coral forms and age. The survival of reefs depends largely on the relative rates of uplift or depression of the sea-bed, of change in sea-level and of coral growth. Slow regional warpings of the ocean floor do not preclude sudden localized movements, up or down, particularly in unstable, volcanic tracts. Moreover, the superstructures of most existing reefs have been affected by fluctuations in sea-level due to recent climatic changes.

Island Life

The two essential characteristics of an island environment are its isolation and its universality. Insularity, especially in the past, caused islands to be havens of refuge and strongholds for men. The universality of islands to-day may be judged from the nodal positions of Ceylon, Singapore, and Hawaii, all of which are on world ocean-liner and air-routes.

Several factors tend to encourage the settlement of islands. The climatic advantages of an oceanic position, especially in moderating the winters of temperate latitudes and in modifying the heat all the year near the tropics, are reflected in the tourist and horticultural industries of islands such as the Bermuda, Madeiras, and Scillies. The possibility of obtaining a livelihood or extra food from the products of the sea encourages growth of population, and in some cases, as in Newfoundland and South Georgia, is, or was, the main attraction for settlers. The multitude of islands in Polynesia and Micronesia support, in the aggregate, a considerable population in what otherwise would be the world's greatest uninhabited space. The twenty-five tiny atolls comprising the Gilbert and Ellice groups support a total population of about 33,000 natives and a few Europeans and Asiatics.

The flora and fauna of islands has long attracted the attention of naturalists. Continental islands usually carry a plant and animal life practically similar, except for a diminution in the number of species, to that of the nearest continent. The 'old' continental island, which was separated from the mainland in more remote geological time, and is now surrounded by deep water, has geological affinities to the mainland, but long isolation has usually given it a peculiar flora and fauna. In Madagascar, for instance, numerous species of its plants are not represented on the mainland while rare forms of animal life, such as lemurs and chameleons, occur. The peculiar birds and plants of Australia and New Zealand probably also show the effects of long isolation. Such islands are, however, relatively rich in life. Even the tiny granite islands of the Seychelles group, which may be an unsubmerged remnant of an old continent, have well over 200 flowering plants, nearly 2,000 species of insects, several reptiles, and many peculiar birds.

True oceanic islands, such as the volcanic peak or coral reef, possess no indigenous life. The relative paucity of species of plants and animals is astonishing. The Maldives, although they are in the track of currents from the mainland, have only eighty-seven plants naturally introduced, and about twice that number of species of animals. The symbols of these and other oceanic islands might well be the turtle, which is amphibious, and the coconut palm, whose hollow fruits float easily.

LIST OF BOOKS

Submarine Relief; Island-arcs; Deep Sea Deposits
Lewis, R. G. 'Orography of the North Sea Bed,' *G. J.*, Oct. 1935 (map)
Robinson, A. H. W. 'Floor of the British seas', *Scot. Geog. Mag.*, 1952, pp. 67–79
Heezen, B. C. *et. al.* The Floors of the Oceans: The North Atlantic, *Geol. Soc. Am. Spec. Paper*, 1959
Menard, H. W. *Marine Geology of the Pacific*, 1964
Betz, F. and Hess, H. H. 'The Floor of the North Pacific Ocean,' *Geog. Rev.*, 1942, pp. 99-116
Jones, O. T. 'Continental slopes and shelves,' *G. J.* Feb., 1941, pp. 80-99
Daly, R. A. *The Floor of the Ocean*, 1942
Shepard, F. P. *Submarine Geology*, 1963; *The Earth beneath the Sea*, 1959
Kuenen, Ph. H. *Marine Geology*, 1950
Johnson, Douglas. *The Origin of Submarine Canyons*, 1939
Sverdrup, H. U. and others. *The Oceans*, 1942
Pettersson, H. *The Ocean Floor*, 1954

Coral Islands
Davis, W. M. *The Coral Reef Problem*, 1928 (full bibliography).
Gardiner, J. S. *Coral Reefs and Atolls*, 1931
Steers, J. A. *The Unstable Earth*, 1950
Stearns, H. T. 'An integration of coral-reef hypotheses,' *Am. Jour. Sci.*, 1946, pp. 772-91
Emery, K. O. and others. *Geology of Bikini and Nearby Atolls*, U.S. Geol. Surv. Prof. Paper 260, 1954
Wiens, H. J. 'Atoll development and morphology,' *Ann. Ass. Am. Geog.*, 1959, pp. 31-54; *Atoll Environment and Ecology*, 1962
Verstappen, H. Th. 'On the geomorphology of raised coral reefs,' *Ann. of Geomorphology* (Berlin), 1960, pp. 1-28
Ladd, H. S. 'Reef building', *Science*, 1961, pp. 703-15
Stoddart, D. R. 'British Honduras cays', *Inst. Brit. Geog.*, 1965, pp. 131-47

Island Life
Gardiner, *op. cit.*, pp. 25-44; Wiens, *op. cit.*, 1962
Wallace, A. R. *Island Life*, 1892
De la Rüe, E. A. *L'homme et les îles*, 1935
Murphy, R. E. 'The economic geography of a Micronesian atoll,' *Ann. A. A. G.*, 1950, pp. 58-83
Johnston, W. B. 'The Cook Islands,' *Malayan Jour. Trop. Geog.*, 1959, pp. 38-57

Section Four

THE LAND SURFACES

CHAPTER TWELVE

THE COMPOSITION OF THE SURFACE OF
THE LITHOSPHERE

JUDGED according to their mode of origin, rocks may be grouped into two main classes, igneous and sedimentary. To these, however, a third class, the metamorphic, must be added.

Sedimentary rocks are estimated to cover some 75 per cent. of the land surface, yet about 95 per cent. of the earth's outer crust consists of igneous rocks. From this it will be seen that the sedimentary layer is, relatively speaking, extremely thin, and over considerable portions of the lithosphere is entirely absent.

The first rocks formed were igneous, but deposits were soon derived from them and laid down beneath water. Sedimentary rocks, therefore, are not necessarily young and soft; their formation dates from the first emergence of the land masses and pre-dates a great many types of igneous rock.

Sedimentary Rocks

Sedimentary rocks may be formed by the sub-aqueous and sub-aerial deposition of erosion products, of organic remains and of chemical precipitates from aqueous solutions.

Chalk consists largely of calcium carbonate abstracted, mainly in warm oceans, by minute marine organisms, cocoliths, ooliths, foraminifera, etc. Larger fossils and shells, whole or fragmentary, also occur in it as in most sedimentary series. A dense accumulation of organisms is also represented in diatomaceous earths; but of greater importance to mankind are the carbonaceous, or coal-yielding, rocks which are formed of the remains of plants, and petroleum, which is derived from plant and animal matter.

The great majority of sedimentary rocks originate from deposition on the continental shelf, or 'epicontinental seas,' which to-day cover about 10 million square miles, and in the past covered more extensive areas. The vast amounts of material carried by streams to the sea may be judged from the estimate that, for each square mile of the earth's surface, 50 tons of matter in solution and over 300 tons of solid matter are removed in this way annually.

The sequence of deposition at the shoreline is of much interest to geographers (Fig. 52). Generally speaking, shingle or pebbly beds lie across the upper beach, and these grade seawards into sands that merge laterally in the deeper water into muds and so into deposits of lime. The deposits merge into each other both laterally and vertically, so that sandy limestones, argillaceous sandstones, and clayey limestones (or marls) occur.

The sequence of deposits at any point of a coastline varies with the source and character of the materials and with the direction and strength of the currents. Extensive tidal flats will form off an estuary such as that of the Chester Dee, while off a steep rocky coast, such as Land's End, beach materials will be comparatively scarce. The velocity of rivers may, for climatic reasons, increase, in which case more material would be carried to the sea. The work of waves and tidal currents in transporting material and in altering coastal deposits is discussed in Chapter Twenty.

In addition, the sequence of deposits locally has usually been altered by relative changes in the level of land and sea. Both progressive and regressive movements of sea-level are often interrupted by stationary periods, and it will be seen that, in the aggregate, a thousand and one events may affect the sequence of sedimentation. The important fact is that a sedimentary series usually consists of alternating layers of conglomerates, sandstones, shales, and limestones, each of which may vary considerably in thickness locally, and may at

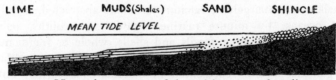

LIME   MUDS(Shales)   SAND   SHINGLE

MEAN TIDE LEVEL

52. Normal sequence of deposition on a shoreline

any point be quite absent. The different strata when exposed at the surface frequently vary sufficiently in their powers of resistance to exert a marked effect on the topography. The elevated summits of Ingleborough, Pen-y-ghent, and the Peak are due to a capping of hard Millstone Grit that preserves the underlying shales (Fig. 60). Or again, the abruptness of Glastonbury Tor originates from a capping of Cotswold Sands that protects the underlying Lias Clays.

After deposition, sediments are consolidated into beds in various ways. The pressure of the overlying deposits converts muds into shales and lime deposits into limestone. The rocks may consolidate on drying, or the loose particles may be cemented together by solutions of calcium carbonate, of silica, and of iron oxides. Thus loose sands will become sandstones and pebbly beds will be cemented into conglomerates. The rocks in general are usually classed as either argillaceous, arenaceous, calcareous, or carbonaceous according to which constituent predominates.

Characteristics of Structure. The mode of origin of most sedimentary rocks ensures that they show distinct signs of bedding or stratification. Consequently most of these rocks will split and weather most easily along planes parallel with their base, as may be seen in many limestones and sandstones.

The same rocks, however, commonly show shorter lines of weakness, known as joints, that lie at right angles to the bedding. The joints, which owe their origin to earth movements or to pressure or to contraction of the rocks on drying, cause the beds to split relatively easily into blocks. It is these vertical planes of weakness that greatly assist erosive agents in breaching outcrops of sedimentary strata.

Owing to earth movements, sedimentary beds seldom retain their original horizontal position, although in some areas the almost horizontal nature of alternating layers of different hardness is reflected in the stepped character of the valley sides. Usually the sedimentary strata are tilted, and the angle at which the general stratification of the rocks inclines from the horizontal is called the dip. The dip, which varies from a few degrees to nearly vertical, has a considerable influence on landforms. Owing to the action of running water, it commonly happens that the highest portions of a tilted series are eroded away first while the less elevated parts form a gently sloping tableland. The resultant land-form is called a *scarpland* or *cuesta*, the steep descent being the scarp or escarpment and the gently inclined plateau adjoining it the dip slope. The strata usually outcrop in narrow bands along the escarpment and occupy larger surface areas on the dip slope. The alternation of hard and soft rocks often assists the formation of a steep escarpment, since erosion of the soft rocks tends to undermine the harder layers above them (Fig. 53a).

Plains that are floored by sedimentary strata may consist largely of a series of scarplands separated by clay vales. In the Thames Basin the major scarplands consist of oolitic limestone and chalk and locally may include in addition a distinct Corallian cuesta.

Occasionally sedimentary rocks dip steeply. A clear example

occurs in the North Downs near Guildford, where the Hog's Back, a rather sudden narrowing and sharpening of the chalk ridge, is due to a sharp increase in the dip of the chalk beds (Fig. 53*b*). At the Needles, Isle of Wight, the dip is practically vertical, and is largely responsible for the shape of the pinnacles along this seashore.

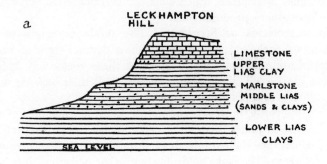

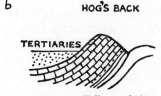

53. Effect of dip on topography

(*a*) The Cotswold escarpment and upper dip slope at Leckhampton Hill, Gloucestershire
(*b*) The North Downs at Hog's Back, Surrey

Where sedimentary strata have been gently folded or domed, the crown of the fold usually becomes eroded away first. In this case relatively abrupt inward-facing scarps may be associated with the outcrop of each of the harder layers. The Weald with its inward-facing chalk rim and Greensand ridge shows the effect of denudation on such a fold (Fig. 58*b*). The Woolhope district near Hereford is a smaller example of a similar feature.

Igneous and Metamorphic Rocks

Igneous or crystalline rocks are formed by solidification from a state of fusion such as may arise from the heat of the earth's interior or from vulcanism. Granite, basalt, and obsidian (volcanic glass) are common examples.

Metamorphic rocks are rocks, either sedimentary or igneous, that have had their original character completely changed by great pressure and heat. Limestone, for example, may be metamorphosed into marble, shale into slate, coal into graphite, and granite into gneiss or schist. The intense pressure involved often makes it possible for metamorphosed sedimentary rocks to be split along lines at right angles to the original stratification. Thus after being subjected to great pressure, fine-grained mud will 'cleave' into thin sheets or slates. These cleavage lines, which are quite independent of bedding planes, offer a line of weakness for the attack of weathering agents. Feldspar, the chief constituent of igneous rocks, is extremely hard, yet it will break readily along two sets of smooth cleavage faces that lie roughly at right angles to each other.

Both metamorphic and igneous rocks are usually well jointed, the

54. Effect of joints on weathering of granite: Great Mountain
Tor, Dartmoor

joints often being arranged at right angles to each other. This is especially noticeable in the many igneous rocks that show a tendency to split up into cuboidal forms and hexagonal columns. Granite frequently weathers into rectangular blocks, as may be seen in the tors of Dartmoor (Fig. 54). Basalt lavas, on the other hand, contract on cooling with joints that favour the formation of hexagonal shapes, the columns of the Giant's Causeway in Antrim and of Fingal's Cave in Staffa being notable examples.

Were it not for joints, crystalline rocks would be extremely resistant to all forms of erosion. As it is, igneous formations usually retain a 'block' shape and frequently show relatively flat summit levels. Generally speaking, igneous rocks form the old hard blocks or 'shield' lands of the world, while sedimentary strata form the newer plains and great folded mountain systems. The five tiny granite 'massifs' of Devon and Cornwall, of which Dartmoor and Bodmin Moor are the chief, have exerted as important an influence on local scenery as have the vast tablelands of Africa, Brazil, Laurentia, and the Deccan on world relief features. In the succeeding chapter it will be shown how the older rocks form a 'girder framework' about which the sedimentary strata are deposited and moulded.

Economic Contrasts of the Rock Groups

In the human mind sedimentary rocks are usually associated with lowlands rather than with the great highland ridges which are equally characteristic of them. Economically these strata are of great importance, as coal and petroleum are found in them. The major areas of sedimentary rocks consist of mud or shales, and on these lives by far the greater proportion of the world's population.

Igneous rocks, on the other hand, are of especial importance to civilization owing to their peculiar association with ore deposits. On the cooling of hot fluids, minerals carried in solution are deposited either as veins or as particles disseminated throughout the whole mass. Primary deposits of, for example, copper, gold, silver, lead, zinc, and tin occur in igneous rocks. Thus the tin and copper mines of Cornwall belong to the granite massifs and their metamorphic aureoles, while the coalfields of Britain belong mainly to the synclinal basins of sedimentary strata. For geological reasons it happens that a locality rich in metals will seldom be rich in coal or oil.

There yet remains to be mentioned a non-marine type of rock deposition of much importance to man. In deserts, vast sand-deposits are laid down sub-aerially to form future arenaceous strata while to windward of them fine argillaceous particles may accumulate as loess.

LIST OF BOOKS

Kirkaldy, J. F. *General Principles of Geology*, 1962
Twenhofel, W. H. *The Principles of Sedimentation*, 1959
Pettijohn, F. J. *Sedimentary Rocks*, 1957
Krumbein, W. C. & Sloss, L. L. *Stratigraphy and Sedimentation*, 1963
Shand, S. J. *Eruptive Rocks*, 1950
Harker, A. *Metamorphism*, 1950

CHAPTER THIRTEEN

CRUSTAL MOVEMENTS

Rapid Movements: Earthquakes

AN earthquake is a sudden trembling or series of rapid vibrations set up in the rocks of the earth's crust by an impulse that originates below the surface. The shocks, as recorded by delicate measuring instruments called seismographs, are of frequent occurrence. In the world as a whole five strong shocks are registered on an average every day, and earthquakes of considerable dimensions occur about once every three weeks. Italy experiences nearly four hundred and fifty shocks in a year, and Japan several a day, few of which, however, are violent enough to be destructive.

The actual horizontal and vertical movement involved at the surface is generally a small fraction of an inch, but the amplitude of the vibrations has been known on rare occasions to exceed three inches. The small vibration involved in a typical disastrous earthquake is exceptionally destructive to tall buildings on the earth's surface for the following reasons: The wave carries the base of a tall structure forward, and this progressive movement passes more slowly up the edifice. The forward movement of the ground is, however, immediately followed by a recoil or reversal. This recoil starts up the building, the top of which is moving forward at the same time as the base is moving backward, with the result that the structure may snap under the sudden strain.

The deformation of the earth's crust that accompanies the tremors occasionally reaches considerable dimensions. After an earthquake in the Disenchantment Bay area of Alaska, barnacle-encrusted rocks were lifted 47 feet above sea-level, while adjacent regions were depressed. California, the chief earthquake zone of North America, experienced a great tremor in 1906 during which fences and pipe lines were displaced up to 20 feet horizontally and 3 feet vertically. These deformations could be traced over a distance of nearly 300 miles. After a strong shock at Messina in 1908, parts of the adjacent coasts of Calabria and Sicily sank by as much as 2 feet. It seems, however, that many of the greatest earth tremors originate beneath the sea. The severe shaking that in 1923 caused the death or wounding of nearly 250,000 people in the Tokyo district, was accompanied by remarkable changes in the floor of Sagami Bay. Some parts of the bay were elevated 700 feet and others were depressed by more than 1,000 feet. Surface alterations of this magnitude are only associated with the most intense earthquake shocks.

The distribution of the major seismic, or earthquake zones is mainly related to two belts: the first circles the Pacific coastline; the second follows the zone of young folded mountains from the East to the West Indies. Among the minor belts is one running southwards from the Mediterranean through East Africa. This distribution shows a very marked connection between chronic seismic areas and zones of surface weakness. Earthquakes are especially common in areas that have experienced or are experiencing: recent intense folding; considerable fracturing; intense surface loading; and vulcanism.

Most of these features are, however, correlated. The Quetta region of Pakistan, the Chilean coast, and the Central American lands have undergone relatively recent earth-folding, but in all these areas intense folding is inevitably accompanied by fracturing. Earth tremors are especially common in Perthshire near the faults of the Central Lowlands and near Inverness, which is situated on the Great Glen fault of the Scottish Highlands. In North America, California is especially subject to them. It is certain that many earthquakes are due to slipping and settling down along lines of fracture.

Regions of excessive surface loading seem disposed to earth tremors, some of which may be due to fractures and slipping on the continental shelf. Charleston, which has suffered from disastrous shocks, adjoins deep coastal deposition. It is noticeable, too, that the loess region of North China, where excessive loading is caused by wind transport, has experienced several severe earthquakes.

Volcanic eruptions are commonly associated with earth tremors, but earthquakes are by no means confined to volcanic regions. Both phenomena are closely connected with lines of weakness; but volcanoes are relatively shallow in origin, whereas, although most earthquakes also originate at less than 40 miles depth in the more rigid crustal zones, about one in five begins at depths of 200 to 450 miles. Below 450 miles, strains seem to be distributed by gentle flow and not by sudden rupture.

Earthquakes, however, are not major geological processes; they are merely the outward sign of them. The vibrations may cause fissures at the surface and considerable landslips, but geologically speaking, their total effect on scenery is temporary and insignificant. The different rates of travel of the various types of waves emanating from an earthquake centre provide a useful means of assessing the density of the interior zones and core of the earth. Yet, it must be admitted that earthquakes attract so much attention largely because of their disastrous effect on human life. Even so, much of the disaster is due to incidental causes, such as fire and 'tidal' waves.

Gradual Changes: Relative Elevation of the Land

Elevated Marine Sediments. The vast expanses of sedimentary *marine* strata that lie almost flat and undisturbed upon lands such as the plains of the Mississippi indicate either a widespread upward movement of the land or a downward movement of the sea. More detailed evidence, pointing to relative elevation of the land, must be sought in coastal districts.

Raised Beaches. In many parts of the Scottish coast raised marine beaches occur at heights of 25 feet, 50 feet, and 100 feet above the present high-water mark. The 25-foot beach is especially conspicuous along stretches of the west coast and notably around Loch Linnhe, Loch Aber, and many of the western islands such as Skye. Numerous settlements in both east and west Scotland have arisen on these flat marine benches; Kirkcaldy, for instance, stretches for three miles along and behind a raised beach. Raised beaches are of fairly common occurrence round the shores of other parts of Britain; near Spittal in Northumberland, around Morecambe Bay, and along the coast of Devonshire and South Wales (Plate 5).

Similar marine benches are common in Norway where the feature, which is of utmost value to settlement on a fiord coast, is called the 'strandflat.' Often the landward side of the raised beach is demarcated by a sea-cliff, and in parts of western Scotland this cliff shuts off the tiny crofting hamlets from communication by land with the interior. Occasionally, up to elevations of 100 feet or so, caves of obvious marine origin may exist in the raised sea cliff. Inland caves of this nature are found at the Gower Peninsula in South Wales, along the coast of Antrim, and the western shore of the island of Jura, Argyllshire. The Gower Peninsula also furnishes examples of how a raised beach may widen out into a wave-cut platform; for example, the flat-topped promontory of Worms Head has every appearance of being levelled by wave action. A typical view of the flat top of a wave-cut platform is shown in Plate 25.

Gradual Warpings. In some parts of the world gradual changes in level (relative to the sea) have been measured over a considerable number of years. According to actual measurements, the northern part of the Baltic coast is rising at about 36 inches a century, the Stockholm district at about 18 inches a century, and the Malmö area at a very small rate. The Baltic coast of Germany seems to be relatively stable, while southern Denmark appears to be sinking slowly. In Labrador the coast is estimated to have risen by amounts varying from 250 to 400 feet since the Glacial Period.

These changes in the relative height of the land surface are local in effect, and denote local movement of the land rather than widespread changes in sea-level. That is to say, the alterations seem to be due to isostasy, such as isostatic recovery after the melting of ice-sheets, rather than to alterations in the height and volume of the oceans, or to eustatic changes as these latter are called. In fact, isostatic adjustments, up or down, often have to be considered with variations of sea-level, which also occur during Ice Ages because of climatic changes and of the growth or melt of ice-caps. In addition, surface warpings are caused by earth-stresses other than isostatic.

Subsidence of the Land

Submerged Forests. Definite evidence of relative subsidence of land areas is necessarily hard to find. During the last great glaciation sea-level was lowered by over 500 feet, but much of the evidence is now drowned to half that extent. The drowning of human structures is rather doubtful evidence, but more reliance can be placed on the presence of submerged forests. Some forests, however, could have arisen in lagoons behind shingle bars at a level slightly below that of the outside sea. If the barrier was washed away the forests would then be submerged. Submerged tree growth is quite common around the British coast, and occurs, for example, at Barry Docks, Swansea Bay, Mount's Bay, Holyhead, Solway, and off Cheshire.

Drowned Valleys. The major reason for postulating a relative submergence of a coastline is found in drowned valleys. Where the present valleys continue beneath the sea as submarine valleys that retain the shape and form of their sub-aerial portions, it seems obvious that the sea has invaded the land. Relative subsidence of the land is, therefore, especially well seen in ria coastlines. There seems no doubt that the floors of rias were cut by running water (Plate 6).

The theory of isostasy and 'floating continents' almost demands these gradual changes in the relative level of land masses. The load, however, that is required to cause warping at the surface is so great that the movements involved are imperceptible, except where the strain leads to a sudden snap and an earthquake occurs. Most of the warping observed to-day can be imputed to the effect of the weight of a former ice-cap. It may well be, however, that excessive deposition, such as is going on at the Mississippi delta, is imperceptibly lowering the coastlines of some districts.

Orogenic Movements

The most startling evidence of relative changes in land- and sea-level is to be found in the fossiliferous limestones and sandstones that mainly compose great fold ranges such as the Himalayas, Alps, and Rockies. The enormous alterations in level involved in these changes were brought about gradually over a vast period of geological time. The movements causing them are called orogenic, or mountain-building, the essential feature being uplift. Many, but not all, orogenic movements are associated with compression of the earth's crust that leads to folding. Folding, however, is a highly comprehensive term that includes a variety of forms which vary in complexity largely according to the intensity of the orogenesis. The main types of folds are shown in Figures 55–57.

Simple Folding. The Jura are relatively gently folded strata, predominantly of limestone. In parts the folds are comparatively shallow and comparatively little eroded, so that to-day the ridges are anticlinal structures. In Pennsylvania, Virginia, and other sections of the Appalachian system large areas also consist of symmetrical folds, similar to but very much older than those in the Jura. In the British

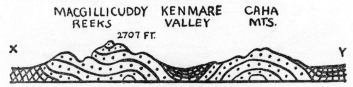

55. Simple symmetrical folds. Section across south-western Ireland from the Lakes of Killarney (*x*) to Bantry Bay (*y*)

(*After Hull*)

The valleys are of Carboniferous Limestone and mountains of Old Red Sandstone (dotted)

Isles, the fold region of south-western Ireland is probably the clearest example of simple folds, the shallow anticlinal structures still forming the higher regions (Fig. 55).

This system of symmetrical folding may be altered by the compression being greater on one side, in which case the fold becomes steeper in that direction. The Hampshire Basin consists of two anticlines, one north and one south of a central syncline. Each of the anticlines is markedly asymmetrical, the north side being the steeper in each case. This can be especially well seen in the steep dip of the chalk ridge that traverses the Isle of Wight (Fig. 56).

The folded strata may, however, also undergo movement along the line of the axes of the folds. Thus an anticline or syncline need not necessarily be horizontal. Some anticlines pitch and may give rise to cigar-shaped hills, while their erosion, as with that of tilted synclinal strata, may cause the formation of canoe-shaped valleys. These topographical features are well developed in the Great Appalachian valley of the U.S.A.

Complex Folding. The relatively simple forms of folding discussed above are less common than the complex structures where intense folding leads to fracturing and overthrusting, with the result that there is a considerable shortening of the horizontal space occupied by many strata. The Jura strata may have been compressed in all about three miles. Yet, the strata of the Appalachians are estimated to have been shortened about 100 miles, those of the Alps by about 150 miles, and those of the Himalayas by about 400 miles. Hence, it is thought that the actual crumpling and

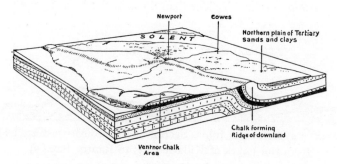

56. Asymmetrical fold in chalk of the Isle of Wight

(*Trueman, 'The Scenery of England and Wales.' Victor Gollancz*) over-thrusting in the chief mountain chains of the world have resulted in a decrease of the former surface area of their folded sedimentary strata by over 750,000 square miles.

In Britain, the Highlands of Scotland show evidence of intense overthrusting. The various layers have been overfolded and overthrust until the structure resembles a tilted pack of cards.

The complex structure of mountains such as the Alps can be best studied in conjunction with their mode of origin. Mountain structures in Britain outside the Scottish Highlands give useful clues to the nature of orogenic movements. In the Weald, a tiny example of an anticlinorium or slightly complicated anticline, the deposits beneath the surface of the dome are at least 4,000 feet thick. In North Devon, the thickness of the Devonian and Carboniferous series of rocks is about 20,000 feet. In the Appalachians there is thought to be 25,000 feet thickness of Palaeozoic sediments, while the same rock groups in the adjacent Mississippi region are only approximately 5,000 feet thick. Thus it appears that in folded mountain ranges there is an exceptionally great thickness of deposits. This is the first clue to orogenesis or mountain-building. The thickness of deposits is explained on the grounds that mountain chains occupy the former sites of geosynclines or great downfolds of the earth's surface. Into the geosynclines, sediments accumulated from off the adjacent land masses. As the accumulation proceeded the syncline gradually sank, so that subsidence kept pace with deposition and the sediments went on increasing in thickness.

The second main clue is the fact that the position and direction of fold mountains show a marked relation to old hard blocks. In other words, the geosynclines were situated at the edge of old hard blocks which have shown a tendency to move. The gradual shifting of an older block, vertically or horizontally or both, results in folding of the sedimentary strata, which folding increases in proportion to the amount of movement. The whole process occupies a vast period of time, and is no more spectacular than is the warping of Scandinavia to-day.

The Alps and other 'Alpine' folds in Eurasia serve to illustrate the theory outlined above. Formerly a wide ocean, called Tethys, stretched between the Eurasian and African land masses. Marine transgressions occasionally formed shallow deposits on the margins of the continents, but sedimentation in the deeper water near shore was continuous. The ultimate result was a considerable thickness of sediment on the continental margins and an enormous thickness of deposits on the adjacent sea-bed. Then, for some unknown reason, orogenesis began to dominate over deposition. The gradual movement at first formed small folds, but increasing elevation extended and complicated the folding (Fig. 57). In other words, the whole orogenesis consisted of a long series of smaller orogenic movements which finally built up the Alpine mountain systems. In the later period of the building, huge recumbent folds were formed and the intense overfolding led to faulting. The upper portions of overfolds were thrust forward long distances until they overlie rock strata far from their original bases. Old crystalline rocks were in parts thrust over younger sedimentary layers until the complexity of the structure almost defies analysis. There occurs, especially in the western Alps, enormous masses of rocks that have no structural continuity with the underlying strata. These overthrust and fractured folds are called 'nappes,' there being, for instance, the Monte Rosa nappe and Great St. Bernard nappe.

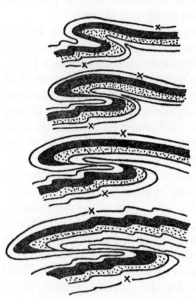

57. Complex folds. Recumbent folding leading to overfolds and formation of nappes

Note the relative position of *xx*

The story of the Appalachians is similar except in detail to that of the Alps. Vast deposits were laid down on the west of a great continental mass of old hard rocks, now called the Piedmont Plateau. The deposits were thickest near to this land mass and thinned out westwards. During the course of the orogenesis the pressure appears to have come from the east and the rocks nearest the edge of the old block were intensely folded. To the west, however, farthest from the area of compression the sedimentary strata remained, as they are to-day, almost horizontal.

It will be readily appreciated that more than one old block may take part in orogenic movements locally. Where two blocks moved toward

Crown Copyright] [*Geological Survey*

5. 25-foot (in foreground) and 100-foot raised beaches at Hilton of Cadboll, Ross-shire

[*J. Dixon-Scott*

6. The drowned estuary of the river Salcombe, near Kingsbridge, Devon

(*By courtesy of the British Council*)

[*Berni, Klosters*

7. Young fold-mountains: the Alps in eastern Switzerland, with ser-rated skyline, pyramidal peaks (Gross-Litzner and Gross-Seehorn), cirques, glaciers and frost-riven slopes

[*Valentine, Dundee*

8. Old fold-mountains: the even summit-levels of the Cairngorms, Scotland

each other, there would be a tendency to ruck-up a complicated mountain system on the advancing edge of each and to leave between them a median area of different relief. Similarly, where a depositional zone was uparched, whole or in part (by subcrustal currents?), gravity could cause sediments to glide down the flanks of the arch, leaving a median core between intensely compressed outer ranges. Either may explain why some vast cordillera consist of high flanking ranges on either side of, what may be, a lofty plateau or a 'basin and range' landscape or even a lowland. Irrespective, however, of such complexities, the essential stages of mountain-building may be roughly summarized as follows:

(a) A downfold in the sea-floor exists near an earth-mass.

(b) Into this hollow thick deposits accumulate; the syncline sinks under their weight, but depression keeps pace with the deposition.

(c) Owing either to movement of the adjacent continental blocks, or to the effect of subcrustal currents beneath the geosyncline itself, orogenic uplift begins.

(d) The orogenic movements continue gradually over a long period of geological time.

(e) Continued uplift leads to much strata shortening and results in overfolding, overthrusting, and the formation of nappes. These are not necessarily very high.

(f) At the same time the great pressure and earth movements involved cause much metamorphism of the sedimentary rocks at the bases of the geosyncline and folds, and the mountain base is thus greatly hardened. In some cases, owing largely to the crustal weakness, lavas may find their way to the earth's surface.

(g) The orogenesis ultimately results in a mountain mass that 'floats' with a root of great depth beneath it (Fig. 6).

(h) Denudation, which from the first had attacked the slowly-rising mountains, continues at a rapid rate and begins to lay bare the core of the mass.

(i) Isostatic uplift, with or without subcrustal up-currents, now becomes active and maintains the elevation of deep-rooted mountain masses.

After long ages of erosion the hard core and metamorphosed roots of the mountain are carved into a block structure. Thus sedimentation is the start of a cycle of orogenesis and an old hard block

N

the conclusion (Plates 7 and 8). In other words, great massifs such as the Laurentian Shield and West Africa are the worn-down stumps of old fold mountains, and on them the trend lines of former mountain ranges are still visible. An excellent example occurs in Anglesey, which, although the oldest part of Wales, is one of the lowest and flattest because its ancient mountains have been almost completely peneplained. In other areas, such as Snowdonia and the Weald, the original upfolds no longer form the highest areas and the relief has been inverted (Fig. 58).

To-day there is a tendency to place less emphasis on epeirogenic or continent-building movements and more on orogenic or mountain-building movements. It is usual to regard continents as produced largely by a series of orogenic movements, each of which added to the land mass. Moreover it is generally held that mountain-building episodes are not confined to particular geological periods but are always going on somewhere in the world.

Mountain-building Movements in Britain. British mountain-building episodes take on a new interest when viewed as part or repercussions of orogenic movements on the Continent. Generally

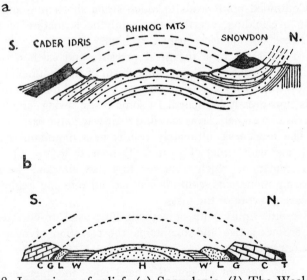

58. Inversions of relief: (*a*) Snowdonia, (*b*) The Weald
T. Tertiary Beds   C. Chalk   G. Gault Clay   L. Lower Greensand
W. Weald Clay   H. Hastings Sands

speaking, three major groups of mountain formations can be recognized in Europe.

1. The Caledonian orogeny, of which the mountains have now been worn down to mere stumps. Portions of this movement remain as low dissected plateaux in north-west Ireland, the Highlands of Scotland, and in the western parts of Scandinavia.

2. The Hercynian or Armorican orogeny, which considerably affected the British Isles south of a line from southern Ireland through the extreme south of Wales and the southern counties of England. Belgium and northern France, especially in Brittany or Armorica, were likewise affected. To the north of this line the ground was rocked by 'earth ground-swell,' which was not sufficient to cause great folding nor to have much effect on the Caledonian blocks. In this orogeny the Pennines were probably uplifted.

3. The Alpine orogeny, during which Britain was too far from the main area of compression to be contorted and elevated into mountains. In the south of Britain the slight fold that crosses the Isle of Wight and the main fold of the Weald are of this age. But the sediments which had been laid down to the south-east of the Caledonian blocks were for the most part only flexured or warped.

Faulting and Fracturing

Another factor which greatly alters the original folded nature of mountains is faulting or fracturing that occurs commonly both in folded mountains and in old hard blocks. Fracturing is probably more in evidence in the old blocks, as in this case the fractures, so common at the base of the original folds, are revealed by erosion. Moreover, any subsequent orogenic movements tend to cause further fracturing in resistant non-foldable rocks.

Whereas local tension in the earth's crust results in the formation of joints in the rocks, regional tension frequently leads to faults and fractures. Faulting usually implies an extension of the surface area, the extension varying with the amount of downthrow. Thrust faulting is, however, often compressive (Fig. 59).

In the vast majority of cases faults make little difference to the surface of a district; a dozen faults may be shown on a geological map, but the surface features may reveal no trace of them. The faults may, however, influence the position of springs and the course of valleys, as river erosion works more rapidly along a line of weakness. Near Oban, Loch Awe drains seawards through a narrow defile,

called the Pass of Brander, which has been cut by ice and running
water along a fracture line. The main direct topographical effects of
faulting are fault scarps and rift valleys.

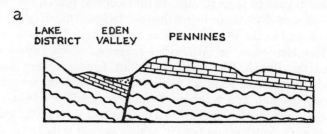

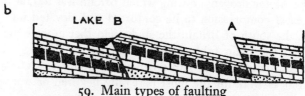

59. Main types of faulting

   (a) Normal fault. Diagrammatic section across the Eden Valley
       and Cross Fell
   (b) Reversed fault. Great Basin, Oregon. A shows initial fault
       scarp; B same scarp after erosion, with a lake formed
       against it

Fault Scarps. Fault scarps, when formed, are usually soon removed by erosion. Often, however, the faulting may later encourage
the formation, or resurrection, of another scarp along the old line
of fracture. Between Giggleswick and Settle there is an excellent
example of a fault directly influencing present surface features. Here
Giggleswick Scar, a step of hard, bare, grey limestone (Carboniferous)
stands above a lower slope of softer Millstone Grit (Fig. 60). The
ground on the south-west has dropped relative to that on the north-
east, but quicker erosion of the softer grits and shales may have
accentuated the 'Scar,' which is actually part of the newly exposed
fault-line. Other well-known scarps along fault-lines occur at the
abrupt western edge of the Pennines overlooking the Eden valley
and at the steep straight face of the Côte d'Or in France. In the
U.S.A., the Great Basin, between the Sierra Nevada and Wasatch
Range, has been fractured into blocks. The blocks, however, are

tilted, so that the faults are 'reversed,' and the Oregon lakes have collected in the fault valleys (Fig. 59b).

Rift Valleys. The greatest topographical features produced by faulting originate either when a tract of country between parallel faults slips down to form a rift valley (Midland Valley of Scotland)

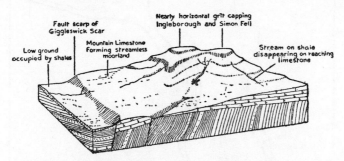

60. Ingleborough and Giggleswick Scar viewed from the south-east. Gaping Ghyll is marked by a cross.

(*Trueman, 'The Scenery of England and Wales.' Victor Gollancz*)

or when a split in the earth's crust widens into a valley with infacing faults (Red Sea). Another famous rift valley occurs along the middle Rhine between Basle and Mainz; here the faults aline a valley about 20 miles wide by 200 miles long. The Vosges and Black Forest have been left upstanding: such mountainous structures, outlined by faults, are known as *horsts* (Fig. 61).

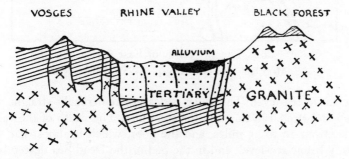

61. The Rhine rift valley; cross section in the latitude of Strasbourg

(*Modified from Joly*)

Shading denotes rocks of Secondary Age

The greatest of all rift valleys is that which persists for nearly 3,500 miles from the Taurus Mountains in Asia Minor almost to the Zambezi (Fig. 62). In Palestine the rift divides the plateau of Judæa from that of Trans-Jordan, both of which stand at 2,000 to 3,000

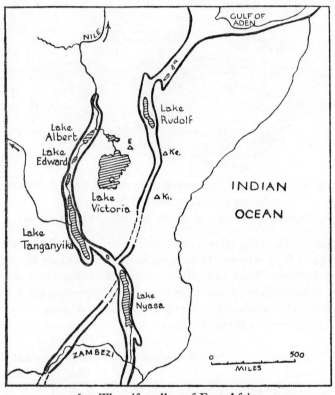

62. The rift valley of East Africa
E. Mt. Elgin   Ke. Mt. Kenya   Ki. Mt. Kilima Njaro

feet above sea-level. The valley sides are remarkably straight and steep and stretch in this way for 400 miles. The width of the valley varies from 10 to 25 miles, and the floor, which is almost flat, lies mostly below sea-level, and in the bed of the Dead Sea descends to 2,600 feet below the surface of the Mediterranean. The Jordan trench passes southwards into the Gulf of Akaba and the Red Sea, which have a typical rift shape. The rift forms the southern edge of the Abyssinian plateau and then continues through East Africa, where much of its floor and slopes is obscured by lava flows. The trench assumes immense proportions along the shores of lakes Nyasa and Tanganyika, where vulcanism is less in evidence. Each lake lies in a rift some 400 miles long by about 50 miles wide, the walls of which are steep or precipitous. The eastern wall consists of one big fault, but the western wall is formed of a series of steps. These sides rise in several parts to over 6,500 feet above the surface of the lakes. The rift, however, has greater dimensions than this, since the floor of Lake Tanganyika (depth 4,708 feet) descends 2,149 feet below sea-level, and that of Nyasa (depth 2,316 feet) 794 feet below sea-level. As a lake, Tanganyika is only exceeded in depth by Lake Baikal. From the northern end of Lake Nyasa a large fault trench extends northwards to the west side of Lake Victoria.

The Asiatic part of this immense rift is due to tear faulting near the Dead Sea and to crustal splitting in the Red Sea. The origin of the African valley is still in doubt. It is probable that earth-movements produced an arch, the keystone of which collapsed and formed the rift. Wegener supports a tensional view that Africa is being split by the tearing away of the East African plateau. More recently it has been postulated that, owing to compression, fracturing occurred and the margins of the area were forced upwards and the central portion thrust downwards; or in other words, that the rift is due as much to elevation of its sides as to the sinking of its floor. Today this compressional theory has been displaced by the arching and keystone-collapse theory, coupled with local vulcanism.

Summary. The correlation between orogenesis, faulting, and erosion is summarized in the following list of ways in which mountains may be formed:

   (i) folding—Alps, Jura, Rockies;
  (ii) Faulting—Pentland Hills, Vosges, Harz, Black Forest;
 (iii) vulcanism—Vesuvius, Etna;
 (iv) circumerosion or eroding away of surrounding strata, thus leaving an elevated mass—Catskills in eastern New York, mesas, and buttes of western U.S.A.

Mountains and Mankind

The effect of mountains on climate and economy has already been discussed; the barrier nature of mountain ranges needs no emphasis here. It is, however, relevant to notice how much plateaux attract settlement in the tropics, especially in lands settled by white people, as in East Africa. The culture of the Incas and Aztecs was typical of the superiority of a plateau environment in hot lands. Above all, it is remarkable how often mountain systems, when wide enough, will co-operate in the formation of national groups. Switzerland, Bolivia, Tibet, and Andorra are nations that have acquired unity in a mountain environment. This aspect is frequently overlooked in the stress laid on the utilization of mountain ranges as political frontiers.

LIST OF BOOKS

On Subject in General
Steers, J. A. *The Unstable Earth*, 1950
Jeffreys, H. *Earthquakes and Mountains*, 1950
Gignoux, M. *Stratigraphic Geology*, 1955
Goguel, J. *Tectonics*, 1962
Hills, E. S. *Elements of Structural Geology*, 1963
De Sitter, L. U. *Structural Geology*, 1964
Holmes, A. *Principles of Physical Geology*, 1965

Earthquakes
Eiby, G. A. *Earthquakes*, 1957
Davison, C. *Great Earthquakes*, 1936
Skrine, C. P. 'The Quetta Earthquake,' *G. J.*, Nov. 1936.
Richter, C. F. *Elementary Seismology*, 1958 (768 pp.)
Gutenberg, B. 'The Energy of Earthquakes', *Q.J. Geol. Soc.*, 1956, pp. 1–14
Bullen, K. E. *Introduction to the Theory of Seismology*, 1963

Changes in Sea-Level; and Land-Level
Baulig, H. *The Changing Sea-Level*, 1935. (Outline of problem.)
King, L. C. 'Pediplanation and Isostasy', *Q.J. Geol. Soc.*, 1955, pp. 353–9
Pugh, J. C. 'Isostatic Readjustment . . .' *ibid*, 1955, pp. 361–74
Fairbridge, R. W. 'The changing level of the sea', *Scientif. Amer.*, 1960, pp. 70–79

Fold Mountains
Collet, L. W. *The Structure of the Alps*, 1936
Ramsay, J. G. 'Stratigraphy, structure and metamorphism in western Alps,'
    Procs. Geol. Ass., 1963, pp. 357–91

For British Isles
Stamp, L. D. *Britain's Structure and Scenery*, 1946
Wills, L. J. *The Physiographical Evolution of Britain*, 1929
British Regional Geologies (H.M.S.O.)

Faults
King, L. C. *South African Scenery*, 1963
Cotton, C. A. 'Tectonic Relief,' *G. J.*, June, 1953, pp. 213-22
Taber, S. 'Fault troughs,' *Jour. Geol.*, 1927, pp. 577-606
Dixey, F. 'Erosion and Tectonics in E. African Rift System', *Q.J. Geol. Soc.*, 1946,
    pp. 339–88; *The East African Rift System*, 1956
Quennell, A. M. 'The Structural and Geomorphic Evolution of the Dead Sea
    Rift', *ibid.*, 1958, pp. 1–24
Girdler, R. W. 'Relationship of the Red Sea to the E. African Rift System', *ibid*,
    1958, pp. 79–105

CHAPTER FOURTEEN

VULCANISM

Volcanoes

A VOLCANO is a vent that links the surface of the earth with
some type of magma reservoir, and that may, in consequence,
act as a channel for the emission of hot gases and molten lava. The
conical mountain so often associated with volcanic vents is by no
means essential, and many great eruptions have occurred on flat land.

Distribution of Volcanoes. Irrespective of those vents erupting
wholly beneath the sea, some 480 active volcanoes have been recorded
during historic times. The distribution of these shows a marked
connection with the insular arcs and young fold mountains of the
Pacific seaboard. In all about 245 active volcanoes occur on islands
in the western Pacific and another 120 on the west side of the Americas.
These include the chain of islands stretching from the Kuriles to
Japan and Formosa, that contains 103 active peaks, and the Andean
chain where at least 40 active and 200 dormant vents are to be found.
The 'fiery girdle' of the Pacific includes the great volcanoes of the
East Indies and stretches eastwards to New Zealand, and may be
continued in Mounts Erebus and Terror in Antarctica.

In contrast to the Pacific, the Atlantic has only 60 active cones,
and these are mostly in Iceland and the Caribbean Sea. The Canaries
and Cape Verdes, however, have active vulcanism, and islands on
the Mid-Atlantic submarine ridge often record eruptions.

Generally speaking there are to-day few active volcanoes in the
central regions of continents. The main exception consists of the 10
active vents and numerous others recently extinct that mark the course
of the great rift system in East Africa.

The close connection of volcanic regions with the seaboard is
emphasized by the distribution in the Mediterranean. But the
correlation to folding breaks down in the case of the Himalayas,
where active cones are not found, and in the case of Iceland, where
recent folding is absent. Vulcanism, however, seems in the main to
follow lines of recent tectonic disturbance, either of a fold or fault
nature. In Java, Iceland, the Galapagos Islands, and East Africa the
vents lie above fractures and at their points of intersection. In other
words, volcanoes occur where the earth's more solid crust is weak or

thin. After what has been said in the previous chapter, it may seem contradictory to talk of a 'thin' crust in an area of young fold mountains. The much folded and fractured nature of an orogenic structure is, however, such as to lead to many lines of weakness that actually favour and are usually accompanied by a considerable intrusion of magma. It is noticeable, too, that the great mountain-building periods have also been periods of intense volcanic activity.

Nature of Volcanic Eruptions. Vulcanism does not imply a continuous layer of molten or fluid rocks beneath the earth's crust, since deep-seated rocks are kept solid by the pressure of the overlying strata. If at any spot the pressure is lessened, the melting-point of the rocks below that area will be lowered. No matter whether the lowering of pressure be accomplished by folding or by fracturing, the effect on the substratum will be comparatively local and may, in a geological sense, be relatively short-lived. (But see pages 32–34.)

Assuming that molten magma forms owing to decrease of pressure beneath a crustal weakness, it is still necessary to explain the expulsion of the lava. In many cases the pressure of the overlying rocks may force the lava upwards; in others the lava may rise because it is lighter than the surrounding rocks. The key to volcanic eruptions, however, lies in the presence of volatiles, especially water vapour. Even in the quietest of eruptions there is a considerable output of gas, and gas rather than lava is the most constant and characteristic product of volcanoes. Steam usually predominates, but nitrogen, hydrogen, carbon dioxide and monoxide, sulphur dioxide, chlorine and many other gases may be given off in large quantities. Experiments and investigations show that the fluidity of magma partly depends on the presence of gas. It is also known that water and carbonic acid are readily dissolved by hot silicate liquids under pressure, but when the pressure is released the dissolved gases escape and blow the liquid up in a frothy mass. All lavas rising to the surface are charged with steam and various gases which strive to escape from solution as the pressure decreases. Where a crustal weakness occurs the volatiles can spread and press upwards taking with them some of the molten lava into which they have diffused. Then a welling-up of volcanic lava occurs, such as has long been observed at Mauna Loa in Hawaii. Some of the released gases burn on contact with air and give out so much heat that the surface lava is kept fluid and the gases beneath continue to escape. Consequently, a volcanic vent of this nature remains active for long successive periods.

Even at 600° C. basaltic lava has been found to be still capable of slight movement, but once it has cooled, and so has lost its gases, the hardened mass must be heated to double that temperature to restore the same degree of fluidity. The hardness of cooled lava has a marked effect on some volcanic eruptions, because, if the escape of gases at the surface temporarily dies away, a hard crust may form over a volcanic vent. This hard crust tends to prevent the future escape of gases, which gradually concentrate in the lava beneath it. The time will come when the pressure of the gases cracks the crust and then the immediate release of pressure leads to a sudden volcanic eruption. In this case, fragments of the solid crust are ejected as well as lava, and the violence of the ejection contrasts vividly with the gentle upwelling of molten rocks typical of the ever-active volcano. Some volcanoes, such as Krakatoa and Vesuvius, erupt with the greatest of violence after lying dormant for long periods; others, such as Stromboli, the 'lighthouse of the Mediterranean,' erupt rhythmically at short intervals with less, but still considerable, violence; others, such as Mauna Loa, are more or less constantly in eruption in a quiet way.

The source of the water vapour and gases in lava is a matter for speculation. The idea that sea-water percolates into the neighbourhood of molten rocks can, with a few possible exceptions, be discounted. The gases usually are part of the rocks themselves having been imprisoned in them when they were originally formed. Thus the sponge-like nature of pumice is due to the escape of gases originally dissolved in the pumice itself. Volcanoes are related not so much to the sea as to the great lines of crustal weakness that happen so often to form the seaward edge of land masses.

Different Types of Volcanic Cones. In addition to gases, of which over 70 per cent. usually consists of water vapour, the most typical products of volcanic action are lava, fragmental material, and dust. Lava varies considerably in fluidity; acid lavas, which frequently contain 65 per cent. to 75 per cent. of silica, are sluggish and highly viscous; basic lavas, which normally contain only 40 per cent. to 50 per cent. of silica, are highly fluid and mobile. The fragmental material consists of parts of the crust formed in the crater and occasionally includes pieces of native rock blown from the top and sides of the vent. The fragments range in size from large angular blocks and rounded masses, called bombs, down to small stony pieces resembling coarse gravel, called lapilli, and to highly pulverized matter. Often the fine dust coats the slopes of a volcanic cone and consolidates into tuff or, if rain falls, solidifies into mud. Much of the fragmental matter named above is scoriaceous, or pumiceous, by which is meant that the rock is filled with cavities. Different volcanoes eject these products in widely different proportions and in widely different ways.

The shape of the topographical features formed depends very largely on the kind of material ejected. Owing to the stiff, viscous nature of acid lavas, the flows consolidate above the point of eruption and tend to be restricted in width, but of considerable thickness. The majority of volcanoes tend to be acid. Notable examples occur in Auvergne, where the Puy de Sarcoui is a typical rounded, dome-shaped, acidic cone; at Vulcano near Lipari in the Mediterranean; in Japan; the Azores; and in the *mamelons* of the Bourbon Islands in the Indian Ocean. Violent eruptions are usually associated with non-fluid lava. That of Krakatoa (1883) is the greatest known in historic times, and after it floating masses of pumice, which forms readily in acid lavas, almost blocked navigation in the Strait of Sunda. At Mont Pelée, in Martinique, the vent seems to have been blocked by a great solidified plug. The accumulated gases eventually escaped (1902) and a vast cloud of hot gas rolled down on the plain beneath, utterly devastating the town of St. Pierre. The solid plug then rose above the crater, but it has since gradually disintegrated.

Basic Lavas, owing to their greater fluidity, move farther on the surface and spread out as relatively thin sheets over a much larger area. The resultant cone is broad and flat or shield-shaped, with slopes seldom exceeding 7°. Examples occur in Hawaii and Iceland. Hawaii consists of five volcanoes of which Mauna Loa and Kea are the greatest volcanoes in the world. They rise 30,000 feet from the ocean bed, and their upper 14,000 feet is above sea-level (Fig. 63). As the sea-floor base of Hawaii has a diameter of 65 miles, the whole mountain feature is extremely flat. The crater that crowns Mauna Loa extends 2 by 3 miles, and is nearly 1,000 feet deep. Here volcanic activity is constant, but the volcano has grown so high that during eruptions lava usually breaks through the sides of the mountain and creeps quietly down them, reaching in places to the sea.

In Iceland similar features occur on a much smaller scale. Kólotta Dyngja, for example, rises to 1,400 feet above its surroundings on a diameter of nearly 4 miles. Its steepest slopes attain only 8° and, as on Hawaii the lava flows are relatively thin, most being between 12 and

30 inches in thickness, but they have been known to reach as much as
75 miles in length.

Some of the smaller volcanic cones consist mainly of lapilli and
scoriaceous fragments. This type seems to be common in regions
where vulcanism is gradually becoming extinct. The gases may break

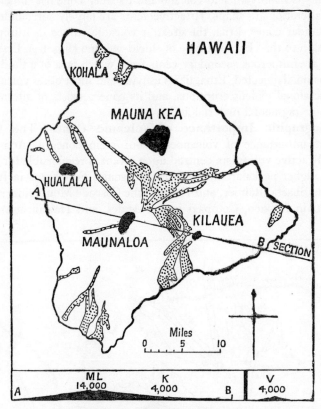

63. The Island of Hawaii

(*Wills, 'Physiographical Evolution of Britain.' Arnold*)

Black shows vents of volcanoes. Historic lava flows are dotted. A–B is
section across Mauna Loa as compared with V, section across Vesuvius
on same scale

up the crust solidified over a vent and hurl the fragments about the
crater. The material in falling lodges in steep angles, especially near
the crater, and consequently the typical cinder cone has a symmetrical
conical shape, its slopes being concave and especially steep near the

summit. Monte Nuovo in the Phlegraean Fields and many of the smaller cones in Auvergne, in Iceland, and near Lassen Peak, California, are of this nature.

Most volcanoes are intermediate in type, yielding lava and fragmental material. These different products cause composite cones. Vesuvius is mainly an acid volcano and its steep cone has undergone many alterations in shape. Its active vents are largely surrounded by small cinder cones. Etna, the greatest volcanic edifice in Europe, is more akin to the Hawaiian type of shield-volcano (Fig. 64). Upon its slopes are numerous secondary vents from which lava of a fluid type is occasionally ejected, Etna, like Vesuvius and most other volcanoes, has occasional violent eruptions, and its cone consists of alternating layers of fragmental material and lava.

Topographic Importance of Volcanic Cones. The topographic importance of volcanoes is our main concern. Active or recently active vulcanism centred upon a vent is responsible for most of the higher peaks around and in the Pacific; Fuji Yama in Japan, Mounts Shasta, Rainier, and Hood in the Cascade Mountains, Popocatepetl in Mexico, Cotopaxi in the Andes, and Hawaii and most

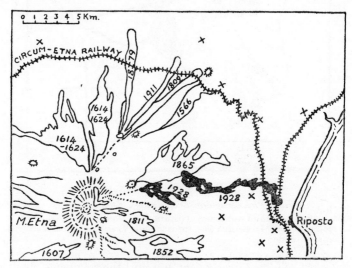

64. The north-east quadrant of Etna

(*After Tyrrell, 'Volcanoes.' Oxford*)

Lavas are indicated by their date of eruption. Latest great eruption (1928) is shown in black

'high' islands exemplify this. Vesuvius and Etna in the Mediterranean, Elbruz in the Caucasus, Ararat in Armenia, and Mounts
Kenya and Kilima Njaro in East Africa serve, however, to illustrate
the elevating and dominating topographic effect of volcanic action in
other parts of the world (Plate 9).

Erosion does not easily destroy the conical shape and circular plan
of extinct volcanic vents. The cone of the Marysville Buttes, California, although a mere skeleton of the former structure, still clearly
retains a typical volcanic shape and has given rise to a radial drainage.
The Cantal massif in Auvergne shows similar features and so, for
that matter, does the summit of the Wrekin in England, which
appears to be a central boss of bedded lavas and tuffs.

It frequently happens that the solidified lava conduit of a volcano
will resist erosion long after the slopes of the cone have been denuded
away. Continued denudation may result in this circular mass of
igneous rock protruding vertically above the surrounding area. The
Devil's Tower, Wyoming, is a volcanic 'plug' rising abruptly some
600 feet. Numerous volcanic 'necks' or 'plugs' occur in Scotland;
among these are North Berwick Law in Haddingtonshire, Dumgoyn
and the sites of Edinburgh and Stirling Castles (Fig. 65). Equally
famous are the *puys* or isolated conical hillocks of Auvergne, many of
which are crowned by a monastery or a chateau.

The craters of island volcanoes may lead on erosion to the formation of markedly circular islands very similar to atolls in outline.
St. Paul in mid-Atlantic, Santorin off Greece, and Deception Island
are well-known examples. Equally interesting is the formation of
submarine plateaux from volcanic peaks. Graham Island, a volcanic
peak formed in 1831 between Sicily and Africa, was raised to 200 feet,
but when the eruptions ceased it was soon reduced by the waves to
a shoal. Falcon Island in the Tonga group has been reduced to a
shoal and rebuilt within living memory.

In several cases the crater of a volcanic cone has become filled with
rain-water and forms a circular lake. Crater Lake in Oregon, which
occupies a crater at over 6,000 feet O.D., has a diameter of 5 miles
and a maximum depth of about 2,000 feet. The lake is surrounded by
steeply sloping volcanic walls that rise in parts to nearly 2,000 feet
above its surface. The basin was formed by the sinking of lava in the
vent, in which a volcanic cone has subsequently developed and
forms an islet in the lake. Other lakes in the craters of extinct volcanoes are Lake Nemi near Rome and Lake Avernus near Naples.

In the Eifel district of the Rhine Highlands, volcanic explosions formed numerous shallow-craters, in which water collects as lakes, called locally 'maare,' of which the Laacher See is one of the largest.

A less distinctive and less common type of lake due to vulcanism originates when a lava stream obstructs the flow of a river. Tiberias

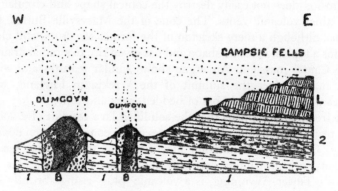

65. Volcanic plugs; Dumgoyn and Dumfoyn, Scotland

(*Tyrrell, 'Volcanoes.' Oxford*)

B. Basalt plugs   L. Basalt Lavas   T. Bed of Tuff   1. Old
Red Sandstone   2. Carboniferous shales and sandstones
Restoration of old volcano shown in dotted lines

in the Jordan rift-valley and Snag Lake in California occupy lava-dammed valleys.

Fissure Eruptions: The Upwelling at Fractures of Basaltic Magma (Fig. 6)

The volcanic phenomena already described are associated with a central vent around which some form of cone usually arises. Another manner in which lava is extruded consists of a welling-up along the length of a fissure. These linear eruptions may flood large areas of country. The flood may be greater at certain points, and it may come from numerous vents, but the ultimate result will be a *plateau* of lava. The products seem to be predominantly basic (or basaltic) in nature, and the fluid lava thins out gradually away from the fissure.

In Washington and Oregon lava-flows of this type cover 220,000 square miles and reach 4,000 feet in thickness. In the north-west Deccan of India similar fissure eruptions cover 230,000 square miles, and probably once extended over twice that area (Fig. 66). In Bombay Presidency the lavas exceed 7,000 feet in thickness, and one boring

[*Dorien Leigh*

9. A volcanic cone. Fuji Yama, Japan

[*W. Lawrence, Dublin*

10. Basalt sills at Pleaskin Head, near the Giant's Causeway, County Antrim

[*Dorien Leigh*

11. Moving dunes in Death Valley, southern California

Crown Copyright] [*Geological Survey*

12. Culbin Sands, Elgin, showing old land surface (on left) swept bare
of advancing sand

revealed twenty-nine distinct flows averaging 40 feet each in thickness.

Notable lava plateaux occur in Antrim, western Scotland, Iceland, and parts of Greenland. The eruptions seem to date mainly from Tertiary times when the areas may have formed part of one lava field. In Antrim the individual basalt flows seldom exceed 50 feet in

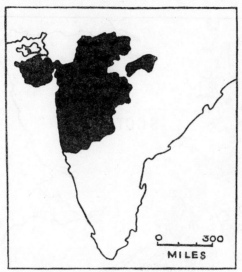

66. Basalt lava plateau of the Deccan, India

thickness and are often separated by the remains of surface soils containing plant and animal remains. Similar evidence of successive upwellings of lava also occurs in other lava fields of this type. The famous columnar structure of the Giant's Causeway and Fingal's Cave appears to be related to the injection of lava rather more recent than the main plateau (Plate 10). In the Western Isles of Scotland, Skye, Rum, and Mull are nearly entirely the product of volcanic outbursts of the Tertiary period. In the Clyde region a lava plateau of an earlier period has been exposed by erosion until it forms many hills and small flat-topped blocks of which Dumgoyn and the Campsie Fells are remnants (Figs. 65 and 67).

Iceland, however, furnishes the only known case of a big fissure-eruption in historic times. At Laki, in 1783, a fissure nearly 20 miles long emitted lava streams that flowed for 40 miles westward and 28

o

miles eastward of the opening. All told, 220 square miles were flooded with lava and one-fifth of the population and the bigger part of the stock of Iceland perished.

When it is added that enormous lava plateaux also occur in South Africa and in the Parana region of South America, it will be realized that this type of vulcanism is the grandest and greatest known. The

67. Chief Tertiary volcanic regions of the British Isles
(*Wills, 'Physiographical Evolution of Britain.' Arnold*)
Basalt-plateaux, dotted; central volcanoes, solid black; main dikes, straight lines

scenic effect has, however, little extraordinary about it. The mighty mountain scarp of the Drakensberg in South Africa is due mainly to fissure eruptions, while the deep canyons of the Snake River, Oregon, have been cut into almost horizontal lava flows. But it is the *plateau* itself that is the dominant topographic effect.

Intrusions of Magma

So far only the extrusion of lava at the earth's surface has been discussed. It frequently happens, however, that topography is affected by magma that intrudes into but does not actually breach the overlying rocks.

Very large bodies of intrusive magma are called batholiths (Fig. 68). These immense masses of igneous rock are extremely common in many parts of the world and abound, for example, in the Laurentian Shield. On exposure by weathering they form rounded hill masses such as the granite batholiths of the Wicklow Mountains, Ireland, and Pike's Peak, U.S.A. These larger magma masses probably form the feeding-ground for other intrusions.

A smaller type of intrusion called a laccolith usually consists of viscous magma that arches up the covering of overlying rocks. After erosion the mass tends to produce an isolated, conical hill of which Traprain Law in Haddingtonshire and the Henry Mountains in Utah, U.S.A., are characteristic.

Laccoliths frequently merge into sills and are commonly associated with dikes. A sill is a stratiform intrusion of igneous material that has been thrust, often along the bedding planes of adjacent strata, to form a horizontal sheet of rock. Sills usually thin out away from the conduit and consequently they vary in thickness from a few inches to hundreds of feet. When tilted and exposed by erosion, sills form

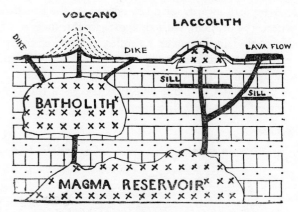

68. Main types of igneous intrusion and
extrusion

Thick upper line shows earth's surface after erosion

scarps, such as that of Salisbury Crags overlooking Edinburgh. The Hudson River crosses a thick sill at the Palisades opposite New York City, and the name tells of the columnar structure of the intrusion. A far greater sill, the Great Whin Sill, occurs in northern England. The underground extent of its dark, blue-grey rock (dolerite) can be traced for about 80 miles from Burton Fell on the western scarp of the Pennines to Bamborough. The lofty crags of High Cup Nick, four miles east of Appleby, the cliffs on which Bamborough Castle stand; the ridge topped by Hadrian's Wall north of the Tyne at Haltwhistle; and the low, rocky Farne Islands illustrate the scenic effect of its outcrops. The sill, which has an average thickness of nearly 100 feet, extends over an area of at least 1,500 square miles and lies nearly horizontally. It is generally parallel to the limestone strata through which it is intruded, but locally it cuts across the beds (Fig. 69). It has been suggested that the little volcanoes of the Phlegraean Fields arise from a sill-like sheet of magma intruded laterally from Vesuvius.

Sills are by no means so commonly exposed at the surface as dikes, which are thin, tabular, parallel-walled masses of igneous material that have been injected at a steep angle into the rocks. Dikes are essentially vertically inclined in contrast to the horizontal disposition of sills. They may form the feeders to a fissure eruption and may extend in thickness to several hundreds of feet and in length to several miles. They are exposed at the surface in greatest numbers near to centres of past volcanic action such as in parts of Western Scotland. A typical 'swarm' of dikes occurs on the south-east coast of Mull where 375 of them, with an aggregate thickness of over 2,500 feet, were counted in just over 12 miles. These relatively narrow, almost vertical intrusions often weather into low 'walls' that cross the country up hill and down dale in a fairly straight line. A notable example is the great Cleveland dike that traverses the North York Moors in a west-north-west direction.

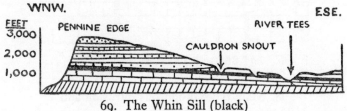

69. The Whin Sill (black)

Diagrammatic section crossing the Tees at Cauldron Snout and just below High Force

The Signs of Decaying Vulcanism

Some writers advocate a 'volcanic cycle': (a) first the fissure eruption producing floods of basalt lava and swarms of dikes; (b) gradually the eruptions are localized at vents and shield volcanoes form; (c) the volcano declines to a partly explosive, partly upwelling type; (d) eventually the activity becomes purely explosive, the main product being steam; (e) finally only steam and gases are discharged.

Evidence of dying vulcanism occurs in a great many parts of the world chiefly in the form of geysers, hot gas exhalations, and hot springs. Geysers are named after the Great Geyser in Iceland, where eruptive springs were first observed and described. The geyser areas of North Island, New Zealand, and of the Yellowstone National Park, U.S.A., where Old Faithful erupts every sixty-five minutes, are, however, more visited. The eruptions of superheated water depend on steam produced in surface fissures by the heat of underlying rocks. Surface water comes in contact with the steam and if the fissure is siphon-shaped an explosion occurs whenever the steam forces its way into the vertical outlet of the fissure. Then some water is forced out at the surface, and the sudden decrease of pressure causes the rest of the overheated water to change into steam. The pipe or fissure is refilled mainly with rain-water, and the process is repeated.

Sometimes the vents give off steam and vapours only. These gas exhalations, mainly of steam, occur in abundance at the 'Valley of Ten Thousand Smokes,' near the Alaskan volcano of Katmai. Here the vents, called *fumaroles*, came into being after the paroxysmal explosion of 1912. Similar exhalations in Italy are termed *solfataras*, after La Solfatara, a gas-vent in the Phlegraean Fields.

The Effect of Vulcanism on Human Life

As vulcanism is associated with igneous rocks, it is of some importance to man as a concentrator or maker of metals and precious stones. The copper deposits of Butte, Montana, occur in a large batholith; the nickel deposits at Sudbury, Ontario, lie in a volcanic intrusion; the diamond mines at Kimberley follow a volcanic pipe or circular dike (see also page 184).

The effect of lava eruptions on farming is, in spite of the physical difficulties and danger involved in central eruptions, often beneficial. Volcanic lavas consist of a mixture of rocks, and eventually, except for the highly acid type, usually weather into fertile soils.

Lava-flows and coarse debris may remain barren for centuries but fine ash-deposits are soon utilized. The deep, fertile, moisture-retaining soils weathered from the basalt plateau of the Deccan form the main cotton lands of India. The volcanic soils of Java yield abundantly, while the closeness of settlement on the lower slopes of Vesuvius depends on the great fertility of the soil. The plains and lower slopes of Etna are equally densely peopled; here, to about 2,500 feet, the fertile volcanic soils induce intensive cultivation. Thence up to 6,000 feet the slopes are clothed with forests of chestnut, oak, and beech, and, towards the upper limits, of pines. From 6,000 feet upwards extends a weird waste of red and black lava, of crags and precipices that culminate in a snow-clad plateau above which rises the terminal cone. To people tilling the slopes of these volcanoes, the eruptive phenomena, unless of some violence, attract little more attention than does a severe thunderstorm in Britain.

The milder forms of vulcanism can be put to direct human uses. In Iceland and New Zealand, housewives utilize hot springs for cooking and laundry. Projects have been undertaken to supply Reykjavik and other towns in Iceland with a hot-water supply from boiling springs. Natural steam-power has long been used in Tuscany, and there is every indication that here and elsewhere underground sources of heat will be increasingly tapped for power purposes.

LIST OF BOOKS

Tyrrell, G. W. *Volcanoes*, 1931. (H.U.L.)
Cotton, C. A. *Volcanoes as Landscape Forms*, 1944
Anderson, T. *Volcanic Studies in Many Lands*, 1903–17. (For photographs.)
Stearns, H. T. *Geology of the Hawaiian Islands*, 1946
Williams, H. 'The history and characteristics of volcanic domes', *Univ. Calif. Geol. Sc.*, 1932, pp. 51–146; 'Calderas', *ibid*, 1941, pp. 235–346
Bullard, F. M. *Volcanoes*, 1962
Rittman, A. *Volcanoes and Their Activity*, 1962
International Vulcanological Association. *Catalogue of Active Volcanoes*, 1951—
Foshag, W. F. & Gonzalez, J. R. 'The birth and development of Paricutin Volcano, Mexico', *U.S. Geol. Surv. Bull. 965*, 1956, pp. 355–489

British Isles

Geikie, A. *The Ancient Volcanoes of Great Britain*, 2 vols., 1897
Mems. Geol. Surv. for western Scotland. Especially Richey, J. E. *The Tertiary Volcanic Districts*, 1961
Wooldridge, S. W., and Morgan, R. S. *An Outline of Geomorphology*, 1959
Wills, L. J. *The Physiographical Evolution of Britain*, 1929
Lake, P., and Rastall, R. H. *Textbook of Geology*, 1941
Charlesworth, J. K., and others. 'Geology of N.E. Ireland,' *Procs. Geol. Ass.* Dec. 1935

CHAPTER FIFTEEN

THE WORK OF THE ATMOSPHERE

THE earth movements already discussed bring into being the lithosphere and cause the major deformations of its surface. The minor topographic features of any landscape are, however, the product mainly of sculpturing agencies. The most universal and yet at the same time the least spectacular of all sculpturing agents is the atmosphere, which produces the surface soils on which and with which other earth-modelling forces work. It is weathering, or the eternal process of rock disintegration and decomposition, that supplies fluvial, glacial, and aeolian agents with much of their load. In a general sense, it is also the atmosphere that, with the exception of sea coasts, controls the predominant type of erosion. The atmosphere works in conjunction with erosive forces to lower elevated portions of the land masses, weathering being usually most rapid where erosion is greatest and least where deposition protects the underlying rocks.

The work of the atmosphere is, however, both disintegrating and eroding. The first is called weathering, while the second, or mechanical aspect, is best discussed as the work of the wind.

Chemical Weathering

The chemical action that leads to the decay of rock surfaces is mainly connected with oxidization, carbonation, and hydration.

Oxidization. Oxygen in the air may combine, especially in the presence of moisture, with certain rock constituents to form oxides. This action is especially common in the case of iron compounds, and it greatly assists the disintegration of iron-bearing rocks.

Carbonation. Water containing carbon dioxide often forms carbonates that are, as a rule, relatively soluble. Ground water of this nature acts most strongly on limestones and on rocks which are cemented by lime (Fig. 145).

Hydration. The mere taking up of water as a chemical constituent may weaken the durability of a rock mass. Some forms of mica, that occur commonly in sandstones, take up water and cause the surface of the rocks to decompose into separate grains. In granite the feldspars disintegrate under the chemical action of ground water, and decompose into relatively insoluble clay minerals, such as kaolin. Generally speaking, igneous rocks are more susceptible to chemical weathering than are sedimentary strata, with the exception of limestones. Rocks consisting largely of silica, especially in its crystalline form of quartz, and of clay minerals are least liable to undergo chemical change. It should be noticed, too, that humid climates with high temperatures considerably increase the solvent action of surface water. The problem of chemical weathering is exceedingly complex, and is discussed further in Chapter Twenty-Four.

Physical Weathering

The physical action of weathering depends in warmer climates on the expansion of rock surfaces on heating and their subsequent contraction on cooling. As rocks are such bad conductors of heat only a very shallow surface layer is affected, and eventually this thin layer tends to break away from the main body of rock. The rise of temperature involved is not that of the air but that of the *rock face*, which becomes much hotter. The splitting action of expansion is considerably stimulated by variations in colour and in durability of the constituents of a rock. Yet even where large homogeneous rock masses, such as granite blocks, are subjected to changes of temperature and humidity the surface layers flake off and *exfoliation* occurs.

In colder climates the physical action of weathering depends mainly on the fact that water expands nearly one-tenth its volume when it freezes. Moisture in pores, crevices, joints, and fissures of rocks expands on freezing, and tends to disrupt the surface of the rock. If freezing is followed by alternate thawing, the moisture works its way farther into the rock, thus loosening successive surface layers. Soon after sunrise many a valley in Alpine Switzerland experiences a daily shower of loosened rubble from the mountain slopes nearby; in Britain the screes or talus slopes of fallen débris at the foot of precipices tell the same story. The vast screes on the precipitous slopes of Snowdon and of Ben Nevis are typical; those alining the lower flanks of Borrowdale and forming the declivities above Wastwater in the Lake District are equally large.

The Work of the Wind: Arid Topographies

The atmosphere, in addition to its infinitely gradual but persistent and almost universal work as a producer of soils, is also capable of considerable mechanical action.

Aeolian work only occurs to an appreciable extent in areas where vegetation is practically absent and where evaporation so much exceeds rainfall that the surface soil is normally dry and incohesive. These regions with dry air and arid soil conditions form the world's chief deserts; they lie mainly between latitudes 20° to 45° in each hemisphere, and, all told, comprise about 30 per cent. of the land surface outside the polar regions. A hot desert cannot be clearly demarcated on either vegetal or climatic bases, and each area is surrounded by a slow gradation of arid conditions. Certain characteristics, however, come to mind; lack of vegetation to bind the soil, bare surfaces of rock or sand; high temperatures and dry, shimmering air; wind-swirls and dust-storms.

Scenic Variations in Hot Deserts. The variations in desert scenery are considerable, but locally the landscape may be monotonous owing to the vast areas concerned and to the relative absence of vegetation which in most other latitudes often tends to hide or subdue surface relief. The surface characteristics of an arid region vary especially with the geological outcrop, with the local balance between wind erosion and wind deposition, and with the part played by water erosion.

Generally speaking, four main types of arid and semi-arid scenery may be distinguished. The sand-desert proper, a maze of billowy sand-dunes and crescentic sand-ridges frequently in a state of change and movement, is called *erg*. A great belt occurs in the Sahara from Mauritania to southern Tunis (El Erg) and again east of the Aïr Massif; the most inaccessible parts of the world's deserts, such as Libya and the Rub 'Al Khali ('abode of emptiness') in Arabia, are largely of this type. Similar scenery in Turkestan is called *Koum*. Large areas are almost completely void of plant-growth, while in others occasional bushes occur in the major hollows.

The rocky, stony desert where bare rock covers large areas and loose sand is largely absent or confined to the major hollows is called *hamada* in the Sahara. This is the fixed desert associated especially with plateaux and elevated outcrops of rock in arid regions. Although its vegetation may be little in some areas, it is generally preferred to the *erg* for caravan routes owing to its firmer surface and the probability of water-holes upon or near it. The Ahaggar and Tibesti highlands in the Sahara are examples.

Some localities seem to represent the transition stage between hamada and erg, these are known as *reg*, and consist of stretches of gravel, pebbles, and stones. An extensive development occurs in the Tanezruft region, west of the Ahaggar plateau. Occasionally the desolate stretches of pebbles look as if they have been polished and varnished.

Some desert and semi-arid localities are dissected by a series of gorges or ravines, called canyons in Arizona and *chebka* in the Sahara, which outline a wilderness of small, steep-sided plateaux. Parts of the Sahara and of the Hadhramaut ('The abode of death') in southern Arabia, show this dissection, but it especially abounds in the Bad Lands of Dakota, U.S.A.

The above varieties of desert scenery are due largely to the relation between geological outcrop and the predominant factor in modelling the landscape of the particular district. The main agents in the formation of desert scenery are weathering, sand corrasion, transportation and deposition by wind, and erosion and deposition by surface water.

Weathering. Mechanical disintegration, owing to the expansion and contraction of alternately heated and cooled surfaces, is at its maximum on the bare rock surfaces of arid regions. Since the rocks are frequently heterogeneous in nature, such as sandstone and conglomerate, the differential expansion of their constituents increases the disintegration of the rock face. More homogeneous rocks, such as granite, are weathered along their lines of weakness, notably the joints, while sedimentary rocks of any kind are attacked along their lines of stratification. The extreme insolation of the dry air causes daily variations of over 150° F. of the rock surface, and the ground may become so heated as to feel burning to the touch. During the heat of the day rocks will occasionally split with pistol-like reports. The small disintegrated particles are swept away by the wind, but, where formed, the coarser rock-waste accumulates at the foot of slopes to form sharp-angled screes.

It is thought that in parts the weathering of harder rocks may be assisted by dew forming at night in the cracks, which would especially happen on the shaded slopes. In a minor degree, too, weathering is assisted by the oxidization of iron compounds and by the presence of salts in the rock, as, owing to the aridity, the salts are not removed by running water. The salts come to the surface through capillary action, and, being hygroscopic, become damp at night and the moisture assists the granulation of the rock surface in the daytime.

Sand Corrasion. Weathering supplies the desert with its sand

and greatly assists in the formation of isolated columns, a type of land-form which exposes a maximum proportion of its surface to atmospheric changes. Desert rocks, especially where soft and arranged in almost horizontal layers beneath a harder crust, will form almost vertical slopes under the influence of weathering and wind erosion.

Desert sands consist mainly of hard mineral grains, in which quartz predominates. These particles are carried by the wind and used as cutting-tools to form a sandblast strong enough to cut through wooden telegraph poles in the Australian desert. The hard particles, once in motion, progress partly by means of saltation or bouncing off the dry ground. They are dashed against projecting surfaces which they proceed to wear away, especially at a short distance above ground-level, where the heaviest particles are travelling. Wind corrasion and deflation gradually lower the rock-surface, but, owing to inequalities in the hardness of the surface layers, some parts are eroded faster than others. Thus, manganese concretions in the Nubian Sandstone of the Sinai desert have been left upstanding as small isolated blocks. Where the crustal layer is harder than the substrata, mushroom- and anvil-shaped rocks, capped by the harder rock, may be formed. These abrupt hillocks, often with overhanging sides, are locally known as *zeugen* or *inselberge*. The variations in this type of hillock, chimney, or pillar are many, but, generally speaking, they are characterized by abrupt faces, at least on the windward side, and by a distinct aspect of isolation.

Sand Deposition and Transportation. The sand being carried by the wind normally tends to accumulate in the hollows on the leeward quarter of areas where weathering is most active. The highest areas are often *hamada* or *reg*; the lower areas normally show on their surfaces one or more of the many varieties of sand deposition. Just as stony desert varies from bare rock to sheets of varnished pebbles, so does sand desert vary from an undulating wilderness of dunes to flat sand-sheets. Perhaps three types of *sand desert* may be distinguished as characteristic of various parts of deserts in the Old World.

The *barchan*, a semi-circular sand-dune with a hollow, crescent-shaped front facing down-wind, occurs, for example, over large areas in Turkestan and Libya. The sand is driven by the wind up the gentle slope and falls over on the concave side; consequently, the dune is highest in the centre and tapers off towards the base of its horns. In size these dunes range up to 90 feet in height, and from 100 to 400 yards between the horns of the crescent. *Barchans* may be scattered singly over a sandy area or they may congregate together in long curving ridges, appearing not unlike a sand-model of a mountain ridge eaten into by a continuous series of deep cirques (Fig. 70). Small dunes of the *barchan* type alter their form with changes in the direction of the wind, but the movement of bigger dunes is largely superficial, and variable winds cannot appreciably affect their lateral

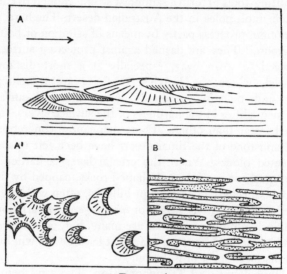

70. Desert dunes
(*After Bagnold and Cornish*)
A. Barchans; B. Seif ridges; A¹, seif-dune ridge deteriorating into barchans

position. A very small moving ridge of the barchan type of dune is shown in Plate 11.

Seif dunes are long, narrow ridges lying parallel with the wind and with crests that rise gently from windward to leeward. A *seif* may be isolated, but frequently it links up with others to form a range or line of dunes, of which the crest may stretch for 40 miles or more at a height of 300 feet above its base. Sand-desert of this type occurs in the southern Thar desert, India and Pakistan, where the monsoon winds are strongest locally (Fig. 70B). In parts of the Libyan desert the *seif* landscape resembles a series of uniform parallel ridges and corridors.

The undulating nature of these shifting sand topographies contrasts markedly with the sand-sheet deserts which occur especially on the Nubian Sandstone country of the Libyan desert. Here the sand-sheets are extremely flat expanses which show no tendency to form dunes; one sheet alone, near Selima, covers 20,000 square miles. The vastness, however, of the Libyan desert (about 1,000 miles from north to south by an average of 800 miles wide) makes these local scenic variations appear so small that one authority describes the whole area as being 'almost uniform in character.'

Loess. Wind deposition is not limited actually to desert areas. The movement of desert sand is largely restricted to the six feet of the atmosphere nearest to the ground, but the very much finer dust may be carried by upward air currents into the atmosphere and be transported far from the place of origin. Dust from the Sahara has given 'red rain' in Britain, and dust storms occur on the flood plain of the Mississippi and Missouri in summer. Indeed, a noticeable feature of dry climates is the dust storm. The microscopic particles of dust carried by wind from arid surfaces accumulate in some areas to form a soil known as loess. The loess of the Mississippi basin seldom exceeds 50 feet in thickness, but along the middle course of the Hwang Ho in north China, thicknesses of 600 feet are known. Here the strong westerly winds in winter deposit a very fine, yellowish dust carried from the deserts of central Asia. The loess has covered the lower land until the higher mountain ranges now overlook a series of broad valley basins. The highly permeable loess is penetrated by innumerable minute vertical tubes, which may mark the roots and stems of successive growths of grass-cover, each growth being smothered in its turn by a new dust surface. These tubes cause the loess to form almost vertical escarpments under the influence of erosion. The Hwang Ho has cut a deep valley bounded by cliffs, while even cart-tracks become worn into narrow ravines as much as 100 feet deep. Along the sunken roads dwellings are excavated out of the loess walls; these storied houses, with wooden doors and windows overlooking the roadway, usually have an upper exit on to the cultivated plateau surface. The landscape of this upper surface appears almost flat, since the ravines are largely hidden from a distance. The loess region of north China covers some 250,000 square miles (cf. France), and has no true counterpart elsewhere. In Europe a belt of loess soils stretches from Picardy to Poland and the Ukraine, but here the wind-borne dust was probably

derived from the sands of the glacial deposits to the north.
Sand on Shore Lines. The transporting work of wind is seen on
many shore lines of the world. Sand, no matter whether formed by
water or weathering, always contains much quartz. These particles
lack cohesion and quickly become movable unless kept permanently
moist by rainfall. The wind, acting on the drier surface layers of sand,
has formed dunes on the coasts of Cornwall and many other parts of
the British seaboard. The Landes region of Gascony in south-west
France has been formed during the last six hundred years by the
sand dunes moving inland under the influence of westerly winds.
Here the movement of the sand has been checked by erecting brush
fences; then the surface was stabilized by planting marram grass
and pines. Moving dunes of considerable size occur in many parts of
the Great Plains of the U.S.A.; here the connection between sand
and wind direction is well seen around Lake Michigan, where the
prevailing winds, being from a westerly quarter, have piled up a
large dune-belt only along the eastern shores of the lake.

In Britain a remarkable area of sand country occurs along the
southern shores of the Moray Firth between Burghead and Nairn.
Here the Culbin estate, which originally comprised some 3,600 acres
and sixteen sizeable farms, has been engulfed by sand, first deposited
by currents and then blown inland by wind. It is said that the
chimneys of the chief farmhouse sometimes reappear above the sand.
Besides moving dunes, many other topographical features typical of
wind-work appear here, including isolated columns capped with a
growth of grass which acts as a resistant surface-layer. Large areas
of the Culbin Sands have to-day been stabilized by 'thatching' with
wattle-hurdles and then by planting with marram grass and pines
(Plate 12).

Work of Surface Water in Arid Lands. There yet remains to
discuss the work of running water as an agent in the modelling of
arid landscapes. It is rather a paradox that upon the dry, loose soils
of arid regions surface water should find the optimum conditions for
erosion. Fluviatile erosion, however, is usually limited by climatic
conditions to the actual stream bed, and the rain-wash of valley sides
and the lubricating effect of rainfall on the soils of valley slopes are
small. Consequently valleys in desert areas usually have abrupt sides;
they are mere gashes in the scenic make-up, and play a minor
part in the modelling of the major landscape. In semi-arid regions
erosion by surface water becomes of much greater significance. The

rainfall, although small, comes as sudden storms which may gully the surface, and, in extreme cases, will form 'bad land' topography. Thus, the soft lacustrine clays in large areas of the western plains of the U.S.A. have been eroded by sudden downpours into a maze of deep gorges that separate abrupt-sided *buttes* and *mesas*.

This deeply dissected type of topography is not unknown in deserts proper; the lower course of the Wadi Hadhramaut in southern Arabia traverses such a country, and parts of the Sahara, chiefly near the edges of the higher hills, are dissected by gorges, here called *chebka*. Generally, however, these ramifying gorges are related to a pluvial climate during the Ice Age (*see* p. 225).

The desert regions of the world are peculiarly areas of inland drainage, and few permanent rivers leave deserts for the sea. The Colorado, with its mighty canyon that is incised through almost horizontal layers of sandstone to a depth of nearly a mile, is a classic exception. The bed of the normal desert wadi is entirely dry for long periods; or it may have a few water-holes towards its upper course; or floods may flow down it for a short distance every so many years before disappearing into a delta of sand and gravel. High mountains in desert areas have a certain amount of running water; in winter, snow rests temporarily on the highest parts of the Ahaggar plateau, and at that season wadis leading southward from the Atlas Mountains often flood. De Martonne gives an instance of a disaster in the Wadi Urirlu south of Ghardaïa in Algeria, where a troop of French soldiers were caught in a flood and twenty-eight of them were drowned. When the rare local showers in deserts fall on the higher, rocky surfaces, the rapid run-off encourages sudden flooding in the steep-walled valley. Normally, the flooded water will travel only a short distance down the wadi before it becomes so laden with silt that its progress is impeded, and it disappears in a fan of débris.

Alluvial and gravel deltas are almost as characteristic of wadis as is the abrupt-sided valley. The loose material forming the fan often has a steepish slope and may be furrowed by gravel channels marking the dried-up courses of streams. At the foot of high hills the deltaic fans will grow outwards till they cover the slopes of a valley or the edges of a basin with a belt of gravel deposits. In the great Tarim basin most of the streams die away in a piedmont zone of gravel from 5 to 40 miles wide. The Turfan basin shows the same features on a smaller scale. Elsewhere a whole basin or valley may be filled with an almost flat bed of pebbly gravels to form what is called in

Mexico a *bolsón*. The oases of west Argentina and north Chile are often centred upon deltaic fans of this nature; here the irrigation channels take the place of streams, and the greatest danger is the very rare flood which will sweep away soils, and litter the cultivated fields with gravel and boulders.

The extensive 'fossil river systems' that are found in the Sahara and Arabia were probably formed during the Ice Age, when surface streams persisted over much longer distances than they can to-day. The beds of these wadis were the work of considerable rivers; former connections with the south are suggested by the presence of degenerate crocodiles in the northern valleys of the Ahaggar. Generally speaking, the wadi and deltaic fan do not occupy a large proportion of desert surfaces, but they are of considerable human importance as in them vegetation and ground water at a reachable depth are most likely to be found.

Human Significance of the Work of Winds

The human significance of the work of wind in modelling landscapes is not easy to assess. The action of the wind in building and moving dunes increases the already inhospitable nature of deserts. More than one civilization has perished probably because wars prevented the people from keeping their irrigation channels free from sand. The fate of the Culbin estate serves to remind us of the buried cities of the deserts of Asia which have attracted the attention of modern explorers. The unfavourable influence of the work of the wind is well seen in the motor routes across the west Sahara which cling wherever possible to the *hamada* areas.

Its hostility is, however, probably of greatest direct human importance in semi-arid areas, such as the 'dust-bowl' of the U.S.A., where years of adequate rainfall may be followed by years of drought. In these prolonged dry periods, the arable crops fail and the loose cultivated soils are attacked by the wind. Dense dust-storms and the piling-up of sand-dunes against abandoned farmsteads then tell that man has seriously underestimated the effect of wind on bare, dried-out surfaces.

Deserts form vast areas of scanty settlement, and their intractable nature is relieved only by the oases that are based largely on sources of underground water-supply. Occasionally the salts, which are especially common in dried-up areas of inland drainage, may be economically useful, as happens in north Chile. In the aggregate,

[*The Times*

13. Upper stretch of the gorge at Cheddar in the Mendips

14. Caves in limestone at Cheddar, showing subterranean stream, stalagmites and stalactite curtains

15. A youthful stream. The Afon Dulas, Montgomery

[*D. J. Davies, Lampeter*
16. A mature river. The Rheidol below Devil's Bridge, Cardigan

however, deserts repel mankind and hinder the spread of land communications.

LIST OF BOOKS

Weathering: See also Chapter XXIV on Soils
Thornbury, W. D.   *Principles of Geomorphology*, 1954
Reiche, P.   *A Survey of Weathering Processes and Products*, 1950
Taber, S. 'Frost heaving,' *Jour. Geol.*, 1929, pp. 428–61
Polynov, B. B. (trans. Muir, A.).   *The Cycle of Weathering*, 1937
Keller, W. D. *Principles of Chemical Weathering*, 1962

Wind as a Landscape Factor, and Arid Topographies
Peel, R. F. 'Aspects of desert geomorphology', *Geog.*, 1960, pp. 241–62
Cotton, C. A. *Climatic Accidents in Landscape-making*, 1942
Davis, W. M. *Geographical Essays*, 1909; reprinted 1954

Sand Waves
Melton, F. A. 'A tentative classification of dunes', *Jour. Geol.*, 1940, pp. 113–73
Olson, J. S. 'Lake Michigan dune development', *ibid.*, 1958, pp. 254–63
Bagnold, R. A. 'Movement of Desert Sand,' *G. J.*, April 1935. *The Physics of Blown Sand and Desert Dunes*, 1941. 'Sand formations in Southern Arabia,' *G. J.*, March, 1951, pp. 78–86
Jennings, J. N. 'On the orientation of parabolic dunes,' *G.J.*, 1957, pp. 474–80

For descriptions of great deserts:
Sahara: Gautier, E. F. *Le Sahara—L'Afrique Noire Occidentale*, 1935
Arabia: works of Doughty, C. M., Lawrence, T. E., and Philby, H. St. J.
Central Asia: works by Sir Aural Stein and Huntington, E.
Kalahari: Passarge, S. *Die Kalahari*, 1904

Loess Soils
Barbour, G. B. 'Recent Observations on the Loess of North China,' *G. J.*, July 1935
Fuller, M. L., and Clapp, F. G. 'Loess and Rock Dwellings of Shensi, China,' *Geog. Rev.*, 1924, pp. 214–26
Yang-Yuet, S. *Le Loess de la Vallée du Rhône*, 1934
Russell, R. J. 'Lower Mississippi Valley Loess', *Geol. Soc. Am.*, 1940, pp. 1–40
Symposium on Loess. *Am. Jour. Sci.*, 1945, pp. 225-303

Underground Water
Sandford, K. S. 'Sources of Water in N.W. Sudan,' *G. J.*, May 1935
UNESCO. *Arid Zone Hydrology*, 1953; *Problems of the Arid Zone*, 1962
For Aeolian Work in Britain
Steers, J. A. 'The Culbin Sands. . . ,' *G. J.*, Dec. 1937
Gresswell, R. K. *Sandy Shores in South Lancashire*, 1953
Landsberg, S. Y. 'The orientation of dunes in Britain', *G.J.*, 1956, pp. 176–89

Note on Climatic Changes and Pediments
During periods of the last Ice Age deserts had wetter climates than at present and these are reflected in their landforms. The flat rock-bench, or pediment, at the base of steep slopes in arid and semi-arid climates, is usually attributed to water-erosion. For pediments *see*
King, L. C. *South African Scenery*, 1963; 'Canons of Landscape Evolution,' *Geol. Soc. Am.*, 1953, pp. 721–52
Tuan, Y. F. *Pediments in Southeastern Arizona*, 1959
Pugh, J. C. 'Fringing Pediments of Nigeria,' *Inst. Brit. Geog.*, 1956, pp. 15–31

P

CHAPTER SIXTEEN

THE WORK OF GROUND WATER

Nature of Ground Water

OF the rain water that falls on to the surface of the land some runs off, some sinks in, and some is evaporated from the soil and transpired by vegetation. In Britain the average ratio of surface run-off to rainfall varies between 25 and 35 per cent., which means that in districts with 28 inches of rainfall a year about 7 to 10 inches would run directly into the streams. The amount of water entering the ground depends on:

(*a*) the abundance and nature of the rainfall,
(*b*) the slope of the surface,
(*c*) the porosity and permeability of the surface rocks,
(*d*) the amount of vegetation cover,
(*e*) the amount of water already in the soil,
(*f*) the dryness of the atmosphere or rate of evaporation.

The significance of these different factors needs little emphasis; suffice it to say that porosity, or the amount of water that can be absorbed, depends on the interstices in the rock. Thus the pore space in loose sand and gravel may be 50 per cent. of the volume, whereas in shales it is usually only 4 per cent., and in some igneous rocks is under 1 per cent. Yet joints may make igneous rocks very permeable.

Position of Ground Water. The downward progress of rain water that has percolated into the soil is eventually halted either by impervious layers or by the fact that the increase of rock pressure with depth has greatly decreased the interstices in the rock. Consequently, beneath most localities there is an underground zone where water is descending and a zone where water is accumulating above a relatively impervious stratum. The first, where air *and* water are in the rock, is called the vadose-water zone, and the junction of this with the underlying saturated layer is called the water-table or level of saturation. The latter, which may be recognized as the point where water begins to seep into wells and borings, is usually within a few hundred feet of the surface. In any region the great bulk of ground water is limited to the upper half-mile of the earth's crust.

Ground water is in constant motion both vertically and horizontally. Locally the level of saturation will rise after heavy rains and will create a sufficient head of water to cause a gradual seep towards drier districts. The movement, however, rarely exceeds a few hundred yards a year. Much depends on the size of the pore spaces; in some close-textured sandstones the flow is exceedingly slow, whereas in fissured limestones it is relatively fast. The horizontal movement of ground water is well illustrated in regions with artesian wells. Most of the water in this case has entered the soil far from the site of the boring, and, in percolating downwards, has eventually become imprisoned between impervious layers.[1] The rise of water up the shaft or above the mouth of the well is dependent on the pressure of the ground water accumulated above the level of the well-head. These conditions obtain in some synclines, such as the London Basin, but are also frequently associated with rocks of varying porosity that dip slightly, as happens over large areas of western Queensland (Fig. 71).

The level of saturation is a very complex problem, as it depends so much on the underlying rock structure. In a district with a fairly

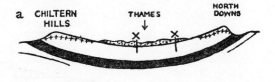

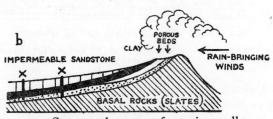

71. Structural nature of artesian wells

(a) Syncline of London Basin. Water-table shown by line of crosses. Black band is Gault Clay
(b) Inclined strata of Great Artesian Basin, Queensland. xx mark mouths of artesian wells

[1] Another theory is that the water is plutonic; that is, that it was incorporated in the rocks when they were laid down.

homogeneous geological structure the water-table tends to follow, in a subdued way, the surface relief. Throughout such an area the water-table rises after rainy spells and falls after droughts, the variations being greatest near the watersheds and least in the valleys (Fig. 72a). On the chalk of Salisbury Plain the level of water in the wells shows seasonal variations up to as much as 120 and 140 feet.

It happens that what geographers call, for example, a 'limestone' scarpland usually consists of alternate layers of sandstones, limestones, and clays (see pages 180 and 182). The more impervious layers form their own water-tables, which may be perched high above the main saturation level of the region. For instance, large areas of the Cotswold uplands would be entirely void of scattered villages and isolated farmsteads were it not for the presence of a thin impermeable layer of Fullers' Earth. When clayey beds of this nature outcrop along the face of a slope they frequently give rise to a series of springs.

a

b

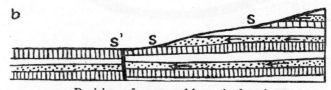

72. Position of water-table and of springs

(a) Springs and water-table in a rainy period (S), and in a dry period (S^1)
(b) Springs due to impervious outcrops (S), and to faulting (S^1)

Some escarpments carry a number of spring-lines of which the highest dry up first in time of drought. Another type of spring occurs where a fault or fissure allows ground water to rise up it to the surface (Fig. 72b): for intimate details of this subject the student is, however, advised to consult the Geological Survey Memoir on the underground water supply of his own district.

Dry Valleys. The water-table is especially interesting to geographers owing to its influence on the formation of the 'dry' valleys that abound on limestone cuestas in England. Along the chalk outcrop of the North Downs most of the springs at the scarp foot are thrown out by the Gault Clay. As the escarpment receded it grew in height, but the level of the Gault outcrop was continually lowered, and, with its fall there occurred a corresponding drop in the water-

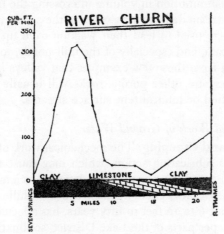

73. Diminution of surface streams in limestone country. The flow of the Churn in a dry autumn

table. Consequently some of the valleys on the chalk upland are now streamless for most of the year.

Over large areas of the Cotswolds, clay beds at or near the surface of the dip-slope have been gradually eroded away, and consequently the water-table falls and the upper courses of the streams sink into the porous limestone, leaving their valleys quite devoid of water except in time of heavy rain. The main rivers now have their perennial sources lower down the slopes, and above the headsprings dry valleys ramify in all directions and incise the tableland with a fossil river-system.

There are other causes of the formation of valleys now streaml ss An appreciable increase of rainfall in various periods in recent geological time is possible. With such an increase the water-table rises and surface-flow and valley-cutting also increase. When the rainfall decreases again the higher valleys become dry. More potent probably was the effect of surface run-off, especially in spring, on the frozen sub-soils of all limestone areas just outside the limits of ice-advance during the last Ice Age. With the onset of a warmer climate, the rainfall again seeps into the porous strata, leaving many valleys dry.

The effect of porous beds on the volume of a river has been measured along the Churn, one of the headstreams of the Thames (Fig. 73). The diminution in volume in crossing the limestones is so great that the stream delivers little water to the parent river. Formerly the local farmers used to lead their walking oxen up and down the bed of the stream, and especially of the mill-ponds, to consolidate it. It will be seen from these few examples that valleys in areas floored with limestones and other porous rocks will usually contain either much diminished or intermittent surface streams.

The Mechanical Work of Ground Water

Slumping and Creeping. The mechanical work of ground water depends on its lubricating action, which may cause rocks either to slip over a moistened layer or to creep downhill as a moistened mass. In the Rhymney valley, Glamorganshire, a downhill creep of soil, at the rate of about 6 to 10 feet in fifty years, has affected a railway. In the higher, steeper parts of the Lake District, solifluxtion or a down-hill creep of soil occurs largely owing to the presence of wet surface soils upon a frozen subsoil in spring. Nivation, or snow-cap erosion, is mentioned in Chapter Nineteen. It is in Siberia, polewards of 50° N. latitude, and in the northern parts of North America that soil-creep becomes a predominant factor in the fashioning of relief. Here, in summer, a surface layer of saturated or half-thawed mud overlying a frozen sub-soil forms the ideal conditions for solifluction.

Landslips. Landslips, as denoting a relatively rapid movement of surface rock, are more common in Britain than is generally supposed. They occur especially along escarpments where steep faces of relatively resistant rock overlie a clay bed. The harder rocks may be undermined by the sapping action of springs issuing from the top of the clays or, after rains, they may become saturated with water and slip downhill with the aid of the lubricating band of clay. The scarps near Ludlow and those on the east side of the vale of Woolhope have experienced great landslips. The Jurassic scarp facing Cheltenham and Gloucester repeatedly experiences the slipping downwards of masses of limestone on the basal bands of clay. Near Cheltenham one slip involved whole fields, which it carried several hundred yards, taking trees and hedges along with it. Near Gloucester a recent slip seriously affected a main road following the summit of the escarpment. The available evidence leaves no doubt that in damp climates, such as that of the British Isles, landslips play a considerable part in the fashioning and recession of scarps.

The Chemical Action of Ground Water

Solution. By far the main work of ground water is accomplished by means of solution. In some localities the large amount of mineral matter in solution in spring water may be inferred from the petrifying action of the streams and the 'fur' deposit left in kettles on boiling. At other places where springs emerge, the shaking of the water and the lowering of its temperature cause a proportion of the dissolved matter to be deposited about the spring-line. At Dursley in Gloucestershire, beds of calc-tufa have been formed of which Berkeley Castle is largely built; near Rome the building-stone called travertine has a similar origin. The importance of the solvent action of underground water may be judged from the fact that each square mile of chalk outcrop in England loses about 140 tons of matter in solution every year.

Ground water only acquires this load with the assistance of carbon dioxide which it takes up to form a very weak acid. Although so weak, the acid acts on limestone to form calcium bicarbonate, which is much more soluble than pure limestone. As dissolution proceeds most quickly along the joints, fissures, and bedding planes, the topography formed depends largely on the abundance in the rock of these lines of weakness.

The Development of Limestone Topographies. Chalk is a limestone but is soft and so often mantled with clay that in spite of its fissured nature underground it only gives rise to a very subdued form of 'limestone topography.' The landscape of the North and South Downs is rolling, and, except on the coast, where wave undercutting goes on unceasingly, is devoid of precipitous slopes. On the uplands many of the smaller valleys are dry or intermittent, and winter-bournes which break forth only after heavy rains are not uncommon. The perennial rivers are widely spaced and flow in large valleys with steep sides and wide, flat bottoms. Swallow holes occur, notably along the river Mole where it is crossing the chalk between Dorking and Leatherhead. Here in dry seasons so much water flows down these holes that the Mole disappears underground for a short distance.

Jurassic and Carboniferous limestones are more fissured than chalk and show distinct bedding planes. As these rocks will stand in precipices, they develop solution characteristics to a high degree. Ground water acting on the Carboniferous limestone of the Mendips has formed numerous sink-holes that lead down to underground rivers and caverns. The subterranean rivers usually emerge from caverns at the base of the scarp, and when, for example, the Axe leaves Wookey Hole it is already a considerable stream (Fig. 75; Plates 13 and 14). The caves at Cheddar, with their wealth of stalagmites and stalactites, are close to the famous gorge. The lower part of the gorge appears to be partly due to the collapse of the roof of an underground river that was running in zigzags as if following a criss-cross of geological joint planes. A smaller but less disputable example occurs at the Ebbor Gorge, near Wells. In this case the valley begins as a dry depression on the plateau slope and deepens gradually until, near the edge of the scarp, it is replaced by a narrow, precipitous gorge, the base of which is choked with the débris of the collapsed roof. No water is visible, but vestiges of caves at the foot of the precipices mark a former stream-level.

In the Pennines well-developed limestone scenery is to be found near Ingleborough and Malham Cove. The peak of Ingleborough consists of a capping of grits and shales above thick Carboniferous limestones, and consequently the streams upon the grits plunge underground on reaching the limestone outcrop (Fig. 60). On the east side of Ingleborough, Alum Pot Hole and Gaping Ghyll are the largest swallow-holes; the latter, which swallows up the Fell Beck, has a vertical shaft 365 feet deep that opens out into a large cavern. On the west side of the hill the Greta stream repeatedly disappears underground as far as Chapel le Dale, where impervious rocks outcrop in the valley. For one short stretch this beck flows in a narrow cavern, the roof of which has collapsed. At this point an underground tributary gushes forth from high up in the side of the gorge and plunges 70 feet into the main stream below. Many similar phenomena in the district emphasize the fact that in limestone lands different sets of streams commonly flow at different levels. Near Ingleborough the flat surface exposures of bare limestone are dissected into irregular blocks (*clints*) separated by a labyrinth of fissures (*grykes*) which attain several feet in width and depth.

A similar outcrop and topography occur above Malham Cove, five miles east of Settle. Here, a river, assisted to keep above ground by clay patches and the natural puddling of its bed, has eroded Gordale, a fine limestone gorge. At the Cove a small head-stream of the Aire issues from the base of a limestone precipice 300 feet high, at the top of which is the dry bed once occupied by the stream.

Further examples in the British Isles are High Tor, above the Derwent at Matlock; the Avon gorge at Clifton; Dovedale; the Wye near Symond's Yat; and Kent's Cavern, Torquay. A more unusual feature occurs along the Shannon in the Central Plain of Ireland, where the limestone underlying the superficial boulder clay has been dissolved in places so that the river widens out into various loughs.

The most highly developed types of limestone scenery are, however, to be found in Yugoslavia and southern France. From Trieste 450 miles south-eastwards into Montenegro stretches the *karst*, a belt of thick, folded limestones that locally reaches 100 miles in width and 8,000 feet in height. The very heavy rainfall, the porous, barren, elevated surfaces, and the well-jointed nature of the rock combine to form the ideal conditions for chemical solution. At the same time the rock is compact enough to stand in deep slopes. Here, in some localities, a definite cycle of erosion has been observed. First the streams work down and enlarge the joints, so allowing the surface water to get underground. Then surface depressions are formed which increase in size and number until all the streams flow underground. Eventually the limestone outcrop is practically dissolved away and the streams reappear on the surface. Details of this cycle are shown diagrammatically in Figure 74, where the limestone rests on impermeable clays. In the first stage, the valleys are almost normal but small solution hollows (*dolines*) are being formed, and faults are encouraging the formation of a large depression (*polje*). In the second stage, the solution hollows are greatly extended and most of the drainage is underground. In the third stage the karst relief is beginning to disappear. In the last stage only a few patches and isolated hillocks (*hums*) remain of the limestone surface, and the drainage flows mainly above ground.

Thus the scenery of the karst is characterized by:

(*a*) numerous *dolines*, or funnel-shaped depressions, many of which lead to underground caverns; as these extend by dissolution, several may unite to form a *polje* or elongated depression. *Poljes* usually drain only underground, but in

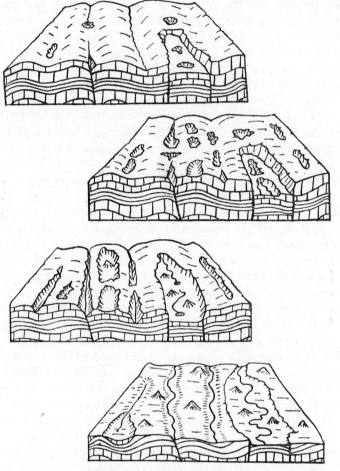

74. Cycle of erosion in a karst land

Thick lines denote faults or fractures

some cases residual clays on their floors hold marshes and
lakes;

(b) subterranean drainage. The courses of the underground
streams are, however, marked by sags at the surface, and
where these collapse the stream is revealed entering and
leaving a gorge by means of an underground channel (Fig.
75). The Narenta, the only large surface river to cross the
region, also flows in a deep gorge;

(c) the dissolution of the flatter limestone surfaces into a maze of sharp pinnacles and fluted ridges.[1] This fretwork of irregular columns, which may exceed 15 feet in height, closely resembles the crevassed surface of an ice-fall.

Another famous limestone region lies on the southern slopes of the Central Massif of France. Here a much-faulted plateau of Jurassic and dolomitic limestone, rising in parts to over 3,000 feet above

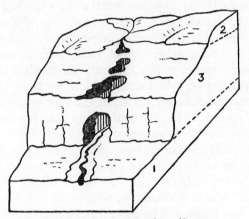

75. Formation of gorge in a limestone
escarpment
(*After Cvijic*)
1. clays   2. sandstone   3. limestone

sea-level, has been dissected into a series of small, steep-sided table-lands called *causses*. The surface of each causse is pitted with sink-holes (*avens*), some of which are 700 feet deep. These extend and unite to form *sotch* or *dolines*, the floors of which may be quite flat save for isolated remnants of limestone. Around these depressions the bare rock forming much of the upland surface is cut by solution furrows, while below it is honeycombed with caves and tunnelled by an intricate network of underground streams.

The Tarn and other rivers that cross the Causses have carved out deep, precipitous gorges. Surface water is practically absent on the plateaux, and in thirty miles the Tarn receives thirty subterranean

[1] Called in France *lapiés*.

tributaries and not one on the surface. The sides of the gorges are in places cut by solution into fantastic towers and pillars, as at the cliffs of Montpellier-le-Vieux.

Ground Water and Human Affairs

It is a commonplace of geography that a water-supply, either from springs or wells, has been in the past one of the chief factors controlling the sites of settlements. Nottingham grew up on Bunter Sandstone that forms a collecting centre for underground water, and which, even to-day, supplies half the city's needs. In fact, all the older towns of the Midlands, such as Coventry, Warwick, Birmingham, and Lichfield arose on water-bearing sandstones that were rich in springs. It is remarkable how in a clay vale (where a supply of good drinking-water is always difficult) the hamlets cluster on the occasional patches of gravel that yield water from shallow wells. Where a spring-line occurs, as along the clay bands of the Weald and of the Lincoln Wolds, the settlements aline themselves upon it. These facts are, however, best based on local surveys and on the study of geological (drift edition) and topographical maps.

In some parts of the world the utilization of ground water permits settlement where otherwise it would be either impossible or impracticable. Much of the recent settlement in Libya depends on deep wells, and desert life in general is bound up with the waterholes. Even in the states of the central U.S.A., such as Illinois, ground water raised by pumping is vital to the watering of stock. On the dry interior plains of Queensland, three thousand bore-holes yield artesian water for stock, and in time of drought prove of especial value by keeping open the stock-routes to the coast.

The quality of ground water is of importance to inhabitants. Deep-seated springs or wells may be warm and highly charged with mineral matter in solution. Some artesian water in Queensland is nearly at the boiling-point and is so impregnated with sodium salts as to render it quite unsuitable for irrigation. Usually the waters of spas, for example, of Bath and Carlsbad, are warm and consequently well charged with dissolved salts.

Also important is the hardness of water from limestone areas, a quality which has caused the water to be avoided for purposes of scouring and bleaching in the textile industry. The Yorkshire woollen manufacturing district virtually ceases northwards near the Aire, which rises on Carboniferous Limestone, and southwards at the

Dearne, which rises on the Coal Measures. The great woollen manu-facturing cities lie on the Colne and Calder systems that derive their soft waters from the Millstone Grit. In the lower Severn Valley the woollen manufacturing area around Stroud obtains an abundant supply of soft water from the Cotswold Sands, the only thick sandy bed in the whole region.

Ground water sometimes affects industrial activities adversely. It becomes a problem in mining and deep quarrying. Miniature rivers are pumped unceasingly from the larger coal pits of the Forest of Dean and many other British coalfields. A powerful underground flow of water was encountered in boring the Severn Tunnel, and only after tremendous expense was this water brought under control.

The effect of limestone topographies on human affairs is not easy to distinguish except in the case of 'karst' districts. These are extremely inhospitable, the arid broken surfaces being useless for agriculture. In the karst the upland settlements cluster around the residual soils and water-supply of the larger poljes; in the Causses the residual clays on the floors of some of the sotches may retain a pond and form a patch of soil for cultivation. Otherwise the popula-tion clings, where possible, to the bottom of the narrow gorges. In Britain the more subdued chalk and limestone topographies, when unmodified by surface deposits, at worst support grassland and moorland that provide pasturage for sheep.

Of minor interest is the great use made of limestone caverns in prehistoric times. Wookey Hole in the Mendips, for instance, and many of the caves in France appear to have been used by troglodyte man since Neolithic times. The main economic uses of caves to-day are, however, for the storage of cheese and wine (Rheims district) and as an attraction for tourists. The Roquefort cheese made of ewe's milk in the Causses owes some of its qualities to maturing at a constant temperature in limestone caves.

LIST OF BOOKS

Ground Water and Landslips
Meinzer, O. E. (ed.) *Hydrology*, 1942
Tolman, C. F. *Ground Water*, 1937
Smith, B. 'Water Supply,' *Geography*, June, 1935
Geological Survey Memoirs on underground water supply
Fox, C. S. *The Geology of Water Supply*, 1949
Sharpe, C. F. S. *Landslides and Related Phenomena*, 1938
Todd, D. K. *Ground Water Hydrology*, 1959

'*Dry*' *Valleys*

This is a very complex problem: for other means of formation of 'dry' valleys, see note in *G. J.*, 1936, p. 477

Bull, A. J. 'The geomorphology of the South Downs,' *Procs. Geol. Assoc.*, 1936, pp. 99–129; 1940, pp. 63–71

Sparks, B. W. and Lewis, W. V. 'Escarpment dry valleys near Pegsdon, Hertfordshire', *Procs. Geol. Assoc.*, 1957, pp. 26–38

Small, R. J. 'The escarpment dry valleys of the Wiltshire Chalk', *Inst. Brit. Geog.*, 1964, pp. 33–52

Limestone Topographies and Chemical Work of Ground Water

Sanders, E. M. 'The Cycle of Erosion in a Karst Region,' *Geog. Rev.*, 1921, pp. 593–604

Cvijic, J. 'Hydrographie souterraine et évolution morphologique du Karst,' *Géographie Alpine*, VI (1918), pp. 375–426. (With 26 figures and 4 photographs.); 'The evolution of lapiés,' *Geog. Rev.*, 1924, pp. 26–49

Also works of E. de Martonne, especially for *Causses*

For England. *Memoirs of the Geological Survey*. Especially *A Guide to Geological Model of Ingleborough*

For underground drainage of Ingleborough, *Procs. Yorkshire Geological Soc.*, XIV and XV

Sweeting, M. M. 'Erosion cycles and limestone caverns in the Ingleborough District,' *G.J.*, March 1950, pp. 63–78; 'The karstlands of Jamaica,' *G.J.*, 1958, pp. 184–99

Corbel, J. 'Erosion en Terrain Calcaire,' *Ann. de Géog.*, 1959, pp. 97–120

Birot, P. 'Problèmes de Morphologie Karstique', *Ann. de Géog.*, 1954, pp. 161–92

Coleman, A. M. & Balchin, W. G. V. 'Origin and development of surface depressions in the Mendip Hills', *Procs. Geol. Assoc.*, 1959, pp. 291–309

Sweeting, M. M., Groom, G. E. *et al.* 'Denudation in limestone regions', *G. J.*, 1965, pp. 34–56

For Caves

Davis, W. M. 'Origin of Limestone Caverns,' *Geol. Soc. Am. Bull.* 41, 1930, pp. 475–628

Swinnerton, A. C. 'Origin of limestone caverns,' *Geol. Soc. Am.*, 1932, pp. 663-93

Gardner, J. H. 'Origin and development of limestone caverns', *ibid*, 1935, pp. 1,255–74

Bretz, J. H. 'Vadose and phreatic features of limestone caverns', *Journ. Geol.*, 1942, pp. 675–811; 'Carlsbad Caverns . . .', *ibid.*, 1949, pp. 447–63

Trombe, F. *Traité de Spéléologie*, 1952

Cullingford, C. H. D. (ed.) *British Caving*, 1962

Moore, G. W. & Nicholas, G. *Speleology: The Study of Caves*, 1964

For illustrations:

Lee, W. T. 'A Visit to Carlsbad Cavern,' *Nat. Geog. Mag.* 45, 1924, pp. 1–40

Casteret, Norbert. *Ten Years under the Earth*, 1939

CHAPTER SEVENTEEN

THE WORK OF RUNNING WATER

FLOWING surface water is the chief agent fashioning the details of the landscape in areas where the annual rainfall much exceeds ten inches and the temperature remains above freezing-point for most of the year. Even over large areas of desert lands, wadis and traces of former river-systems occur. The hearts of deserts and regions with ice-caps and permanent frost alone lack traceable signs of the action of running water. Topographical details formed by surface streams predominate in all the lands most densely settled by mankind, and consequently they are of supreme importance to the geographer.

The Work of Flowing Surface Water

Flowing water accomplishes three main types of work: first, it erodes and dissolves the land surface in contact with it; second, it transports the products of its erosion and solution; third, it deposits this material.

Erosion. Running water of itself has little erosive power, but once it has become charged with débris it begins to corrade the surface over which it is flowing. When the material carried is fine, the minor currents and eddies of the stream continually thrust it against the sides and bottom of the channel. It will be seen that the stream's erosive powers increase in proportion to its velocity *and* the amount of material carried in suspension.

Streams, however, also accomplish erosion by pushing angular fragments and rolling roundish pebbles along their beds; the impact and friction of these larger fragments assist in the deepening of the channel. At the foot of small rapids and waterfalls along the course of fast streams, erosion by pebbles weakens or wears away even a bed of solid rock. Here, pebbles and rock fragments that fall into a slight inequality of the stream bed may be turned round and round by the force of the stream. The rock fragments become more rounded in the process, but at the same time the slight hollow is worn deeper until it becomes a circular hole. In some cases the whole bed may be honeycombed with 'pot-holes,' and these may eventually unite and cause the bed-rock to be lowered. Pot-holes and similar phenomena of a less marked nature occur in Britain along most mountainous streams; the classic example in France is at Pont des Oules near Bellegarde (Ain), where the Valserine, just before its junction with the Rhône, enters a gorge, formed mainly by the uniting of pot-holes. Pot-holes of great size with the large rounded boulders actually rotating under the influence of falling water, may be seen at the 'glacier garden' at Lucerne.[1]

It happens, however, that débris moving as traction load along the bed probably seldom possesses erosive power equal to that of matter in suspension. Bottom load often acts partly as a protective agent, especially in sluggish rivers that are aggrading their beds.

Transport. The load of a tributary stream is derived partly from the action of weathering and rain-wash on the slopes of its valley and partly from the erosion of its own bed. The type of material and methods of transport are supremely illustrated by the Mississippi, which annually carries to the sea about

340 million tons of matter in suspension
136 million tons of matter in solution
40 million tons of matter by sliding and pushing along its bed.

Much has already been said on solution; the question of suspension is of more general importance. The amount of material carried in suspension depends on the volume and velocity of the stream and *the size of the particles forming the load*. The finer the material the greater is the capacity of the stream's minor eddies to keep it in suspension. The turbulence of a stream is a great factor in transporting fine material (Fig. 76). If the load is fine and the turbulence

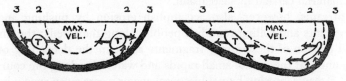

76. Areas of maximum turbulence in a mature stream

1. Axial region of high velocity and moderate turbulence
2. Flanking region of moderate velocity and high turbulence
3. Lateral region of low velocity and low turbulence.
Arrows show main direction of diffusion of suspended matter from chief areas of turbulence (T)

[1] Although at Lucerne the water fell through crevasses in the glacier on to pebbles resting on the rocky bed, these 'glacier mills,' as they are called, form pot-holes in the same way as a stream does. Pot-holes may be formed similarly by waves on a coast.

[G. P. Abraham, Keswick

17. A lake delta. Buttermere and Crummock Water separated by the delta of the Sail Beck

[Greek Topographical Service

18. Delta of the Vardar at the head of the Gulf of Salonika

Crown Copyright] [*Geological Survey*

19. Rejuvenation. Incised meander of the Rheidol Gorge, two miles
north of Devil's Bridge, Cardigan

Crown Copyright] [*Geological Survey*

20. Incised meander of the Wye near Chepstow, Monmouth

great, not only will more material be carried but it will be transported much greater distances before being deposited. Yet much depends on the nature of the rocks; sluggish streams crossing clays are usually well laden with suspended matter, whereas swift streams, such as the Dart and Teign, crossing granite, are relatively clear.

Velocity cannot be divorced from the type of material available, but experiments show that with pebbles of the same shape and composition, the diameter of pebble that a stream moves varies as the square of its velocity. That is, a river twice as fast as another will move pebbles four times greater in diameter. The actual volume of homogeneous débris carried varies as about the sixth power of the velocity but if the débris is mixed in size and nature the quantity moved varies only as the third or fourth power of the speed.

The combined result of erosion and transport is gradually to lower the drainage basin of the river. The Mississippi basin is being lowered at the average rate of about 1 foot in 4,000 years; this rate, however, is greater than that of most of the world's large rivers owing to the high proportion of soft rocks in the drainage area.

When discussing the destructive work of rivers, much emphasis should be placed on the great increase in volume and speed during flood-time. Then for a few days a year the stream accomplishes the work of a much greater river. Annually the Nile below Khartoum rises as much as 25 feet; in spring the lower Yenisei rises 30 feet or more and floods an area 35 miles wide; parts of the Amazon rise nearly 50 feet and turn vast areas into an inland sea. During these times the river accomplishes in a few days work that it otherwise would never have performed, or at least not within a long space of time. Typical flood-time achievements are the severing of meander necks, the bursting of levees and changing of course, the transport of vast quantities of material from regions normally undergoing deposition, and the deposition of silt far from the normal channel of the river. Some English streams appear quite incapable of having carved their large valleys unless seen in time of flood, when a raging torrent covers the whole valley-bottom. If it is realized that this happens for a few days each year, or on an average one day in about one hundred, the importance of river spates will be better appreciated.

Deposition. There is a tendency for some load to be deposited wherever the velocity of flowing water is diminished, as, for example, where:

Q

(a) the volume decreases owing to the river entering an arid region;
(b) the slope of the bed decreases considerably;
(c) the channel widens and shallows appreciably;
(d) the flow is obstructed on entering a lake or the sea.

Such accidents may happen to a river at any point in its course, but deposition occurs most commonly near to the sea, where the load is usually greatest and the current slowest. The youthful stream builds up small sandbanks in the lee of boulders, the mature stream constructs alluvial patches upon the inside banks of its meanders, but in the bed of the sluggish river deposition goes on almost incessantly.

Many features of river deposition cannot, however, be adequately explained solely by the dropping of sediment when the rate of flow is checked. The diffusion of fine matter by turbulence plays an important part in deposition upon the side of a river's bed, especially in the formation of levees and in the accumulation of silt across the mouth of severed meanders thereby forming ox-bow lakes (Fig. 76).

Local Factors influencing the Work of Rivers

Relief. The course of a river is usually divided into youthful, mature, and 'old-age' sections according to whether the current is swift, vigorous, and sluggish respectively. The relief control is, however, seldom as simple as these three stages suggest. A torrent plunging down the side of a glaciated valley into a lake passes from youth straight into old age, and it is obvious that any section of a river's course may show sudden variations in speed. Indeed, each stream and its valley must be studied on their own merits, and for the student it is a question of compiling notes and diagrams from field-work and map study.

Climate. The shape of river-valleys is influenced by climate, since rain-wash is the beginning of river-erosion. The fact that the river really controls only the depth of the valley is illustrated by the canyons in comparatively rainless climates.

Geology. Rock structure exerts a greater local influence than climate, as it affects the shape of the valley and the slope of its bed. Except in dry climates, clays seldom stand in steep slopes, whereas porous limestones and sandstones frequently develop bluffs and gorges. Where a river has incised its valley through alternating hard and soft horizontal strata, the valley-slopes may be stepped. Further,

if the river flows along an outcrop of tilted strata, the valley formed
will be asymmetrical or steeper on one side than the other, as happens
along the Lea, north of London and the Thames at Radley.

The outcrop of resistant strata across a valley often results in
gorges, rapids, and waterfalls. The Shabluka gorge and five cataracts
of the Nile below Khartoum mark the points where the river leaves
the sandstones and crosses resistant crystalline rocks. The fine gorge
of the Yangtze Kiang above Ichang, and the narrowing of the
Danube at the Iron Gates, also reflect the outcrop of more resistant
strata.

Waterfalls are to-day of great importance owing to their potential
hydro-electric power. They may be formed in many ways, by 'hanging' tributaries in glaciated valleys, by obstruction of river courses, by
stream capture, by the undercutting action of waves on cliffs, by the
dissolution of joint planes in limestone districts, and by the formation
of fault scarps. By far the chief causes, however, are *differences in
hardness of the river's bed.*

At Niagara the river plunges 160 feet over thick limestones underlain by soft shales. The beds lie almost horizontally, and the undercutting of the shales leads to the collapse of the harder limestone. As
a result the falls recede upstream, and a gorge nearly seven miles
long marks the extent of the recession. The Victoria Falls on the
Zambezi are caused by a belt of igneous rock.

In England, many waterfalls show the same control. High Force
in Teesdale drops over the vertical face of the Whin Sill 70 feet into
a small whirlpool (Fig. 77). Farther upstream, at Cauldron Snout, the
Tees tumbles down the stepped outcrop of the same sill. Small

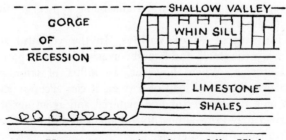

77. Upstream recession of waterfalls: High
Force, Teesdale

Tumbled blocks of igneous rock and limestone lie in the bed
of the gorge

waterfalls are common in the Yorkshire dales. The Aysgarth Falls in Wensleydale and the Lumb Falls near Hebden Bridge (west of Halifax) occur where thick beds of grit overlie softer shales and limestones.

It will be noticed that some falls tend to increase in height with recession (*cf.* Niagara), while rapids formed by harder layers that dip gently upstream gradually disappear as the stream deepens its bed.

The Shaping of River Valleys

The Deepening of River Valleys. The deepening of the valley is characteristic of the work of swift-flowing streams. The downward erosion usually proceeds rapidly enough to form a deep, narrow, steep-sided valley that is markedly V-shaped in cross-section (Fig. 78). Rain-wash on the valley sides may be heavy, but the valley remains steep-sided, as heavy rainfall also increases the erosive powers of the permanent channel of water.

The channel and valley of the youthful stream are relatively

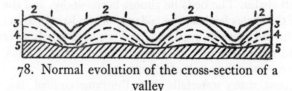

78. Normal evolution of the cross-section of a
valley

Numbers mark successive stages

straight, and, where changes in direction do occur, they are right-angled rather than curved. The stream-bed itself shows many breaks of slope, rapid stretches and waterfalls being typical of this stage of growth (Plate 15).

The Widening of River Valleys. In the youthful stage the volume of a river is small, and lateral corrasion is less important than downward erosion. When, however, an influx of tributaries has increased the volume and load of a river, it can accomplish greater work both as a transporter of fine material and as an agent of corrasion. A river may grow almost to full size and still retain sufficient vigour to transport and to erode.

The characteristic feature of the work of mature streams is the widening of the valley-*floor*. In such rivers, the bottom current is, as usual, retarded by friction with the bed while the unimpeded axial surface flow (Fig. 76) acquires an exceptional velocity. Such rivers begin to wind and to impinge on the sides of their valleys. The windings once started are accentuated as the river's current increasingly undercuts the bank on which it impinges and simultaneously tends to deposit silt on the inside bank of the bend.

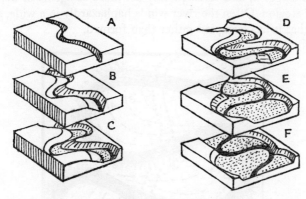

79. Widening of a valley-floor

The windings grow in size until eventually the river's course becomes a series of meanders (Fig. 79). Wherever the river impinges on the valley side it cuts it back into a steep bank or cliff and the valley-floor is widened by the extent of this back-cutting. Hence there is formed a winding valley in which steep slopes on the one side oppose gentle slopes on the other, and these features alternate as the river hugs first one side and then the other side of its valley.

Along the middle Trent most of the riverine villages are situated on the cliffs of the undercut bank. Typical meanders also occur along the course of the Wye above Ross.

This meandering and cutting back is only part of the process whereby river valleys are widened. The meanders change their positions, the general tendency being to *grow outwards* and to *migrate downstream*. It will be seen that the current gradually wears away the upstream side of spurs projecting into meanders, and deposits sediment on the downstream side (Fig. 80). Gradually spurs are sharpened, then shortened, and eventually they may be worn completely away. Spurs in all these stages of erosion are to be seen along the Ribble just west of Blackburn.

The final result of the lateral corrasion and down-stream migration

of meanders is that the valley-floor is widened to at least the width of
the meander belt. The river now winds over a wide, flat floor, and the
valley, bounded on both sides by steep slopes, is no longer V-shaped
in section (Plate 16).

The Shallowing of River Valleys. Near to sea-level most rivers
become sluggish, and consequently act more as agents of deposition
than of erosion. Often the river winds haphazardly in a wide, shallow
valley that it has occupied rather than formed.

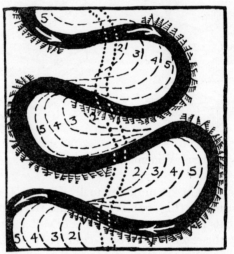

80. Downstream migration of meanders
(*After W. M. Davis*)

The following are the characteristics of this stage:

1. A relatively wide *flood-plain* adjoins the river. This plain, which
is formed by deposition in time of flood, usually extends beyond the
width of the meander belt. It may lie only just above sea-level, since
the *bed* of the stream, which is the base-level to which the whole
river-system works, may be at, or actually below sea-level. The flats
or flood-plains of the Mississippi are exceptionally extensive; at
Dubuque they are over one mile wide, near St. Louis they increase
in width to 10 miles, and at Memphis to 35 miles.[1] British rivers,
although small, show the feature quite clearly, as, for instance, do
the Dee near Chester, the Trent near Gainsborough, and the Mersey
near Warrington.

[1] About 1,000, 700, and 500 miles from the sea respectively.

2. *The channel of the river is liable to undergo frequent changes in direction*, irrespective of the general downstream migration of meanders. In time of flood the necks of meanders may be severed to form semi-circular back-waters known variously as ox-bow lakes, bayous, mort-lakes and cut-offs. Evidence of such lakes abound along the lower Evenlode, the Cuckmere, and the lower Trent, where the flood-plain is seamed with depressions of abandoned channels. The evidence of parish boundaries occasionally emphasizes the relative rapidity of these changes in course.

In some cases the sluggish stream aggrades its bed so much that it bifurcates or splits into numerous channels, and becomes a maze of braided water-courses. Notable examples occur along the rift valley section of the Rhine, along the Platte River in Nebraska, and the Danube in the plain of Hungary. The frequency with which the main channel alters its position in the valley is occasionally aided by the bursting of levees in flood-time.

3. *Natural levees* are built up in time of flood when the river overflows its banks. As soon as water escapes from the rushing current of the main channel its velocity is suddenly checked on coming in contact with the shallower, slower-moving water on the flood-plain. Owing to the decrease in speed, and to the diffusion of fine matter by turbulence, much sediment is deposited on the banks near to the main stream. The natural levees formed in this way are low and generally slope gently away from the river; those of the Mississippi rise nearly 20 feet above the flood-plain but reach in parts 2 or 3 miles in width. These low mounds do not diminish liability to flood, as the river aggrades its bed at the same time. Gradually the height of both levees and river-bed increases until the river is flowing well above the adjacent flood-plain. To prevent serious floods the inhabitants usually strengthen and heighten the natural levees, which along the Hwang Ho have been raised by 10 to 25 feet. In 1852 the Hwang Ho burst its banks, and shifted its mouth from south to north of the Shantung Peninsula, a distance of 300 miles. In 1938, during war, the river was deliberately diverted to the south and was not turned north again until 1947. In England along the lower Trent, the embankments, largely artificial, are 6 to 10 feet high in places, and in time of flood they rise as low, elongated islets above miles of submerged flood-plain.

4. *Deferred junctions*. The tendency of sluggish streams to aggrade their beds and to build up levees often causes their floors to be

convex rather than flat in section. As a result, tributaries rising on or flowing across the flood-plain find difficulty in effecting a junction with the main stream. Along the lower Trent streams rising on the wide flood-plain at first flow *away* from the parent river, and one stream near South Muskham flows for about 5 miles parallel with the Trent before joining it. More famous is the Yazoo that parallels the Mississippi for about 200 miles, and the Bahr Yûsuf that flows parallel with the Nile for almost the same distance.

Deltas

The greatest deposition of material occurs at the mouth of a river where deltas may be formed. Deltas, however, are not necessarily confined to sea-level.

Nearly all torrents rushing down a steep slope on to a flat-bottomed valley build up alluvial fans where their velocity is suddenly checked at the foot of the slope. These alluvial fans abound in glaciated regions where tributary streams are hanging, and in the district of Valais, Switzerland, there are 295 of them with a village on each.

Where swift streams descend on to a semi-arid plain, the rapid decrease in volume may cause the streams to disappear into fans of their own sediment. The rivers draining the western slopes of the Sierra Nevada in California have built up a series of fans or alluvial cones that unite to form of kind of piedmont terrace. In Europe, the greatest example is the Plateau of Lannemezan, a vast fan-shaped cone of débris laid down north of the Pyrenees by flood waters in glacial times.

Deltas formed in lakes by streams are very similar to alluvial fans in shape and origin. The Lake District has many examples. The Aira Beck has built a considerable delta into Ullswater, while Buttermere and Crummock Water have been formed from one lake (Plate 17). The two main ways of lake-filling can be studied in this district; namely, the growth of head deltas where the main river enters (Grasmere; Derwentwater), and the division by fast-growing lateral deltas. Many of the smaller lakes have become very shallow, while some, such as Rosthwaite Basin in Borrowdale and Langdale basin west of Ambleside, have been completely filled.

The largest deltas occur at the seaboard. On entering the sea, rivers first deposit their traction load of gravel and coarse sand, then the suspended load of fine silt, and finally the matter in solution.[1] Rivers,

---

[1] Contact with salt water flocculates some of the coarser clays in suspension.

however, debouch on to widely different coastlines. Thus, along a coastline that has undergone submergence, the sediment from rivers may sink into deep submarine estuaries—as off the mouths of the Hudson, Loire, and Seine—and consequently deposition seldom accomplishes more than the building up of silt banks at the head of the estuary. In bottle-shaped estuaries, strong ebb-tides and long-shore drift may carry silt away from a river's mouth and prevent delta formation. But deltas usually form wherever rivers with much sediment enter relatively shallow water, no matter how high the tides. The deltaic Ganges has a tidal range of 15 feet and the Colorado of 30 feet. In partly enclosed seas, deltas form readily as the small tides render wave- and current-action less effective in keeping the deposition below sea-level.

The growth of a delta is commonly marked by three stages. Firstly, the estuary or inlet into which the river flows is turned into a lagoon by the growth of a spit across its mouth. This is well seen in the haffs of the Baltic and notably in the Frisches Haff at the mouth of the Vistula. The haffs are bounded seawards by low, narrow spits that are broken usually only at the end away from the direction of the prevailing waves and sea-currents; landwards the lagoon shallows rapidly towards the mouth of the river, which may already be splitting into distributaries. The *étangs* of the coast of France are similar to haffs.

Secondly, the lagoon gradually silts up to form a marsh. This process is seen in the 'Green Marsh' on the seaward edge of the English Fens, in the forested swamps (*sundarbans*) of the Ganges-Brahmaputra mouth, and in the 60-miles wide zone of mud-flats and mangrove swamps of the Niger delta near Lagos.

Thirdly, the action of currents and the growth of vegetation raise the salt marshes above sea-level and eventually they become solid land.

The shape of deltas may vary considerably from the △ (Greek 'delta') from which the name arose. The triangular form is caused by the stream repeatedly changing its course about a fixed pivotal point, the mouth of the river. The coarser material carried by the river is pushed outwards by the main channel to form a bank which gradually grows in length and height. At the same time much of the finer material tends to be deposited at the margins of the river's current and there forms banks. These banks are gradually raised to flood-level and the river runs between them. In other words, the phenomena connected with the building of levees and aggradation of an old stream's bed are repeated beneath sea-level. If, as happens frequently, the banks break, the river leaves its channel and begins to build up a new stretch of bed and embankments beneath the sea.

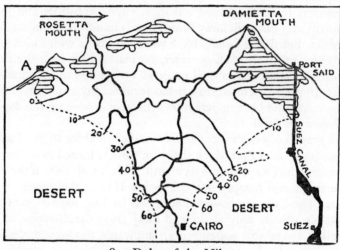

81. Delta of the Nile

Arrow shows direction of longshore drift. Contours given in feet

A delta grows outwards chiefly at the mouths of its main distributaries The Nile delta has retained its triangular shape partly because, owing to irrigation needs, the main channels carry practically no water into the Mediterranean except in time of flood. As a result, the delta grows seawards at less than one-quarter mile in a century, whereas parts of the Rhône and Po deltas extend seawards at 2 or 3 miles a century. The main embankments of the chief distributaries of the Mississippi increase in length about 8 miles in the same time.

The delta of the Nile is a monotonous, triangular plain stretching for 155 miles between Alexandria and Port Said and for 100 miles from north to south (Fig. 81). In contrast, the deltas of the Mississippi, Po, and Ebro are lobed or pronged owing to the greater strength or greater volume of the main distributaries (Fig. 82). The Mississippi has three chief distributaries which have built up their beds far out into the Gulf of Mexico before each splits fanwise into several branches. A further example of a forked delta is shown in Plate 18.

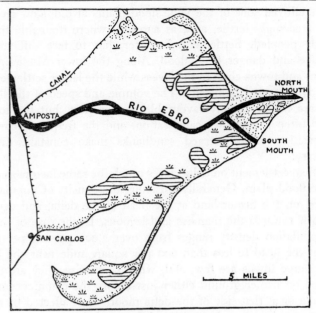

82. Delta of the Ebro

The prevailing sea currents are southwards. The canals roughly mark the
inner edge of deltaic land

Suggestions on Human Geography

The valleys of youthful streams often afford the only practicable
routeways into countries difficult of access. The railways across the
Canadian Rockies and Swiss Alps follow valleys furrowing opposing slopes of the main range and usually tunnel through the narrow
divide between the streams. The narrow, V-shaped valleys, however,
seldom contain sufficient flat-land on their floors to accommodate
easily road or rail communications. Nor is there often sufficient
space to encourage riverine settlements. The streams are usually
useless for transport (except lumber) but some compensation for this
may be derived from the development of hydro-electricity.

The valleys of mature streams afford space for riverside settlements which arise on the steep undercut bank or on bluffs of harder
rocks. The valley-floor is useful agriculturally, and in England was
in the past invaluable for irrigated water-meadows. The stream,
where large enough, favours transport and, at the same time, its
valley-floor affords sufficient space for land communications.

The alluvial soils of the wide flood-plains of old, slow rivers are often amazingly fertile, and, as a result, where drainable they are densely peopled. Settlement, however, has to face difficulties of drainage and danger from flood. Along the lower Mississippi the villages and towns cling to the levees while the larger settlements are strongly protected artificially. The volume and speed of the sluggish river are usually highly favourable to transport, but the tortuous course often necessitates canalization, and the frequent changes in the position of submerged sandbanks make constant dredging necessary.

Human settlement on deltas faces much the same factors as prevail on the flood-plain. Generally speaking, the density of population is greatest on the firmer land at the head of the delta, and decreases seawards towards the marshes and lagoons. In the case of the Nile the population density ranges from over 2,000 persons per square mile at the head to less than 600 per square mile near the lagoon strip. Large towns are few, and, when they arise, they are usually situated on firmer ground either just before the commencement of the delta or at that side of the delta mouth least affected by silting. The west to east longshore drift of the south Mediterranean carries the silt of the Nile away from Alexandria; in the same way the east to west drift prevailing along the north coast carries the silt of the Rhône away from Marseilles.

LIST OF BOOKS

Alexander, H. S. 'Pothole erosion,' *Jour. Geol.*, 1932, pp. 305-37
Leighly, J. 'Turbulence and the transportation of rock débris by streams,' *Geog. Rev.*, 1934, pp. 453-64
Davis, W. M. 'The geographical cycle,' *G. J.*, Nov. 1899. 'The development of river meanders,' *Geol. Mag.*, X (1903). 'The Seine, Meuse, and the Moselle,' *Nat. Geog. Mag.*, VII, pp. 181-202, 228-38
Russell, I. C. *River Development*, 1909
Smith, B. 'Some recent changes in the course of the Trent,' *G. J.*, May 1910
Clapp, F. G. 'The Hwang-Ho, Yellow River,' *Geog. Rev.*, XII, 1922, pp. 1-18
Melton, F. A. 'An empirical classification of flood plain streams,' *Geog. Rev.*, 1936, pp. 593-609
Pardé, Maurice. *Fleuves et Rivières*, 1955
For changes in bed of channel of mature river: Mark Twain's books on the Mississippi
Todd, O. J. 'The Yellow River reharnessed,' *Geog. Rev.*, 1949, pp. 38-56

For human and physical aspects of deltas:
Lozach, J. *Le Delta du Nil*, 1935
Sykes, G. *The Colorado Delta*, 1937
de Martonne, E. *Les Grandes Régions de la France. Région Méditerranéenne*

Russell, R. J. 'Geomorphology of the Rhône delta,' *Annals Ass. American Geog.*, 1942, pp. 149-254. *Physiography of lower Mississippi River delta*, Louisiana Geol. Surv. Bull. 8, 1936; 'Aspects of alluvial morphology', *The Earth* (studies presented to J. B. Hol), 1957, pp. 163–74

Fisk, H. N. *Geological Investigation of the Alluvial Valley of the Lower Mississippi River*, 1944; 'Loess and Quaternary geology of the lower Mississippi Valley', *Jour. Geol.*, 1951, 333–56

Mackay, J. R. The Mackenzie Delta Area, *Canada. Dept. Mines, Mem.* 8, 1963

For Waterfalls:

Kensit, H. E. M. 'The World's Great Cataracts,' *Canadian G. J.*, Sept. 1934

Molyneux, A. J. C. 'Physical History of the Victoria Falls,' *G. J.*, Jan, 1905

Rashleigh, E. C. *Among the Waterfalls of the World*, 1935

For English waterfalls, see especially:

Trueman, E. A. *The Scenery of England and Wales*, 1938

Avebury, Lord. *Scenery of England*, 1912

For recent investigations on river-flow:

Leopold, L. B. and Maddock, T. 'The Hydraulic Geometry of Stream Channels', *U.S. Geol. Surv. Prof. Paper*, 252, 1953

Wolman, M. G. and Leopold, L. B. 'River Flood Plains', *ibid.*, 282–C, 1957; 'River Channel Patterns', *ibid.*, 292–B, 1957

Leopold, L. B. 'Downstream change of velocity in rivers', *Am. Journ. Sc.*, 1953. pp. 606–24

Dury, G. H. 'Contribution to a general theory of meandering streams', *Am. Journ, Sc.*, 1954, pp. 193-224; 'Tests of a general theory of misfit streams', *Inst. Brit. Geog.*, 1958, pp. 105-18

Sundborg, A. 'The river Klarälven: a study of Fluvial Processes', *Geografiska Annaler*, 1956, pp. 125-316 (with full bibliography)

Pardé, M. *Sur La Puissance des Crues en diverse Parties du Monde*, 1961

Leopold, L. B., Wolman, M. G. and Miller, J. P. *Fluvial Processes in Geomorphology*, 1964 (with recent bibliography)

CHAPTER EIGHTEEN

THE DEVELOPMENT OF RIVER-SYSTEMS

The Headward Extension of Rivers

AT their initiation, streams flow downhill according to the relief; they may occupy folds or depressions, but these are not made by the running water. In time the streams erode their beds downwards and begin to adjust their courses to the underlying geological structure. Those flowing on soft material tend, with the aid of rainwash, to enlarge their valleys relatively quickly and to grow at the expense of streams on more resistant rocks. Gradually the rivers etch in the details of the topography.

This adjustment of rivers to the resistance of the surface rocks is achieved not only by downward erosion, but also by headward extension of the valleys. The extension of a valley beyond the source of the stream is mainly, but not entirely, the work of rain-wash and weathering. Headward erosion about the source of a stream is often well developed on a scarp face, especially if it consists of alternating pervious and impervious layers. In this case the scarp becomes stepped, each step corresponding with a spring-line. The retreat of the impervious layers (clays) causes the overlying layers (limestones and sands) to collapse, and so to recede (Fig. 53a).

This headward regression and undercutting works to some extent about the source of any spring. Most springs arise in a small hollow which owes its existence to two factors.

First, a stream deepens its bed right from its source and consequently enlarges its point of issue. As the exit of the spring grows in depth, the undercutting action frequently causes small pieces of earth and rock to tumble into the running water, and so to be removed. Gradually the slope immediately above the exit is steepened and laid increasingly bare to the effects of weathering and rainwash. But at the same time the whole slope above the spring is being attacked by the work of the atmosphere and of rain-water, and this general lowering of the slope facilitates the very localized undercutting action of the stream. Gradually the small hollow or break of slope about the source of the spring grows in depth and retreats into the hillside.

There is, however, a second factor at work; the definite erosive action of the spring is often aided by the lubricating effect of water in the soil. A springline is often a zone of moist soils, and solifluction, or a downhill creep of damp earth, often assists in the formation of the hollow or of the flattish step commonly found about the source of a stream.

The headward extension of a river's source is thus mainly the work of rain-wash, weathering, and solifluction, combined with the erosive powers of the spring. The rate at which a valley is extended varies considerably according to the local climatic and geological conditions. Much will depend on:

(a) the amount of rain-wash (or rainfall);
(b) the size and persistency of the spring;
(c) the resistance of the rocks; and
(d) the steepness of the slope above the spring.

Watershed Regression. The headward extension of river valleys is the key to the development of drainage systems. Most upland areas in moist climates have streams draining down their slopes, and the valleys of these streams extend headwards into the hillsides. The watersheds between the streams are lowered by rain-wash and weathering at the same time as they are being narrowed by the headward growth of the stream valleys. Consequently, in time the divide may become insignificant (Fig. 83). It is rare, however, for both slopes of a divide to be denuded at the same rate. The rainfall of opposing slopes is seldom equal; the streams flowing down one slope are usually faster or larger than those of the other, or, again, the rocks of each slope are rarely exactly equal in their powers of resistance to erosion.[1] As a result of these inequalities, most divides shift in position laterally as they become lower.

When the divide shifts laterally, the one stream gains territory, and consequently grows in volume, at the expense of the other. This shifting of the divide is known as watershed regression and is the normal method whereby a fast-flowing river enlarges its drainage basin at the expense of its slower neighbours. The process has been studied recently in the Lowther Hills in southern Scotland. Here the northward slopes are drained by the Clyde and Tweed, which are

[1] Plate 3 shows the steep face of the windward slopes in Oahu, a trade-wind island. Here the heavy rain-wash of slopes facing the north-east winds is causing the divide to retreat rapidly into the basin of the Nuuanu river flowing in the opposite direction (seen in the distance through the gap in the watershed).

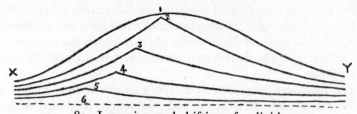

83. Lowering and shifting of a divide

6 shows complete removal of the divide and stage when *x* drains into *y*

mature rivers meandering relatively slowly in a flood-plain. The southern slopes are drained by vigorous, youthful streams that tumble in deep, gorge-like valleys down to the Annan and Nith (Fig. 84). These swift streams are gaining ground at the expense of the slower Clyde and Tweed because the southern slope is rainier

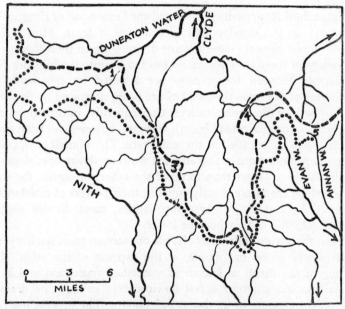

84. Main drainage of the Lowther Hills

(*After Lebon*)

Dashes mark present watershed, and dots probable position of former watershed.
1. Crawick Pass (river-capture) 2. Mennock Pass (watershed migration)
3. Dalveen Pass (river-capture)
4. Beattock Summit (river-capture)
5. Devil's Beef Tub Pass (watershed migration)

than the northern and consists of less resistant rocks. The regression of the watershed is especially marked at the 'Devil's Beef Tub,' a gigantic steep sided hollow where the headsprings of the Annan are cutting back into the basin of the Tweed.

River-capture

Watershed migration will in time lead to the capture of the upper parts of a river's territory, but the process is extremely slow as the erosional factors are working into *main* divides. There is a more spectacular and more definite method of beheading a stream, known as river-capture or river-piracy.

Not infrequently the valley of one stream grows headwards—not opposite but at an angle to the course of an adjacent river. The time may come when the regressing valley cuts into that of the other stream and actually captures the section of the stream above that point. Thus, river-piracy entails the beheading of part of a river by another stream that is flowing at an angle to it.

River systems develop mainly by the captures effected in this way by their subsequent tributaries. As the river adjusts itself to the geological structure of its basin, secondary tributaries develop along lines of weakness, such as the bands of softer rocks in the Weald, and the lines of faults in Wales. These secondary or subsequent streams usually occupy hollows that lie parallel with the main divide but at right-angles to the main rivers. The development of subsequent tributaries is well seen in young scarplands; here, the consequent streams flow down the dip slope, and the sub-sequent flow at the foot of the scarp, entering the consequents at an acute angle (Fig. 85). Subsequent streams which occupy the same line of weakness are usually divided from each other by a relatively low and flat watershed. Opposing streams work headwards into the low watershed until it becomes so indefinite that it may actually drain to either stream. This indecisive stage prevails until the faster downward erosion of the swifter stream causes the whole watershed area to drain to it. This change often occurs when heavy rains have flooded or water-logged the low water-parting, for at such a time the opposing sources are, in a sense, joined. Once the swifter subsequent has captured the drainage of the watershed it proceeds to annex the opposing valley at a faster rate, since the feeble division between the two streams steadily diminishes in height at the same time as it retreats. Thus far, the process outlined above is in reality

R

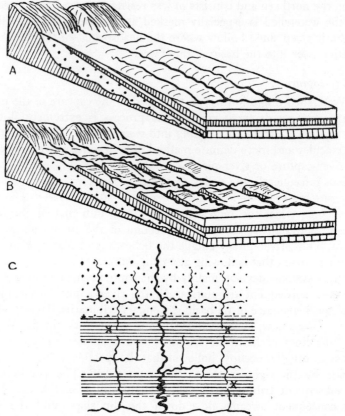

85. Development of streams on tilted strata of unequal hardness

A. Consequent drainage. B. Headward growth; formation of escarpments; strong development of subsequent streams. C. Beheading by subsequent streams; development of obsequent streams; wind-gaps (x) across escarpments

the watershed regression of a minor divide. But the time comes when the stronger subsequent, having captured all the valley of the opposing stream, eats its way into the main valley of the consequent river into which the captured stream used to flow. Then the consequent river is beheaded and enters the pirate stream at an abrupt angle, *the elbow of the capture.*

Although the above description is of rivers in scarpland topographies it applies in principle to the majority of cases of river-capture. It occasionally happens, however, that a main river is beheaded by direct capture. Thus, a fast headstream can work back into the valley of a main stream flowing at a considerably higher level than itself (Fig. 86). In this case, the watershed continually narrows until it becomes so thin that the higher stream begins to seep into the pirate. During floods, the major river overflows into the valley of the pirate stream; then, in time, the flood water forms a

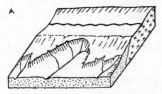 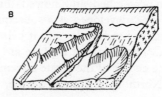

86. Headward erosion of a spring

A. Valley of mature river on highland block (crosses) being approached by headward erosion of youthful stream on softer strata.   B. Capture completed, and incision, due to increase in velocity and volume, is commencing

definite channel that gradually deepens until it drains the main stream into the pirate. Such an overflow junction, known as the Casiquiare Channel, joins the Orinoco to the Rio Negro, a tributary of the Amazon; this channel is 227 miles long and drops to the Rio Negro at an average rate of $3\frac{1}{2}$ inches per mile.

In areas floored by limestones, porous sandstones, sands and gravels, the final act of capture may take place underground and only appears above-ground later when a surface channel develops. There are some known examples of river-capture that has just been effected; the capture of the Rio Fenix in the Andes is so recent that it was rediverted from the Pacific to the Atlantic by a trench dug by six men (E. de Martonne). In the majority of cases, however, the evidence of river-capture is based on:

(*a*) an *elbow of capture* or a sudden change in direction, not explicable on structural grounds;

(*b*) a marked change in the shape of the valley below this elbow of capture; and

(*c*) a dry or almost streamless hollow that joins the valley of one stream to the valley of another.

Very rarely the existence of similar gravel deposits in the valleys concerned will prove their former connection.

In **Northumberland** the land slopes generally from the watershed of the Cheviots and Pennines eastwards to the North Sea. The South

Tyne and Coquet rivers rise near the main watershed and flow almost directly to the sea (Fig. 87). The Wansbeck and Blyth, however, rise midway down the slope, and, beyond their present sources, dry valleys link up with the headstreams of the North Tyne. It is probable that formerly the Wansbeck and Blyth rose near the main watershed and flowed direct to the sea. The North Tyne, however, developed as a subsequent of the South Tyne and worked its valley headwards until it beheaded the Blyth and then the Wansbeck. The points of capture are marked by the characteristic elbow in the course of the pirate stream. Above the elbow near Bellingham, the North Tyne flows in a wide, well-graded valley; below the elbow

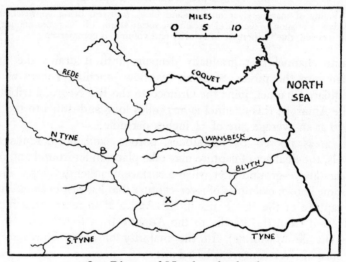

87. Rivers of Northumberland

B. Bellingham   X. a well-marked wind-gap

it quickens its rate of flow and almost immediately enters a deep, narrow gorge, the cliffs of which rise 100 feet sheer in one spot. The fast river threads several other narrow stretches before entering the South Tyne near Hexham. The broken and irregular gradient of this section of its course contrasts with the smooth gradient above Bellingham. The former is the bed of a pirate subsequent stream; the latter is the bed of the captured headstream of the Wansbeck, an older consequent river (Fig. 90).

The Lowther Hills contain examples of river-capture as well as of simple watershed migration. The Evan Water, the longest head-stream of the Annan, has pushed the source of the Clydesburn, a headstream of the Glasgow Clyde, back about four miles, thereby capturing most of its tributaries. The points of capture are marked by 'fish-hook' or right-angled bends (Fig. 84).

The Weald. In south-eastern England the present escarpments and drainage of the Weald are the result of water erosion on gently folded sedimentary rocks. The chalk and associated rocks have been denuded from the dome of the anticline and, on its periphery, have been modelled into concentric ridges and valleys. The softer beds are the Weald Clay and Gault Clay which now floor the vales; the more resistant strata are the chalk and greensand which now form the main scarps (Fig. 88). The consequents are the Stour, Medway, Darent, Mole, and Wey, draining from the centre of the region to the north; and the Cuckmere, Ouse, Arun, and Adur, draining to the south. To-day these cross the various escarpments in deep valleys, which is to be expected, as the rivers are *older* than the ridges.[1] The subsequent tributaries developed mainly on or at the foot of the Gault Clay and of the Weald Clay. The Rother, the Len

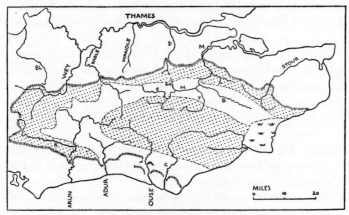

88. Drainage of the Weald

Bl. Blackwater D. Darent B. Beult E. Eden T. Teise L. Len M. Medway R. Rother C. Cuckmere Sh. Shode. Outer dotted area marks Lower Greensand Series, and inner dotted area the Hastings Beds (*see* Fig. 58)

[1] Prof. Wooldridge stresses that the Weald, in intermediate stages of its history, underwent peneplanation and, later, marine planation except in the High Weald. At the close of this stage, the present main rivers arose near the higher centre, and the development of the present drainage began. Hence, the streams of to-day are not the *original* consequents of the Wealden *dome*, but are consequent upon a later phase of its history. The main streams were superimposed by marine planation.

(a tributary of the Medway), the Great Stour and the East Stour flow mainly on the Gault Clay; the Eden and Beult, headstreams of the Medway, have developed on the Weald Clay. The relative ease with which subsequents could work headwards along these clays and capture first the opposing subsequent streams and then the parent consequent rivers, has led to much river-piracy. A subsequent stream working westwards from the main Teise-Medway river, captured the headstreams of the Darent and turned them into the headstreams of the Medway. In the same manner the Wey has beheaded the Blackwater and has, in its turn, been partly captured by the Arun. The Mole has beheaded the Wandle, and the points of capture are marked by large wind-gaps at Merstham and Caterham. Numerous other wind-gaps in the North and South Downs occur above the present sources of beheaded consequent streams.

Figure 88 also shows the Shode, an example of an obsequent stream, or one that develops on the newly formed scarp-face, down which it flows in an opposite direction to the main drainage.

The Cotswolds. A vast literature has grown up on the possibilities of the headstreams of the Thames rising on the flanks of the Welsh Mountains. It is supposed that these tributaries were beheaded by the Avon and Severn at the time when the latter rivers were working headwards along the foot of the limestone escarpment. The evidence, however, is not conclusive. That the present escarpment once extended farther out over the Vale of Avon and of Severn is shown by the presence of isolated limestone hills rising above the vale as far as five miles from the escarpment (Fig. 89). The main evidence pointing to the beheading of the larger Cotswold rivers is:

(a) Each big dip-stream flowing to the Thames rises directly opposite to a small stream rushing down the scarp face to the drainage of the Severn.

(b) The valley of each big dip-stream is continued across the scarp edge by a depression or wind-gap. The gap at Lyne's Barn above Winchcombe prolongs the valley of a headstream of the Windrush; that used by the main Oxford-Cheltenham road near Andoversford is a continuation of the valley of the Coln.

(c) The present dip-streams—Evenlode, Windrush, Coln, and Churn—are too small for their huge valleys; they are 'misfits,' and their relatively small volume may be partly due to loss of water by beheading.

The above evidence and other topographical details all suggest that the streams were once much bigger and rose farther to the west. How much farther west the escarpment extended it is impossible to say, but a matter of 15 miles or so would increase the volume of the rivers by one-third and would, if allowance is made for heavy floods in glacial and pluvial times, do much to explain the 'misfits.'

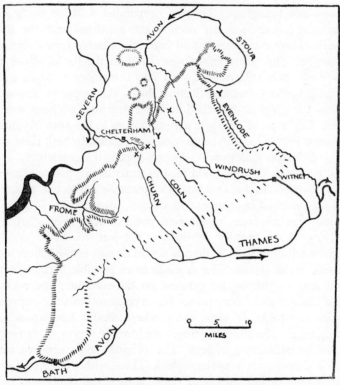

89. Drainage of the Cotswold Hills

Showing weak divides (*y*), notable wind-gaps (*x*), and intermittent streams (broken line). It should be noticed that in pre-glacial times the Warwick Avon flowed to the Wash and not to the Severn

There is, however, a probable example of river-capture near Stroud. Here the headstreams of the river Frome flow southwards for some miles before turning abruptly westwards into the wide, deep valley of the master stream. Only a low divide separates this elbow from the source of the Thames at Thames Head. The gravels found in the valleys of each have, however, so far failed to yield evidence of a former junction. The rapid headward erosion of the Frome is explained by the fact that the scarp face here is the rainiest district in the Cotswolds, and is the only locality with a thick, water-yielding bed of sands.

River Development in Areas of Folded Rocks

Where the topography consists of parallel folds at fairly close intervals, the main streams occupy the synclines and the minor tributaries furrow the slopes and enter the main streams almost at right angles. The minor tributaries rising on the flanks of the synclines work their valleys headward into the adjacent ridges and so eventually capture streams flowing down the opposite slopes. In this way the ridge is breached and the drainage assumes a trellis pattern. This type of drainage pattern occurs in the Jura Mountains (south-east France) and in south-western Ireland. In both these areas most of the major valleys are definitely synclinal in structure.

It is, however, very common for the strata in the valleys of mountainous districts to be anticlinal in structure while the mountain ridges are *synclinal* in structure. Snowdon, for example, is in reality a shallow syncline from a geological point of view (Fig. 58a). This so-called *inversion of relief* depends on the fact that in folded areas the minor tributaries often breach the folds and erode valleys along the crests of the arches. This is made more possible because water-erosion and weathering are greatest on the summits, and rocks in upfolds are by nature more vulnerable to erosion than the compressed rocks in downfolds. In some cases, where erosion has exposed the softer substrata of a crest, the ridges are lowered at a much faster rate than the neighbouring valleys. The original folds are eventually denuded almost as low as the valleys. Then, with the help of some outside factor such as tilting or river-capture, the rivers may erode the upfolds still lower until they become the valleys. The original synclines now form the higher areas, and the relief, from a geological and topographical point of view, has been inverted.

Interruptions in the Cycle of Erosion

A river is constantly grading its bed; therefore in theory it would eventually become perfectly graded and would reach its base-level of erosion. In reality, however, the cycle of erosion requires so great a length of time that it is rarely fully accomplished. The most common interruption in the progress of the cycle is a relative uplift in the drainage basin of the stream caused either by:

(a) a sinking of sea-level, or
(b) an elevation of the land mass.

Either of these changes will cause the stream to be rejuvenated. A lowering of sea-level (eustatic change) would be world-wide and is probably unusual except during Ice Ages, whereas a movement of a land-mass would be more local and is probably of frequent occurrence.

Rejuvenation. If an area is elevated, the rivers draining it quicken their rate of flow and begin to erode their beds to the new base-level. It is found that the regradation normally works upstream from the former base-level. The rate of upstream recession varies with the character of the rocks. In the eastern U.S.A. the rivers have eroded their beds more quickly on the soft coastal rocks than on those farther inland, and the difference in the rates of erosion has helped to form a break of slope, known as the 'Fall Line.'[1] Evidence of rejuvenation is not, however, based only on a break of slope that

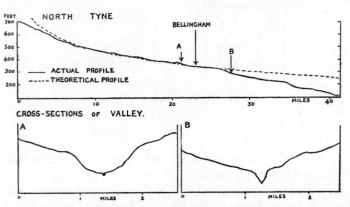

90. Profile and cross-sections of the valley of the North Tyne, showing break of slope and change in shape of valley below Bellingham

(After R. F. Peel)

Vertical scale of cross-sections is exaggerated six times

[1] The break of slope or Fall Line is in reality a zone which consists of the exposed face of an old tilted peneplain. *See* Renner, G. T. 'The physiographic interpretation of the Fall Line', *Geog. Rev.*, 1927, pp. 276-86

marks a new cycle of erosion. More reliance is to be placed on the incision or entrenchment of the river-bed. On rejuvenation, the river begins to cut a new valley inside the old, and even a fast stream, if made more youthful, will erode a narrower bed within its already narrow valley (Fig. 90).

The result of rejuvenation on meandering streams is of much interest. Where the flow of the meandering stream is quickened considerably and where the rocks are relatively soft, the meanders become entrenched between steep, symmetrical sides. If, however, the rejuvenation is slight and the rocks of the valley-floor are resistant, the downward erosion proceeds slowly and the river has time to shift its windings and to accomplish some lateral erosion (Fig. 91A). In this case the valley is wide, with spurs and amphitheatres, and the actual valley itself is entrenched. Hence, the stream is said to be

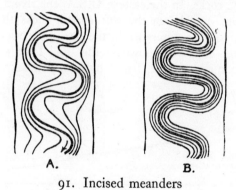

A.   **B.**

91. Incised meanders

(After Rich)

A. Ingrown   B. Entrenched

ingrown rather than entrenched. In both cases the meanders will be deeply incised (Plates 19 and 20).

Among the finest examples of incised meanders in Europe are those of the Wye below Ross, of the Dee at Llangollen, of the Wear at Durham and of the lower Seine near Rouen.

The meanders of the Monmouthshire Wye between Ross and Chepstow are exceptionally large. Some of the curves have an amplitude of three miles and a depth of several hundred feet. On the high plateau above the present Wye, at about 900 feet O.D., there are traces of the meanders formed by the river before it was rejuvenated. With rejuvenation, the river proceeded to erode into the floor of its valley. The old meanders of the higher cycle of erosion were abandoned, and to-day are nearly 400 feet above the river. They may be seen at the villages of Redbrook and Newland, the latter being partly on a gravel patch of an abandoned curve (Fig. 92). Another abandoned meander at a lower stage of erosion occurs immediately below the castle at St. Briavels. Thus, the Wye has gradually eroded its bed

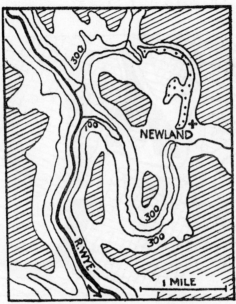

92. Abandoned hanging meander of the
Herefordshire Wye at Newland

Deposits of former channel are shown by dots

downwards, forming successively lower valleys within its former shallow valley. This valley-in-valley form is typical of rejuvenation.

At Llangollen the Dee has incised its bed, leaving old meanders high and dry above it. In this case, however, some of the meanders were plugged by glacial drift which partly forced the river to straighten its course. The abandoned meanders of the former erosional cycle are shown in Figure 93.

The entrenchment and abandonment of meanders also occur along the Evenlode and the Windrush, the longest headstreams of the Thames. Near Hanborough, many abandoned meanders of the

Evenlode occur, while the present small stream has also numerous ox-bow lakes and cut-offs. The Windrush, just below Burford in Oxfordshire, has cut across the narrow necks of two meanders of the more mature and larger stream that once occupied the valley.

It is difficult to determine whether the rejuvenation of any of the above streams is due to the elevation of the land mass. In many the rejuvenation arose from increase of volume because of climatic changes, especially during the Ice Age. Moreover, it is possible that

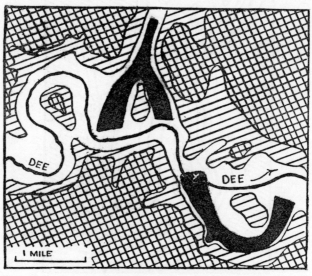

93. Valley of the Dee at Llangollen (*x*), showing abandoned meanders and severed necks of spurs

(*After Wills*)

Black marks glacial drift that assisted in the diversion of the river

sections of a river's course may be rejuvenated by an increase of volume due to the capture of another stream, and by the wearing away of a hard barrier across the stream bed which checked erosion above it. The whole question of uplift and of rejuvenation is extremely complex, and in the above discussion no more has been attempted than to draw attention to the topographical evidence, which usually includes one or more of the following features:

(*a*) a valley cut within a valley;

(b) a slight break of slope where the new cycle of erosion is attacking the old;

(c) entrenched and ingrown meanders;

(d) abandoned meanders hanging above a stream which shows a tendency to straighten its course;

(e) river terraces.

River Terraces. Rejuvenation frequently results in the formation of river terraces. If a river that is depositing gravel and alluvium is rejuvenated, it proceeds to cut a new valley in the deposits of the old. During this downcutting, patches of the deposits may escape erosion and may be left perched on the lip of the new valley. It is possible for the stream to reach again the stage of maturity at which deposition occurs and then to be rejuvenated afresh. In this way a series of terraces may be formed on the sides of the valley. The terraces should be of both alluvium and of gravel, but the alluvium deposits are so easily removed by rainwash that usually only the gravel patches remain. On the lower Thames near London the alluvium and flood plain of the present river is below 50 feet O.D., and reaches a width of 3 miles opposite Woolwich and Barking. The lower, or Taplow terrace, lies mainly between 50 and 100 feet above the river; its gravel patches can be traced at Hyde Park, from Paddington to Holborn and from St. Paul's to the river Lea (Fig. 94b). The upper gravel terrace is about 50 feet higher than the Taplow; it has survived chiefly south of the present Thames, where it caps several small hills. Patches of this upper terrace occur at Clapham Common, Wandsworth, Roehampton, and Richmond Park.

The terraces of the upper Thames near Oxford are named after villages situated upon them (Fig. 94a). There is evidence of four main periods of deposition that were interspersed with three long periods of downcutting. The oldest or highest stage of deposition is marked by the Hanborough terrace, which is about 100 feet above the present river. Its formation was followed by a long period of erosion, at the end of which the river again began to deposit extensive gravels on the valley floor. The remnants of this aggradation form the Wolvercote or 40-foot terrace. The next cycle of erosion terminated in the period of deposition that formed the 20-foot or Summertown-Radley terrace. Subsequently the river was rejuvenated, and proceeded to erode its bed nearly 30 feet below its present level; this channel is now buried beneath the extensive alluvium that floors the wide plain on the north and west of Oxford. The city itself rose on the 20-foot terrace, and spread northwards along it. Only in recent years has the built-up area of north Oxford expanded beyond the gravels of this terrace on to the wetter alluvial soils of the valley bottom. The Cam,

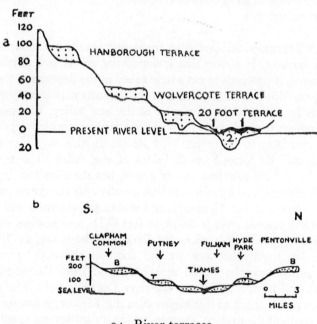

94. River terraces

(*a*) The upper Thames drainage in the Oxford district
     1. Alluvium   2. Buried river channel. (*After K. S. Sandford*)
(*b*) The lower Thames at London. Black shows alluvium of present river.
     B. Boyn Hill or 100-foot terrace. T. Taplow or 50-foot terrace

the lower Severn, and the Warwickshire Avon all have gravel terraces that have been carefully investigated. As yet, however, it has proved difficult to correlate these with the terraces of the upper Thames.

River Courses Largely Independent of Surface Structure

Superimposed Drainage. During a normal cycle of erosion, the original surface rocks are denuded away over large areas and the drainage adjusts itself to the underlying structure. This has been so in the Weald and most other districts of Britain. There are, however, several areas where the influence of the original surface rocks has continued to predominate over that of the newly exposed strata. In these districts the direction of the present rivers was determined by the former rock cover, and is to-day independent of the nature of the strata being traversed. In other words, the drainage has been superimposed on the present rocks by strata that have since disappeared.

The course of the river Wye shows a remarkable indifference to the rock structure (Fig. 95). The probable reason is that the river

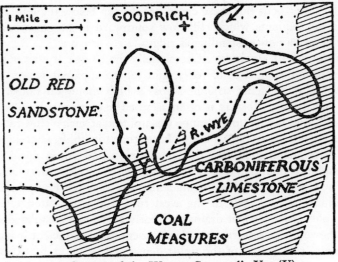

95. Course of the Wye at Symond's Yat (Y)

formerly ran over a homogeneous layer of young rocks. Having eroded its bed through this layer it began to incise itself into the more complicated underlying strata. But the stream was forced to retain the bed it had cut in the original surface strata, and so to-day it crosses widely different rocks and ignores their differences.

The rivers of South Wales, which now cross the various outcrops of the coal basin, probably originated on a southward-sloping layer of younger rocks. The clearest example in Britain of superimposed drainage is, however, the Lake District. Here, the present river system originated on layers of Carboniferous Limestone and younger strata that covered a dome of old hard rocks. The drainage radiated fanwise from the crest of the dome. In time the rivers breached the younger rocks and began to etch their beds into the old core. They had to keep to their original direction; and to-day, although the younger strata have disappeared except from the circumference of the dome, the main rivers retain their radial pattern.

At seaboards, superimposed drainage may form upon newly-emerged marine platforms. The new consequents are superimposed by the gentle slope which is independent of the underlying strata. In south-western Ireland the new consequents flowed southward over a marine platform either laid or cut upon strata with east-west folds (Fig. 55). Soon subaerial erosion etched out east-west lateral valleys in the softer beds but these could not deepen below the level of the main consequents which still flow southward in gorges across the harder strata now upstanding as ridges.

Antecedent Drainage. Occasionally the course of a river shows a strong independence of relief, and for no apparent reason heads straight for and cuts straight across a high mountain range. This peculiar drainage feature is often ascribed to earth movements. It is thought that the mountain range was uplifted at such a slow rate that the river eroded its bed as fast or faster than the uplift. Consequently the river maintained its original direction and cut a deep valley through the mountain region that was being elevated athwart its track.

The Ogden River, which cuts across the Wasatch Range, and the Columbia River, both in the western U.S.A., are often cited as examples of antecedent drainage. Nearly all known postulated cases are, however, necessarily based on obscure or insufficient evidence. The localized nature of the uplift demanded is not easily explained away, and often the breaching of a mountain range may be more satisfactorily explained by river-capture or by the disappearance of former surface strata. Yet, slow changes of a mountain-building nature must be admitted, and there is nothing inherently impossible in a river being able to erode its bed as fast as the land is uplifted. It should be emphasized, however, that the idea of antecedent drainage is based on the *localized* uplift of land across the path of the stream, and not on the uplift of the whole basin, which would result in rejuvenation. Such localized uplift and resultant antecedence appears to have happened on the River Salzach in the Austrian Tirol, in the Arun and many other rivers across the Himalayas and in parts of the Colorado.

LIST OF BOOKS

General
Cotton, C. A. *Geomorphology*, 1942. *Landscape*, 1948
Wooldridge, S. W. and Morgan, R. S. *Outline of Geomorphology*, 1959
Sparks, B. W. *Geomorphology*, 1960

Wooldridge, S. W. and Linton, D. L. *Structure, Surface and Drainage in South-East England,* 1955
Thornbury, W. D. *Principles of Geomorphology,* 1954
Horton, R. E. 'Erosional development of streams,' *Geol. Soc. Am.,* 1945, pp. 275–370
Zernitz, E. R. 'Drainage patterns and their significance', *Jour. Geol.* 1932, pp. 498–521
Schattner, I. *The Lower Jordan Valley,* 1962
Leopold, L. B. *et al. Fluvial Processes in Geomorphology,* 1964

River Capture and Watershed Migration

Crosby, I. B. 'Methods of Stream Piracy,' *Journal of Geology,* July and Aug, 1937
Lebon, J. H. 'On the Watershed Migration . . . of the Lowther Hills,' *S. G. M.,* 1935, pp. 7–13
Merrick, E. 'On the Formation of the River Tyne Drainage Area,' *Geol. Mag.,* 1915, pp. 294–304, 353–60
Bury, H. 'The Denudation of the Western Weald,' *Q. J. G. S.,* 1910, pp. 640–92
Edmunds, F. H. *The Wealden District.* H.M.S.O., 1954. British Regional Geology
Linton, D. L. ''The Origin of the Wessex Rivers,' *S.G.M.,* 1932, pp. 162–75; 'The Sussex rivers', *Geog.,* 1956, pp. 233–47
Sissons, J. B. 'Erosion surfaces and drainage systems in northern England', *Inst. Brit. Geog.,* 1960, pp. 23–38
Yates, E. M. 'The development of the Rhine', *ibid.,* 1963, pp. 65–81

Rejuvenation

Rich, J. L. 'Certain Types of Stream Valleys and Their Meaning,' *Journ. of Geol.,* 1914, pp. 469–97. (Deals with entrenched and ingrown meanders.)
Miller, A. A. 'The Entrenched Meanders of the Herefordshire Wye,' *G. J.,* Feb. 1935
Morley Davies, A. 'The Abandonment of Entrenched Meanders,' *Procs. Geol. Assoc.,* 1923, pp. 81–96
Sherlock, R. L. *London and the Thames Valley.* H.M.S.O., 1960. British Regional Geology, No. 2
Sandford, K. S. 'The River Gravels of the Oxford District,' *Q. J. G. S.,* 1924, pp. 113–70; 1925, pp. 62–86
Hare, F. K. 'The Geomorphology of Part of the Middle Thames,' *Procs. Geol. Assoc.,* 1947, pp. 294–339
Brown, E. H. 'The River Ystwyth, Cardiganshire, A Geomorphological Analysis,' *Procs. Geol. Assoc.,* 1952, pp. 244–69
Sealy, K. R. & C. E. 'The Terraces of the Middle Thames,' *Procs. Geol. Assoc.,* 1956, pp. 369–89
Kidson, C. 'The denudation chronology of the River Exe', *Inst. Brit. Geog.,* 1962, pp. 43–66

Superimposed Drainage

Marr, J. E. *The Geology of the Lake District,* 1916
Mitchell, G. H. 'Geological history of the Lake District', *Procs. Yorks. Geol. Soc.,* 1956, pp. 407–63
Jukes, J. B. 'On the formation of some river valleys in the South of Ireland,' *Q.J.G.S.,* XVIII, 1862, pp. 378–403
Miller, A. A. 'River development in southern Ireland,' *Proc. R. Irish Acad.* 45, 1939, pp. 321–54

Antecedent Drainage

King, L. C. *South African Scenery,* 1963
Coleman, A. 'The terraces and antecedence of a part of the River Salzach,' *Inst. Brit. Geog.,* 1958, pp. 119–34
Wager, L. R. 'The Arun River drainage and the rise of the Himalaya', *G. J.,* 1937, pp. 239–49

CHAPTER NINETEEN

THE WORK OF SNOW AND ICE

Snow-line and Existing Ice-fields

LARGE areas of the world never see snow, and even in the British Isles upon the coastal strips of Cornwall and Devon snow-cover is rare enough for children to be taken to see drifts on the tors. Patches of snow may linger most of the year on the flanks of Ben Nevis and Helvellyn, but the permanent snow-line is not attained in Britain. The height of the snow-line in any locality depends mainly upon the summer temperature and the annual amount of snowfall. Heavier snowfall may counteract the warmer temperature of sunnier slopes; thus many ranges in the Himalayas and Swiss Alps have so much more snow on their southern slopes that the snow-line is lower on the southern side in spite of its higher temperatures.

In the equatorial peaks of the Andes and East Africa the snow-line remains at some 16,000 to 18,000 feet; in middle latitudes it sinks to about 6,500 feet in the western Pyrenees, and to about 9,000 feet in the Alps. Antarctica contains great ice-fields at sea-level, whereas those in the Arctic Circle are usually at heights of over 2,500 feet. The Greenland ice-cap rises to over 10,000 feet and probably attains a maximum thickness of about 10,800 feet. Here the whole topography is hidden beneath ice except near the margins of the ice-field, where glaciers and lobes of ice creep down to the sea. In Antarctica the snowfall is smaller than in Greenland and the higher hills are by no means covered; yet the more continental conditions give rise to an ice-sheet 13,000 feet thick locally, that has spread out as a solid wall over the Ross Sea.

These lands within 30° of either pole nourish most of the world's permanent snow-fields. But the five million square miles of Antarctica, the half-million square miles of Greenland, and the much smaller ice-caps in Iceland and Spitzbergen are only the shrunken remnants of the vast ice-sheets of the Quaternary period. Studies of existing ice-fields and glaciers, although important and interesting in themselves, are of even greater importance since they supply the key to the wider work of ice in the Ice Age. To-day six million square miles are covered by land-ice or over one-third the area covered in the great glacial invasions (Plate 21). This ice if melted and returned to the oceans would raise their level by about 250 feet, if allowance is also made for the isostatic adjustments of the land-masses freed from ice-caps.

Glaciers and Moving Ice

Falling snow accumulates deepest in sheltered and shaded hollows where it escapes the full force of wind and sun. Here, if the hollow is above the melting zone in summer, snow collects gradually year after year. Eventually the weight of the accumulated snow, coupled with the effect of a certain amount of thawing and regelation, changes the lower layers into a kind of granular ice. The pressure of further snowfall causes this ice to begin to move outwards, and under the action of gravity, downwards. There may be formed a wide lobe of ice or a more restricted valley glacier; much depends on the amount of snowfall and the topography of the area of accumulation. A ridge previously dissected by normal fluviatile erosion tends to develop valley glaciers, while a plateau tends to favour a larger ice-field.

In the Alps, where there are nearly 2,000 glaciers, the commonest form is a wide lobe of ice that can scarcely be distinguished from the accumulation basin. The largest glacier, the Great Aletsch, is about 16 miles long, nearly 1 mile wide, and in parts 2,500 feet thick. In Norway the Jostedalsbreen ice-field, which has accumulated on a plateau, covers 580 square miles and gives rise to a score of large glaciers. Somewhat the reverse happens in Alaska, where several mountain glaciers unite upon the flat plain at the base of a ridge to form a piedmont ice-sheet. This, the Malaspina Glacier, extends over an area of 70 miles by 20 miles and has extensive forests growing upon the débris that has collected upon its edges. One of its many feeders, the Seward Glacier, stretches for 50 miles, and is never less than 3 miles wide.

Conditions where snow- and ice-fields increase in extent until they cover the whole land, except for the highest and steepest parts, are not found to-day outside polar regions.

Movement of Ice. The nature of the movement of ice is not fully understood, but it is known that a glacier tends to move more slowly at its edges owing to the friction of the rocks of its bed. The Mer de Glace (in Switzerland) moves about 20 to 27 inches a day at its centre during the warmer months and some 8 inches per day less at its sides. The Rhône Glacier in parts moves 318 feet a year, while glaciers slipping from the Greenland ice-cap towards sea-level progress up to 100 feet a day. The significant point is that the motion is extremely slow.

Many processes are at work in ice motion. The ice may become plastic under pressure and move downhill as a viscous fluid. Or melting and re-freezing of the ice crystals at their points of pressure may permit movement, since the force of gravity would ensure that each partly melted granule would move towards sea-level before it re-froze. On the other hand, ice, especially in its crevassed surface layers, may move partly as a solid by fracturing and then by slipping and shearing. The plastic nature of ice appears to increase with depth, and the mass may become plastic where the pressure is great enough (say, beneath 150-250 feet thickness), but it may remain as a solid at the surface, especially in very cold regions. 'It is a solid close to its melting temperature, which is delicately adjusted to small changes in temperature and pressure.'[1]

The Work of Moving Ice

Erosion. Pressure is of great significance in the erosional work of moving ice. Ice has weight and carrying power, and without these its erosive powers would be insignificant. The enormous pressure, both downwards and to a lesser extent outwards, of several hundred feet thickness of moving ice abrades the sides and bed of the valley. The abrasion is made possible by the presence of rock fragments and finer débris that become frozen into the ice mass. These act as grinding and cutting agents; the abundance of fine rock-flour formed and the striations on rocks in glaciated districts bear witness to their efficiency. As a result of abrasion and of the freezing of any loose materials into the base and sides of the ice, the hollow occupied tends to become flat-bottomed, and its sides lose their irregularities. Thus the valley becomes more U-shaped in section and straighter in profile.

When the slope of the bed of moving ice is increased, the ice develops crevasses at its surface and forms a kind of ice-fall. The increase of slope leads to an increase of speed, but not to a corresponding increase of erosion in that particular part of the glacier, as the crevassing causes a decrease of weight on the glacial bed. *At the foot of the slope* the ice is again compressed, its pressure increases, and owing partly to the weight of the ice-fall above, it exerts greater downward pressure than it does on level surfaces elsewhere. Consequently at the foot of each increase of slope a glacier tends to erode a depression in

[1] H. Bradford Washburn, Jr., *G. J.*, 1936, p. 493

its bed, while at the same time the slope itself undergoes relatively little erosion (Fig. 96). So it often happens that the bed of a glacier is stepped, and when the ice melts, these steps may be occupied by ribbon-shaped lakes. Where the deepening at the foot of gradients did not proceed far enough to gouge-out a true hollow, glaciated valleys usually consist of an alternation of narrow, steeply-graded gorges and wider, flat-floored basins. The stepped nature may, of course, have been present before glaciation because either of rejuvenation or of rock-structure, but it was much emphasized by glacial action.

One school of glaciologists holds that ice is mainly a protective agent. In their view, a step could be formed by ice retreat during inter-glacial periods; the 'step' would be covered and protected by ice while the area below it would be eroded away by running water. The general belief, however, is that ice does erode, but only slowly. The slowness of glacial erosion compared with stream erosion is, however, partly compensated by the relatively greater volume (or surface contact) of ice masses.

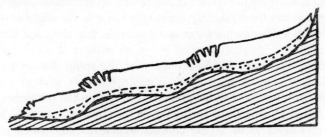

96. Longitudinal profile of a glacier valley

Broken line shows course of the pre-glacial river. Dotted area is part of bed eroded away by the ice

Transport. Of all erosional agents, ice can carry the greatest load. It happens that snow-fields occur where weathering, due to water freezing in crevices and to diurnal temperature changes, is greatest. Showers of loose rocks and stones fall daily on to the edge of the snow-field. Much of this material is soon covered with snow and becomes incorporated into the ice forming the *névé*, or gathering ground. In a similar way, débris ranging from dust to large boulders falls on the sides of a glacier and is piled up to form lateral moraines. Where tributary glaciers come into the main ice body their lateral deposits streak the surface of the central parts of the moving ice, or they may be hidden beneath the surface until exposed by melting near the snout of the glacier (Plate 22). The importance of rock fragments falling from neighbouring slopes can be appreciated in Greenland, where the ice-cap is bare of deposits save where exposed ridges (called *nunataks*) provide the ice below their slopes with rock débris.

The collection of débris from the adjacent mountain slopes and the matter acquired from the sides and base of its bed, usually cause a glacier to be well loaded with material both in its mass and on its surface. The accumulation is usually greatest under the surface near to the edges of the moving glacier.

Deposition. When the ice mass reaches its melting-point the transported matter is deposited usually in the form of hummocks that consist largely of boulders, irregular rock fragments, fine clay, or rock-flour. These huge dumps, or terminal moraines, are often arcuate to the snout of the glacier and are usually channelled by streams issuing from the base of the ice mass. The melting-point even of a short cirque glacier is marked by an enormous collection of débris that has been piled up, especially towards the edges of the ice. Rivulets of swift-flowing water carry away the fine clays and move the smaller débris downstream, all the time making it more spherical in shape. The large blocks remain near the snout, but in times of rapid thaw a spate may push unstable boulders a short distance downstream by sheer water pressure assisted by the undermining of their bases. During these spring floods the streams become so loaded with rock-flour as to appear chocolate-coloured, and small pebbles are rolled rapidly along the stream's bed.

Quaternary Ice-sheets and Glaciers

The Ice Age: Its Cause. Serious objections can be raised against all the hypotheses put forward to account for the decrease in temperature and the increase of snowfall that led to the great ice-sheets. Theories based upon changes in the earth's orbit, when a long winter in aphelion would not be balanced by a short hot summer in perihelion, demand regular periodical Ice Ages, whereas geological evidence shows that cold periods form the exception rather than the rule. Similar objections can be brought against theories postulating variations in the tilt of the earth's axis. Other hypotheses assume that the carbon dioxide and impurities in the atmosphere increased sufficiently to absorb enough heat from the sun's rays to cause an Ice Age in lands that were already cold. The achievement of these conditions, even granting great volcanic outbursts, cannot be conceived. Other theories utilize variations in the sun's heat owing to the development of sun-spots; or variations in its radiation because of its passage through cosmic dust; and others, the elevation of certain land masses so as to lower their temperature and to interfere seriously with the normal world distribution of temperature (*see* Note, p. 294). For our purpose, however, it is sufficient to notice:

(*a*) that the Ice Age lasted over a long period of time (roughly 1,500,000 to 10,000 B.C.), and its conclusion is so recent that its topographical effects are still clearly visible;

(*b*) that during this period the ice-sheets advanced and retreated several times;

(*c*) that considerable changes in the relative level of land and sea occurred;

(*d*) that then, as now, glaciers and ice-sheets persisted far beyond the permanent snow-line, and glaciation was by no means confined to the regions of snow accumulation;

(*e*) that at the outer edge of the ice (and especially during periods of ice retreat) running water was the great transporting and eroding agent. Great floods occurred each spring, and this happened even in warmer areas that underwent nivation (or periglaciation) as distinct from glaciation. Consequently, landforms in areas of glacial deposition frequently bear the mark of running water (Plate 21).

Extent of Ice-sheets. Nearly one-half of North America (or 4 million sq. miles) and nearly one-third of Europe were covered by ice during one or the other of the glacial advances (Fig. 97). In America five major glacial advances are recognizable, in the Alps four, while in north-west Europe seldom more than three or four can be distinguished. The American ice-sheets advanced mainly from two great accumulating centres on the Laurentian Shield and extended as far south as St. Louis on the Mississippi. In Europe, the Scandinavian Highlands formed the main centre of ice dispersion.

At the same time the permanent snow-line descended nearer sea-level in mountainous regions throughout the world. In the Alps it extended farther downwards for altitudes varying from 650 to 1,600 feet, and in the Andes it fell 3,000 feet below its present height. Mountainous regions such as the Vosges, Snowdonia, Cumbria, and the Scottish Highlands attained the permanent snow-line.

In the British Isles the general movement of ice was outwards from western Scotland, but each block of mountains in the west

97. Extent of main Quaternary ice-sheets in the northern hemisphere

Showing land area not glaciated (black); unglaciated land, to-day and in the past, with a frozen subsoil (shaded); approximate maximum limits of land-ice (dotted line); approximate maximum limits of sea-ice (broken line)

formed its own small ice-cap. In addition, an ice-sheet from Scandinavia invaded parts of the east coast lands. Roughly speaking, the glaciation extended as far south as a line from London to Bristol (Fig. 98).

Changes made by Glacial Erosion

Effect upon Plateau Surfaces. The major effect of an ice-sheet

98. Direction of ice-flow and limits of glaciation in the British Isles
(*Wright, 'The Quaternary Ice-Age.' Macmillan*)

upon a plateau surface is further peneplanation. The summit areas
are smoothed and scraped bare of soil, while valleys transverse to the
ice-flow tend to become filled with transported material (Fig. 99).
Valleys, however, that run parallel with the main direction of ice
advance are deepened, and usually at a greater rate than the summit
levels are lowered. The topographical characteristics of a glaciated

peneplain are seen in the Laurentian Shield of Canada, in the Highlands of Scotland, and on a minute scale in Anglesey. Smoothed,
barren summits, an immature drainage system, and frequent lakes
are common features. Laurentia is the paradise of the canoeist, and
excels even Finland, with 35,000 lakes, in bodies of water (Plate 23).

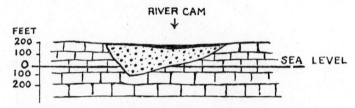

99. Valley filled with glacial débris: the valley of the Cam at
Littlebury

(Modified from Whitaker)
Vertical and horizontal scale are the same

In Scotland, the lochs are mainly ribbon-shaped and more connected
in origin with glacial valleys than with the irregular deposition of drift.
The deep lochs often occur along fault lines, such as in the Great
Glen, where the ice erosion worked faster along a line of weakness.
Yet, in other instances, ice moving across the grain of the relief filled
the consequent valleys with débris and utilized and enlarged the
hitherto insignificant valleys of the subsequent streams. The lower
Tay, Clyde, Tyne, and Humber flow over a considerable thickness of
glacial deposits that fill their pre-glacial beds.

Effect of Ice Erosion on Mountains and Valleys: *Cirques*.
In areas of ridged relief and of steep slopes the work of ice results
mainly in the formation of cirques and of U-shaped valleys. Such
formations are associated largely with valley glaciers. A cirque (or
corrie or cwm) is the armchair-shaped hollow in which the *névé* ice
accumulates. Patches of snow which collect in hollows on hillsides
may form incipient cirques, as the snow-patch continually moistens
the ground beneath and around it, so causing solifluction or a downhill creep of the moistened surface soil upon a frozen subsoil. This
nivation erosion is especially common when temperature changes
about freezing-point are frequent. The frost action further breaks up
the surface layers and melt-water lubricates the downhill creep. At
the same time weathering extends the headwalls of the hollow backwards. But these nivation hollows are minor features.

In the case of cirques, where sufficient snow collects to form ice, melt-water occasionally trickles down the enclosing walls and passes beneath the edge of the ice surface (Fig. 100). This penetration by water is often achieved by means of a gap that commonly exists between the *névé* and rock basin.[1] As a result, at least the edges of the hollow are affected by the action of water expansion on freezing. The disintegrated rock is removed either by slipping into the ice mass or by being gripped and frozen into it. Hence a normal cirque has its headwalls and its bed, at least in its subglacial edges, steepened by the action of freeze and thaw. The fact that the ice is usually thickest and pressure greatest in the centre of the ice-field accentuates the basin nature of the hollow, as also does, in some instances, a slight rotational or pivotal movement of the whole ice-mass.

Nothing is more typical of a glaciated mountainside than a series of cirques incised at a fairly even height along the length of the slope. The cirques may grow until they are separated by only narrow steep ridges, called *arêtes*. Mountain ridges may be cut through in this way

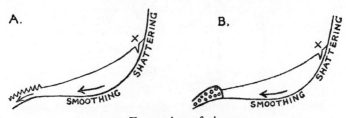

100. Formation of cirques
(After Wright and Lewis)
A. Cirque as the source of a valley glacier
B. Cirque glacier, with moraine at its lower edge
x marks position of *bergschrund*

until only pyramidal or Matterhorn-shaped peaks are left upstanding. Many cirques in areas now free of ice contain small circular lakes; in Scotland, Lochan Nan Cat near Loch Tay; in Cumbria, Red Tarn above Ullswater, Blea Water, and Stickle Tarn; in Snowdonia, Glaslyn and Marchlyn Mawr are examples (Plate 24).

U-shaped Valleys and Truncated Spurs. The tongue of ice moving downwards from the cirque tends to pluck, abrade and smooth

[1] A narrow gap, seldom above an inch or two wide, actually at the rock face, is called the *randkluft*; a larger crevasse, that in summer marks the separation zone of stationary snow-field and moving ice, is called the *bergschrund*. In winter the *randkluft* and *bergschrund* fill with snow. The latter is absent in some glaciers.

its valley-sides. The bases of the spurs of pre-glacial river valleys are truncated, the valley becomes U-shaped and is straightened locally. In many cases to-day the tributary streams hang or drop abruptly

101. Origin of U-shaped valleys and truncated spurs
(Modified from W. M. Davis)
A mountainous region before and after glaciation

down the valley sides on to the flat valley floor. Hanging streams are not confined to glaciated districts, but the great majority are formed by glacial action. The moving glacier steepens the sides of its bed until they are almost precipitous. When the ice disappears, the tributary streams from the neighbouring heights must now pass over an abrupt change of slope on entering the main U-shaped valley (Fig. 101). Characteristic glacial valleys with hanging streams are Nant Ffrancon in Snowdonia, Borrowdale in Cumbria, and the great Lauterbrunnen valley near Interlaken in the Swiss Bernese Oberland.

Ribbon-shaped Lakes. The power of valley ice to wear away its bed at the base of a change of slope often gives rise to the ribbon-shaped lakes so typical of glaciated mountain regions. These lakes may be partly dammed by morainic deposits, but occasionally they are true rock basins and occupy hollows that have been gouged out below the local level of the valley, and in some cases below the level of the sea. In Skye, Loch Coruisk, which is surrounded by naked ice-smoothed and striated rock, has a depth of nearly 100 feet below the level of the sea, from which it is separated by only a low narrow neck of solid rock. The floor of Loch Morar lies nearly 1,000 feet below sea-level; other examples of over-deepening occur in North Italy, where the floors of lakes Como, Maggiore, and Garda all descend to 689 feet or more below the level of the Mediterranean Sea (*see* Table on p. 334). Even after allowing for a large fall in sea-level during glacial times, it seems that the power of ice to erode below sea-level must be admitted.

Glacial Overflow Channels. Occasionally valley glaciers and ice-sheets overflowed the hollow they occupied and sent a tongue of ice across a neighbouring divide. These tongues often moved in a direction different from that of the main ice movement, and their channels subsequently appear as transverse valleys which are at a relatively high level. The mountain ridge between lochs Morar and Nevis narrows and sinks to only 200 feet O.D. at Glen Tarbet, which is a typical overflow glacier channel. Another example occurs in the Pass of Brander, where an overflow glacier worked along a fracture line to originate the deep-sided valley that links Loch Awe to the coast near Oban.

Changes made by Glacial Deposition

The transporting and depositing work of ice is mainly to be seen on lowlands and in piedmont districts. On pre-glacial plateaux, the filling-up of river valleys not utilized by the ice advance has created a new drainage of an indeterminate nature, with frequent rapids, low, marshy divides, and many streams. This is excellently shown at the Moor of Rannoch on the upper Tay and in the Lochinver neighbourhood near Ben More in the north-west Highlands of Scotland.

Moraines and Glacial Drift. The major deposition, however, took place where the ice melted either at the glacier's snout or at the edge of the ice-sheet. The type of material and manner of deposition depends largely upon the nature and movement of the ice and the load it acquired from the country it had traversed. In the case of a large lobe of ice the terminal moraine may spread as a series of arcuate ridges stretching over a great width of country, as does the

Salpausselka in Finland. Valley glaciers usually deposit a heterogeneous collection of striated and non-stratified débris in the form of
a hummocky ridge or series of hummocks (Fig. 102). The drift
left by melting ice-sheets is usually finer in character and is
often deposited over an extensive area and is not restricted to
marked belts. This widespread type of deposition is often called
ground moraine; it causes scenery that varies from the highly

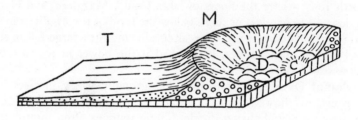

102. General sequence of deposition at the snout of an
ice-tongue or valley-glacier

C. rock basin	D. drumlin zone
M. moraine zone	T. outwash plain

irregular topography found in parts of New Jersey and Montana
U.S.A., to the hummocky landscapes of the Solway Plain and Holderness. These districts are riddled with lakes and marshes that collect
in surface irregularities. In Holderness, much drainage has been
carried out, but one 'mere' still remains near Hornsea and the many
'carrlands' mark the sites of the peaty beds of drained lakes.

    In East Anglia the mantle of boulder clay was probably deposited
by ice-sheets from the north, as the drift is chalky in nature. The
evenness of the drift and rarity of terminal moraines may be put down
to the fact that the region was covered by the front of an ice-sheet
that melted slowly as a stationary block of ice. The area was, however, invaded by an ice-sheet from Scandinavia, and it seems that this
'North Sea Ice-sheet' deposited the Cromer Ridge, the greatest
terminal moraine in Britain. The ridge consists of a belt of hummocky
hills, largely of sands and gravels, covering some 5 miles in width and
15 miles in length and rising to over 300 feet. Very much smaller is
the chief terminal moraine in the Vale of York, that consists of a
broad ridge about 50 feet high extending from the Yorkshire Wolds
to the Pennine slopes between the Wharfe and the Ure. It is cut by

the Derwent at Stamford Bridge and by the Ouse at York. Lateral moraines border Llyn Llydau on Snowdon and many similar morainic deposits occur in the Lake District.

Of more interest than of topographical importance is the transportation by ice-sheets of large blocks of rock or erratics. These boulders may be large enough to interfere permanently with agriculture, as happens in parts of New England, U.S.A., but in northeast Scotland they have been dragged from the land and dumped into ditches. Their geological structure often throws much light on the movement of the ice-sheets. The boulders of the peculiar rock from Ailsa Craig, an island in the Firth of Clyde, are found scattered in western Wales and eastern Ireland.

Roches Moutonnées. Small outcrops of rock and great boulders when traversed by an ice-sheet become smoothed and rounded only on the side facing up hill, while on their lee side, with the aid of freeze-and-thaw, the bare rock is plucked rough and jagged. In some mountainous areas, such as Snowdonia, the polished rock surfaces abound and are called *roches moutonnées*.[1] Occasionally glacial erosion and deposition work together to form distinctive topographical features. In Scotland a landform, known as 'crag and tail,' occurs where upstanding volcanic plugs have undergone this process, the plug forming the crag and the sloping bank of glacial débris behind it the tail.

Drumlins. Elongated mounds shaped like a whale or upturned keel of a boat occur frequently near the margins of an ice-sheet. They consist mainly of glacial clay, and are usually less than half a mile long and 300 feet high (Fig. 103). Their formation is very obscure. One suggestion is that they correspond to patches of débris-laden ice in the base of a moving ice-sheet. The dirtier ice would tend to move more slowly than the surrounding ice, and so would be dragged into an elongated shape previous to the thawing of the ice-sheet. Drumlins abound in the lower Tweed valley, where they almost govern the local pattern of the hedges, roads, and streams. They are equally common in the Solway lowlands and in the Central Plain of Ireland, where their alinement reflects the direction of advance of the ice-sheets. In the U.S.A., the Boston district is rich in drumlins, Bunker Hill being a well-known example.

[1] A term given by de Saussure about 1790 because when seen downhill in mass the glistening, rippling rock surfaces fancifully resembled contemporary wigs smoothed down with mutton tallow.

Eskers: the Work of Glacial Water. Débris carried by ice is eventually subjected to the influence of flowing water, especially in the form of seasonal floods. Moreover, probably at most times of the year a certain amount of water would be moving within

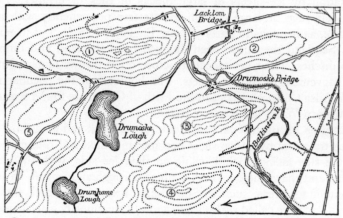

103. Contour map of drumlin-topography, in the neighbourhood of Ballintra, Co. Donegal

(*Wright, 'The Quaternary Ice-Age.' Macmillan*)

Five drumlins are shown at a scale of 1 inch to 600 yards and contour interval of 25 feet

and beneath an ice-sheet at some distance from its front. This sub-glacial water is believed to have caused the long, narrow winding ridges, composed chiefly of sand and gravel, that are known as eskers or ösar. Examples abound in Finland, where roads and railways utilize them to escape the many lakes. The finest examples in the British Isles occur in central Ireland, where the ridges ramify like a stream system and frequently run across irregularities of the ground (Figs. 104 and 105). Some eskers are probably formed by a stream that is rushing along under pressure in a narrow channel at the base of the ice. If the stream deposited débris upon its restricted bed, the accumulation would gradually assume the shape of the narrow winding ice-channel. This would especially occur where the flow of the stream was obstructed or checked in any way. The argument is supported by the way in which most eskers aline themselves parallel with the most probable flow of glacial water.

Most eskers, however, seem partly to be formed where waters issuing from sub-glacial or even intra-glacial channels at the edge of

[C. A. Lindberg

21. South-western margin of the Greenland Ice-cap
(*By courtesy of the National Geographic Society*)

[C. A. Lindberg

22. Glacier and fiord south of Scoresby Sound, eastern Greenland
(*By courtesy of the National Geographic Society*)

[National Film Board of Canada

23. Plateau after glaciation by continental ice-sheet: the Laurentian Shield north of Sioux Lookout, Ontario, showing maze of lakes between eskers and moraines

[Aerofilms Ltd.

24. Mountains after glaciation: the English Lake District, showing cirque lakes (Small Water to left; Blea Water in centre, with a small moraine across lower lip of the rock-basin), arêtes, U-shaped valleys and head of ribbon-shaped lake (Hawes Water)

the ice-sheet had their velocity checked on leaving the ice. As the streams burst free from the ice they deposited débris, which accumulated as a mound banked up against the ice-front. Frequently the deposits were laid down in a shallow lake formed by melting at the

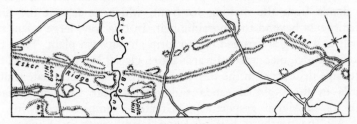

104. Portion of an esker near Tyrell's Pass, central plain of Ireland
(*Wright, 'The Quaternary Ice-Age.' Macmillan*)

ice-fronts (Plate 21). If the ice retreated gradually, the deposits grew in length to form an elongated mound. When the retreat was slow the mound grew wider, and some eskers are like a string of beads showing alternating periods of slow and rapid ice-retreat. It is highly probable that where the ice-sheet ended in standing water, during periods of little or no retreat, the sub-glacial streams formed deltas, and consequently the 'beaded eskers' consist largely of a series of

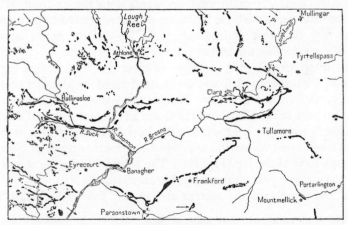

105. The main eskers west of Tyrell's Pass, central plain of Ireland
(*After Sollas, from Wright, 'The Quaternary Ice-Age.' Macmillan*)
Arrows show direction of ice-motion. Area is 50 miles by 30 miles

T

half-formed deltas. A notable beaded esker extends between Wolverhampton and Newport, Shropshire.[1]

The Moraine Zone and Outwash Plain. Beyond the area where eskers and drumlins are common lies the drift or moraine zone, already discussed (Fig. 102). These terminal deposits are acted upon by floods from the melting ice-front, which re-deposit the finer constituents beyond the morainic belt. In this way, large quantities of silt and sand are spread out to form an outwash plain. Thus the Cromer Ridge merges southwards into a glacial loam region, and the extensive drift areas in the Severn and Avon valley terminate southwards in fluvio-glacial sands and gravels near Cheltenham and Moreton-in-Marsh respectively. Greater examples occur in the U.S.A.; the northern half of Long Island, with its hummocky, irregular topography, consists of part of the terminal moraine of a Hudson valley glacier, while the flat gently sloping southern part is an outwash plain. The great morainic belt of the Laurentian ice-sheets lies roughly south of the Great Lakes, but beyond it stretches a vaster area of fluvio-glacial loams that form the Corn Belt, the richest outwash plain in the world.

Diversion of the Pre-Glacial River Drainage. The depositing and eroding work of ice is reflected in the changes brought about in the courses of rivers near to the melting front of the ice-sheet. The northward-flowing streams of the north German Plain were dammed by ice and their waters diverted westwards along the ice-front. When the ice finally retreated, the moraines deposited tended to perpetuate these new east-west drainage channels. To-day the newer courses are of considerable use in the country's canal system.

An interesting English example of river diversion occurs in the Vale of Pickering, which was formerly drained by the Derwent to the North Sea at Filey Bay. The ice from the north covered the western part of the Cleveland Hills, and in the east dammed a series of lakes between the ice-front and the Yorkshire Moors. One of these lakes overflowed across the watershed southwards through Newton Dale to Lake Pickering (Fig. 106). The town of Pickering stands on the large gravel delta formed by this overflow.

At the same time the eastern end of the Vale of Pickering was

---

[1] The word 'kame' has caused much confusion. This feature seems to be usually applied to a cuesta-shaped ridge formed by the confluence of closely spaced deltas of sub-glacial streams; for example, at Carstairs in south Scotland. It seems a halt-stage in the formation of eskers and, as such, might be better included in that group.

blocked by an advancing ice-sheet, and the great lake formed over-flowed westwards into the Vale of York. By the time the ice had retreated, this overflow channel had been deepened into the fine gorge which may be seen near Kirkham Abbey. In the meanwhile the

106. Glacial Lake Pickering and the river Derwent

(*After Kendall, from Wills, 'Physiographical Evolution of Britain.' Arnold*)
Cross-hatching shows unglaciated area; black marks the York moraines.
Dotted line shows pre-glacial drainage.

ice-sheet had deposited a line of low moraines near to the coast, and consequently the Derwent retained its new western direction and now flows directly away from the sea to join the Yorkshire Ouse. Further examples of river diversion due to glacial action occur along the river Leven draining Lake Windermere and along the Severn, the upper portions of which probably formed the upper Dee until they were diverted through the gorge at Ironbridge.

Glaciation and Human Affairs

The ice-sheets denuded large areas of their surface soils, but those areas often happen to have been among the least desirable for settlement. Many other districts were given either a boulder-ridden or an irregular marsh-frequented topography as happened near Lossiemouth, near Newton Stewart, and to a lesser degree in Holderness.

Against these disadvantages must be placed the attractiveness of drift-covered lands such as East Anglia and the Vale of York. The drift usually becomes, with careful drainage, a good farming soil; moreover, the soils were carried southwards, where the climate is better suited for farming.

The fluvio-glacial sands and gravels are widely quarried in England, for example, near Cheltenham and London.

The interference with pre-glacial drainage is usually more of a help than hindrance to mankind. In Scotland communications are facilitated by old overflow channels of ice-sheets and of ice-dammed lakes; these channels are often transverse to the grain of the relief, and so help to counteract the hindering effect of the lochs on communications in certain directions. In the case of the Clyde and Tyne the glacial débris flooring their beds can be easily excavated, whereas the new channel of the Wear is over limestone, which greatly handicaps efforts to deepen the waterway.

It seems as if the British Isles were more fortunate than many other glaciated countries. The vast areas of ground moraine forming the Baltic heights are largely *geest* or sandy, heather-clad moorlands much dotted with marsh. In Finland, where lakes cover about 11 per cent. of the whole country, land communications are hindered, the climate is deteriorated, and farming land restricted. The Great Lakes of North America cover nearly 95,000 square miles[1] of what would be very flat farming land, and this loss alone does much to counterbalance the beneficial effect of the great outwash plain in the Corn Belt. Further details of glacial lakes are given in Chapter Twenty-two.

Besides the value of lakes for water-supply and attracting tourists, glaciation provides opportunities for the generation of hydroelectricity. Suitable falls abound in most glaciated regions—hanging

[1] Lake Superior alone is nearly as big as Ireland.

valleys are especially easy to harness, lakes regularize the water-flow, and, in some areas, melting snow maintains the water-level in summer. In Norway the falls possess a potential output of 12 million horse-power, and in Finland a potential $2\frac{1}{2}$ million horse-power; in Switzerland hanging-streams with single falls up to 1,000 yards are unusually common. In these lands nature offers ample compensation for the denudation of districts that in any case would be of little agricultural value.

LIST OF BOOKS

General
The *Journal of Glaciology*, London, 1947 onwards
Cotton, C. A. *Climatic Accidents in Landscape-making*, 1947
Flint, R. F. *Glacial and Pleistocene Geology*, 1957
Woldstedt, P. *Das Eiszeitalter*, 1954-8
Tricart, J. *Géomorphologie des Régions Froides*, 1963
Brooks, C. E. P. *The Evolution of Climate*, 1925 (climatic aspect of Ice Age); *Climate through the Ages*, 1950
Shapley, H. (ed.) *Climatic Change*, 1953
Daly, R. A. *The Changing World of the Ice Age*, 1934 (reprinted 1963)
Wright, W. B. *The Quaternary Ice Age*, 1936
Charlesworth, J. K. *The Quaternary Era*, 2 vols., 1957 (1,700 pp.)

Glacial Erosion
Nye, J. F. 'The mechanics of glacier flow,' *Jour. Glaciology*, 1952, pp. 82-93
Sharp, R. P. 'Glacier flow: a review,' *Geol. Soc. Am.*, 1954, pp. 821-38
Bailey, E. B. 'Geology of Neighbourhood of Fort William,' *Procs. Geol. Ass.*, 22, 1911, pp. 179-203
Hobbs, W. H. 'The Cycle of Mountain Glaciation,' *G.J.*, Feb. and March 1910
Odell, N. E. 'The Mountains of Northern Labrador,' *G. J.*, Sept. and Oct. 1933. 'The Glaciers . . . of the Franz Josef Fjord Region of N.E. Greenland,' *G. J.*, Sept. 1937
Ray, L. L. 'Some minor features of valley glaciers and valley glaciation,' *Jour. Geol.*, 43, 1935, 297-322
Lewis, W. V. 'A melt-water hypothesis of cirque formation,' *Geol. Mag.*, June 1938, pp. 249-66. 'Snow patch erosion in Iceland,' *G.J.*, Aug. 1939. 'Valley steps and glacial valley erosion,' *Inst. Brit. Geog.*, 1947, pp. 19-44
Cotton, C. A. 'The Longitudinal Profiles of Glaciated Valleys,' *Jour. Geol.*, 1941, pp. 113-28
Linton, D. L. 'Watershed Breaching by Ice in Scotland,' *Inst. Brit. Geog.*, 1949, pp. 1-16; 'Radiating valleys in glaciated lands,' *The Earth* (studies presented to J. B. Hol), 1957, pp. 83-98; 'The forms of glacial erosion', *Inst. Brit. Geog.*, 1963, pp. 1-28
Linton, D. L. & Moisley, H. A. 'The Origin of Loch Lomond,' *Scot. Geog. Mag.*, 1960, pp. 26-37
Lewis, W. V. (ed.) *Norwegian Cirque Glaciers*, 1960

Glacial Deposition. Drumlins, Eskers and Kames
Gravenor, C. P. 'The Origin of Drumlins,' *Am. Jour. Sci.*, 1953, pp. 674-81
Flint, R. F. 'The Origin of the Irish Eskers,' *Geog. Rev.*, 20, 1930, pp. 615-30
Charlesworth, J. K. 'The Glacial Retreat from Central and Southern Ireland,' *Q. J. G. S.*, 84, 1928, pp. 293 *et seq*. 'The Eskers of Ireland,' *Geography*, 1931 pp. 21-27

Cook, J. H. 'Kame-complexes', *Am. Jour. Sci.*, 1946, pp. 573–83
Holmes, C. D. 'Kames', *ibid.*, 1947, pp. 240–49
Sissons, J. B. 'The Deglaciation of part of East Lothian,' *Inst. Brit. Geog.*, 1958, pp. 59–77
Reed, B. *et al.* 'Some aspects of drumlin geometry', *Am. Jour. Sci.*, 1962, pp. 200–10
Hoppe, G. 'Glacial morphology and inland ice recession in northern Sweden', *Geog. Annaler*, 1960, pp. 193–212

River Diversion, Glacial Lakes, and Lake Overflow Channels or Spillways
Kendall, P. F. 'A system of Glacier lakes in the Cleveland Hills,' *Q. J. G. S.*, 58, 1902, pp. 471–569
Kendall, P. F., and Bailey, E. B. 'The Glaciation of East Lothian,' *Trans. Roy. Soc. Edin.*, 46, 1908, pp. 1–31
Wills, L. J. 'Late-Glacial . . . changes in the Lower Dee Valley,' *Q. J. G. S.*, 68, 1912, pp. 180–86. 'The Development of the Severn Valley in the neighbourhood of Ironbridge and Bridgnorth,' *Q.J.G.S.*, 80, 1924, pp. 274–314; 'The Pleistocene Development of the Severn,' *Q.J.G.S.*, 1938, pp. 161–242
Peel, R. F. 'Two Northumbrian spillways,' *Inst. Brit. Geog.*, 1951, pp. 73–89; 'The profiles of glacial drainage channels,' *G.J.* Dec. 1956, pp. 483–87
Clayton, K. M. & Brown, J. C. 'The Glacial Deposits around Hertford,' *Procs. Geol. Ass.*, 1958, pp. 103–19
Soons, J. M. 'Glacial Retreat Stages in Kinross-shire,' *Scot. Geog. Mag.*, 1960, pp. 46–57
Sissons, J. B. 'Glacial drainage channels', *ibid.*, 1960, pp. 131–46; 1961, pp. 15–36
Gregory, K. J. 'Proglacial Lake Eskdale', *Inst. Brit. Geog.*, 1965, pp. 149–62

For Lakes (see references at end of Chapter Twenty-two)
Murray, J., and Pullar, L. 'Bathymetrical Survey of Fresh-Water Lochs of Scotland,' *G. J.*, 1900–5 (with photographs and magnificent large scale maps)

For Study of Special District (for example, Cumbria)
Mem. Geol. Survey. *Carlisle*, 1926. *Maryport*, 1930
Raistrick, A. 'The Glaciation of Borrowdale,' *Procs. Yorks. Geol. Soc.*, 20, 1925, pp. 155–181
Hay, T. 'Glaciology of the Ullswater Area,' *G. J.*, Aug. 1934, pp. 136–148
Hollingworth, S. E. 'The influence of glaciation on the topography of the Lake District,' *Jour. Inst. of Water Engineers*, 1951, pp. 485-96; The Geology of the Lake District,' *Procs. Geol. Ass.*, 1954, pp. 385-402
Linton, D. L. 'Radiating valleys in glaciated lands,' *op. cit.*, 1957

Note. *For Climatic Change*
Brooks, Shapley, Flint (1957) and
Schwarzbach, M. *Climates of the Past*, 1963
UNESCO *Changes of Climate*, 1963
Butzer, K. W. *Environment and Archaeology*, 1965

Today an increase of carbon dioxide is known to increase air temperatures as it absorbs solar heat re-radiated as long waves from the earth's surface. Thus only a decrease of carbon dioxide would lower air temperature. Sun-spots are no longer considered relevant to Ice Ages but longer term decreases in the sun's radiant energy are favoured as a possible cause by some astronomers. For a summary and recent references see Beckinsale, R. P. 'Climatic Change', in *Essays in Geography*, edited by J. B. Whittow and P. D. Wood, 1965, pp. 1–38

CHAPTER TWENTY

THE WORK OF THE SEA: COASTAL TOPOGRAPHIES

Erosive Action of Waves

THE destructive power of high waves is perhaps the most im-
pressive of all forms of erosion. When a large wave breaks, a
great weight of water is hurled against the shore face chiefly above
sea-level. During storms, the force of breakers around Britain
reaches between 8 and 25 tons to the square yard. The spray during
severe gales has broken open a door at Unst, in the Shetland Islands,
at a height of 195 feet above the sea, and has smashed the windows of
Dunnet Head Lighthouse, 300 feet above sea-level.[1] The breaking
face of gale-driven waves at North Beach, Florida, shifted a solid
block of concrete weighing 2 tons some 12 feet horizontally and
turned it over on its side. A mass of breakwater at Wick, weighing
1,300 tons was moved out of position, and the structure built in its
place, a mass weighing twice as much, eventually suffered the same
fate. It is, indeed, difficult to imagine the power even of normal waves.

The force of the wave that strikes a coastline usually depends
largely on the strength and 'fetch' of the wind causing the wave, and
the depth of the water off-shore. The longer the fetch, or distance of
open ocean over which the wind has travelled, the more powerful are
the waves it causes, so that a gale with a short fetch may accomplish
less erosion than a slower wind with a greater fetch. The depth of the
water controls the breaking-point of the wave motion. When deep
water comes right up to the face of a cliff, the waves break against
the cliff-face and hurl themselves against it with great force and to a
considerable height. Thus, on sections of the coast of Hawaii where
the spray reaches 50 feet or more above sea-level, small pits or alcoves
and signs of a marine bench occur at these levels. Where the shore
shelves gradually the waves break farther out to sea and the breakers
lose much of their force in travelling up the shelving beach.

Wave erosion often depends, however, on more than the direct
force of the water. When a breaker strikes a cliff face, the air in the
crevices of the rocks is imprisoned and is subjected to a sudden
increase of pressure. Consequently, the air tries to expand and sub-
jects the rock enclosing it to an intense strain which eventually
assists in enlarging the irregularities of the cliff-face.

[1] At Dunnet Head waves crashing on the cliffs have swept pebbles to this height.

Furthermore, wave erosion is greatest when waves find suitable tools with which to work. Waves not armed with pebbles and shingle accomplish tremendous destructive feats, but similar waves with the aid of a load work more quickly and efficiently. The load, torn from the shore in the form of pebbles and gravel, is hurled continuously backwards and forwards especially against the base of the cliffs at, or just below, high-water mark.

Coastal erosion, however, is mainly confined to a narrow belt between high- and low-water mark. The lower limit of effective wave action is normally very much less than 100 feet below the surface; the upper limit is normally only a few feet above high-tide mark. This zone of erosion is relatively narrow where the shore is steep and relatively wide on a flat, shallow coastline. Yet, in the former, erosion is more definite and usually more effective.

Marine Benches and Sea Cliffs. The most common topographical features formed by wave erosion are the marine bench and sea cliff (Plate 25). Wherever waves break strongly on a coast they tend, by means of their destructive force and undertow, to wear down the land between high- and low-tide marks. Eventually a niche is cut in the shoreline which, under the action of pebbles and spray, gradually widens into a narrow bench bounded by a cliff on the landward side. As the bench widens, the cliff usually grows in height, and consequently becomes more liable to undercutting by the work of the waves (Fig. 107).

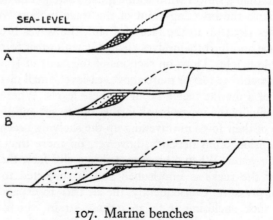

SEA-LEVEL

A

B

C

107. Marine benches
(*After D. W. Johnson*)
A B C show successive stages in formation

Influence of Geological Structure on Coastal Topography.
The nature of the cliff formed depends much on the dip of the strata.
Where strata dip towards the sea, under-mining at the base of the
rocks frequently causes landslips. On the other hand, strata that dip
steeply landwards are less liable to collapse or to slip, and often
produce overhanging or beetling cliffs.

The outline of a wave-beaten coast also depends to a considerable
extent upon the resistance of the rocks exposed to the sea. Where the
rocks are fairly homogeneous, a relatively straight coastline is to be
expected, as is seen at Beachy Head and Holderness. Where the rock
strata differ in hardness, cliffed headlands frequently adjoin bays with
gently sloping shores. Torquay headland illustrates how the outline
of a coast varies with the resistance of the strata (Fig. 108). The cliffs
north and south of Oddicombe are composed of Devonian limestone,
whereas Oddicombe Bay itself is of softer New Red Sandstone. At
Anstey's Cove the limestones dip sharply and form steep cliffs.
Black Head is a projection of igneous rock, just as Hope's Nose is of
limestone, but Tor Bay has been eroded by the waves out of softer
slates and sandstones.

Variations in the resistance of any one rock outcrop lead to minor
coastal topographical features. Breaking waves seek out every crevice
and every line of weakness open to their attack. Fissures may be
gradually enlarged into blow-holes and caverns. Where a narrow
promontory is exposed to wave attack on both faces, caverns may be
worn into its opposing sides. When these caves join, the promontory
is pierced by a wave-tunnel that will in time convert it into a natural
bridge. This has happened at London Bridge, near Torquay, and at
Durdle Door, near Lulworth. When the arch of a natural bridge
collapses, a stack or pinnacle is left standing isolated from the cliff-
face; in this way, Old Harry of Swanage has been detached from the
chalk cliffs.

At Lulworth Cove, Dorset, the sea has completely breached a
narrow coastal band of limestones and eroded a circular bay on the
softer shales inland. Here, the incessant attack of the waves gradually
enlarges the greater fissures of the limestone into blow-holes. These
blow-holes gradually become tunnels which grow landwards until
the narrow limestone band is completely pierced. This process is
well seen at Stair Hole, just west of Lulworth Cove, where three
tunnels have pierced the limestone rampart. In time, the arches of
the tunnels collapse and the waves gain ready access to the Weald

Clay inland, which, being soft, is eroded relatively rapidly away. At Lulworth Cove, the narrow entrance where the sea breaches the limestone band now gives access to a circular harbour nearly a quarter of a mile in diameter.

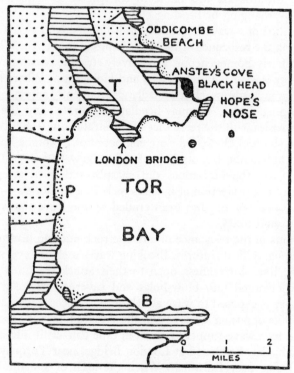

108. Influence of geological structure on a coast-line: Torquay Headland and Tor Bay

Black marks igneous intrusion; shading, limestones; dots (inland), New Red Sandstone; white, slates and grits
T. Torquay   P. Paignton   B. Brixham

Where the rocks of a cliff-face have well-developed bedding- and joint-planes, both lines of weakness facilitate the destructive action of the waves and often give rise to rectangular rock shapes. The famous Castle Rock near Lynton owes it origin to wave action working along joint- and bedding-planes to form rectangular blocks suggestive of a ruin.

Wave erosion is greatest where a coastline composed of soft

material is exposed to strong waves and currents. Thus, the greatest coastal erosion around the shores of Great Britain occurs at the cliffs of Holderness, parts of which recede at an average rate of $4\frac{1}{2}$ to 7 feet a year.[1] Nearly thirty small towns and villages have disappeared since Roman times, and several more are threatened with destruction to-day.

Constructive Action of Waves and Currents

Beaches. The débris derived from the land by means of wave erosion is deposited on a zone of the shore lying between and just below high- and low-tide marks. The detritus is sorted out in much the same way as a river sorts out its load; the large pebbles are moved least, the shingle somewhat farther, while the finer sand and silt is carried farthest out to sea (see pages 179 and 180). As a result, a coastal slope of material torn from the land is laid down below high-water mark, and this débris forms, with the wave-cut terrace, a wide platform. The platform, or marine bench, may be wide enough to preserve the coastline from further erosion, as much of the force of the waves will be spent in travelling up its shelving beach.

Any alteration above normal high-water mark on a gently shelving shoreline usually takes place during gales. Beach or storm ridges can be formed by waves driven inland by strong winds. In the shingle foreland of the Crumbles, near Eastbourne, there are about sixty ridges roughly parallel with the shore that faces Pevensey Bay. These storm-ridges are not easily effaced by subsequent wave action, especially where deposition is extending the shore seawards.

Spits and Bars. The constructive work of the sea is seldom so simple as suggested above, and the ebb and flow of a tide is seldom merely an up-and-down movement. Such factors as tidal drift, the direction of prevailing waves, and river-borne sediment usually enter into the question of marine deposition.

Where strong prevailing waves approach a shoreline they transport sediment derived from the land in the direction of their advance, as is notably the case with the north to south currents along the east coast of Britain. Strong winds nearly always cause a longshore drift of breakers which becomes of great importance as a transporting agent where the prevailing winds have a long fetch or where they reinforce the action of tidal currents. Frequently the influx of river water off estuaries affords an additional load for the longshore drift.

[1] The cliffs consist mainly of boulder clay, loose sand, and gravel.

Currents connected with the general circulation of seas and oceans are extremely gentle, and have, at most, a very small influence on deposition. Thus, at the mouth of the Nile, where bars have been formed stretching eastwards across the mouths of the lagoons, coastal deposition is probably mainly related to wave approach and not to currents.

Frequently, at various localities along a shoreline, the transporting power of waves or of the longshore drift is obstructed in some way, and at these points deposition occurs. This tends to happen where river currents enter the sea, where strong winds blow across the prevailing direction of the longshore drift, where tidal movements neutralize each other, and where the shape of the coastline forms areas of sheltered water. The deposition often takes the form of beaches at the heads of bays, and of spits or bars across mouths of bays and estuaries. The construction of spits has been much studied in both southern and eastern England.

Along the north shores of the English Channel the prevailing wave direction, and consequently the longshore drift, is eastwards. Thus, quartzite pebbles from Devon have been found on beaches in Kent, and at Folkestone harbour a large pier was necessary to stop the drift of shingle from the west. In many localities where this eastward drift is checked, spits and bars are formed.

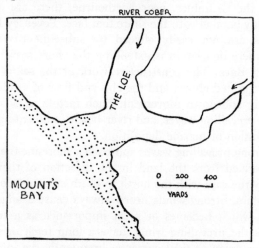

109. Loe Bar, near Helston, Cornwall

The river water escapes by an artificial tunnel cut
through the north bank of the mouth of the Loe

At Loe, a bar has been built up across the mouth of the 'Pool' (Fig. 109). These 'bay-bars,' as they are called, frequently occur where river-currents impinge on a longshore drift and where the drift skirts the relatively motionless water in a sheltered bay. The bars often consist largely of river sediment on their landward side and of shingle on their seaward edge. Similar deposits almost enclose the *haffs* of the Baltic coast and the *étangs* of Languedoc. It is usual for the bar to be broken at the end away from the direction of approach of the prevailing waves.

Near Weymouth a long bar of shingle, called Chesil Bank, links Portland Bill to the mainland (Fig. 110). Except at its extremities, a narrow channel of shallow water separates this bar from the coastline proper. The bank seems to have arisen where powerful south-westerly waves—coming from the open Atlantic—meet shingle drifted from the west and held up by Portland promontory. The great height of the bank, which rises steadily from 20 feet in the north-west to over 40 feet near Portland, is a measure of the power of the waves, and its

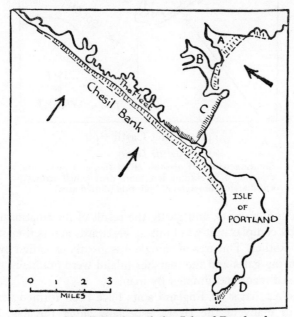

110. Chesil Bank and the Isle of Portland

Arrows show winds of greatest effect
A. and B. Small bay-bars
C. line of cliff   D. raised beach

direction is related to that of the dominant wave-fronts (Plate 26).
Hurst Castle Spit, a more complex type of structure, has been
formed at the mouth of the Solent. This shingle growth continues the
direction of the shoreline for a short distance and then turns eastwards under the influence of prevailing south-westerly waves
(Fig. 111). The spit normally grows eastwards, but when north-
easterly winds prevail for a short while, lateral extensions pointing
northwards are formed. The coast of New England, for example near
Cape Cod, affords numerous instances of curved and pronged spits.
Dungeness, in Kent, is partly the result of the constructive work

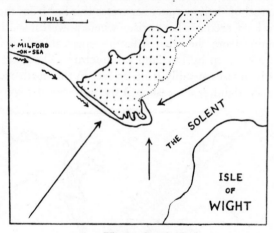

111. Hurst Castle Spit

(*After Lewis*)

Showing winds of greatest effect (large arrows); the
direction of prevailing longshore drift (small arrows);
salt marsh submerged at high tide (dotted area)

of waves and currents, and partly the result of reclamation by man.
Probably a complex spit was built up eastwards across the estuary of
the river Rother. The area of shingle was greatly extended by storm-
ridges during gales and the marshes inland were gradually filled up
by river sediment and drained by man.

On the east coast of England spits have been formed across the
mouths of many of the rivers, especially along the East Anglian
coastline. Here marine erosion wears back the promontories, and
longshore currents deposit the débris across the mouths of the
estuaries. In this way, the rivers are either deflected southwards by a

shingle and sand spit or else are completely dammed. Orfordness is a spit some eleven miles long, consisting mainly of shingle ridges, that has been formed across the mouth of the river Alde. In 1165 the end of the spit was opposite Orford Castle, but by 1897 it had grown 5½ miles farther southward, getting gradually narrower as it extended. Subsequently its thin extremity was eroded away (Fig. 112).

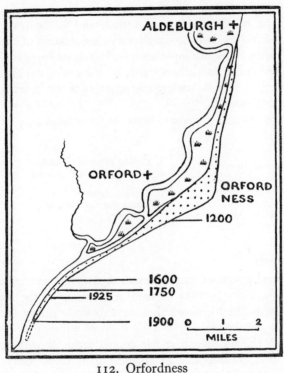

112. Orfordness
(*After Steers*)
Showing extent of shingle spit (dotted) at various dates

The great spit across the mouth of the Yare also formerly extended much farther southwards, but the Yarmouth burghers used to make cuts across it, and later the southern extension was washed away. Like the other spits of the East Anglian coastline, it appears to be formed by the 'combined action of waves and of longshore drift' (J. A. Steers). The waves with the longest fetch come from a northerly quarter, and probably cause southward beach-drifting.

In 'Broadland' north of Great Yarmouth most of the lakes are medieval peat-diggings, since drowned, but the area in which the peat accumulated is protected from the sea by a coastal bar. Farther north at Blakeney and Scolt the bars lie a mile or so offshore and grow mainly westward under the push of prevalent north-easterly waves. At Scolt, sand-dunes cover the shingle.

Coastal Topographies

The shape of coastlines is so varied, and is the result of such complicated factors, that no satisfactory classification of coastal types has been made. Some attempts are based solely on topography, others partly on structure, and others partly on the geological history of the coast—whether it has undergone emergence or submergence or remained stationary.

For topographical purposes, however, coastlines may be grouped simply into:

(a) Flat coastlines
  { i. Coastal plain of emergence
  { ii. Coastal plain of deposition

and

(b) Steep or Mountainous
    coastlines
  { i. Longitudinal
  { ii. Transverse or oblique
  { iii. Glaciated
  { iv. Fault
  { v. Volcanic.

Flat Coastlines

(i) **Coastal plain of Emergence.** A negative movement of the sea, due either to elevation of land or lowering of the water-level, usually reveals the foreshore that used to slope gently towards the edge of the continental shelf. This exposed submarine bench then forms, with the old beach, a flat coastal strip. It will be noticed, however, that considerable emergence of a steep, rocky coastline would only reveal a narrow, submarine bench, whereas a slight emergence of a shallow coastline would expose a large expanse of gently sloping sea-bed. Definite examples are difficult to establish, as, frequently, the relative emergence is followed by a slight submergence, or the coastline may warp so that emergence is dominant only in certain localities. The coastal strip of the eastern U.S.A. in South Carolina, Georgia, and Florida is the emerged portion of the continental shelf. This flat coastal plain attains in places a width of 200 miles, and is bounded on its landward side by the Fall Line. The east coast of India south of Madras is probably also of this nature.

[J. Dixon-Scott

25. Coastal erosion. The north cliffs near Camborne, Cornwall
(*By courtesy of the British Council*)

[*The Times*

26. Marine deposition. Chesil Bank, with Portland Bill in the distance

27. Lake dammed by land-slips: Lac des Brenets or Chaillexon, a
widening of the obstructed valley of the River Doubs
(*By courtesy of Swiss National Tourist Office*)

28. A volcanic crater lake: Lake Duluti in northern Tanzania. Notice
circular shape
(*By courtesy of Public Relations Department, Tanzania*)

(ii) **Deltaic Coastline.** Where deposition is heavy and continuous, a flat coastal plain will often be formed. The deltaic coastline, the chief of this type, has already been described.

Steep or Mountainous Coastlines

Quite distinct from the flat, coastal-plain shorelines are hilly or mountainous coastlines, the outline of which depends mainly on the direction of the relief-lines with regard to the coast, and on the agencies that have formed the relief of the land. Frequently these shorelines have undergone submergence.

(i) **Longitudinal Coastlines** occur where mountain ranges run parallel with the shoreline. They are characterized by a fringe of long, narrow islands and inlets with narrow mouths that open into long, narrow bays. All the landforms are alined parallel with the coast and with the inland mountains. The classic example is in Dalmatia

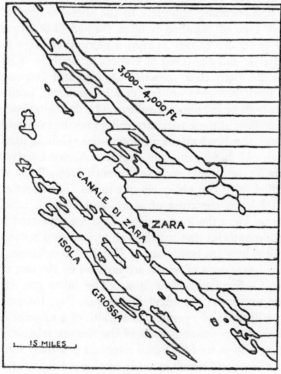

113. The Dalmatian coast near Zara

U

—by which name the type is often known—where, from Fiume to Dubrovnik, a region of narrow folded ranges and valleys has been drowned on its western margin. Inland the mountains rise sharply from the coast, while seawards is a broken edge of islands and channels and sheltered bays, all alined from north-west to south-east (Fig. 113). The long, narrow islands are arranged in rows, and those farthest from the coast, being most submerged, are the smallest. The rows of islands are divided from each other by narrow lanes of water (*canali*) which communicate with one another by transverse gaps formed by river erosion across the coastal ranges before they were drowned.

A similar structural example occurs in southern Ireland, though it is not developed to the same extent. Here, Cork harbour consists of a typical double basin and narrow entrance, the result of the submergence of longitudinal ranges and valleys.

(ii) **Transverse or Oblique Coastlines.** Where mountain folds and relief lines running obliquely to the shoreline are submerged, a distinctive type of coastline is often formed. The drowned river valleys are generally known as *rias*, a name that originates in the submerged river-mouths (rias) of Galicia in north-west Spain (Fig. 114). Here the many river estuaries broaden and deepen gradually seawards. Their branches are few and insignificant, occurring only along softer beds or in tributary valleys. The drowning and widening by marine erosion of these tributary valleys results in an island fringe that prolongs the head of the promontories. Galicia itself consists largely of an old, hard peneplain deeply dissected by river erosion, but rias are by no means confined to such areas. They also occur where parallel folds run obliquely out to sea. The finest example in the British Isles is the coast of south-west Ireland, which is of a folded nature. Here the long, narrow bays are separated from each other by promontories that usually rise to well over 2,000 feet (Fig. 115). These ridges are composed of Old Red Sandstone, while the troughs, the lower ends of which are invaded by the sea, are floored with softer Carboniferous Limestone. The inlets gradually widen and deepen seawards; the small Dunmanus Bay, for example, increases, in a distance of 15 miles, to a width of 4 miles at its mouth. As in Galicia, the containing walls of the rias are relatively straight, and tributary valleys are few. Small rivers are gradually silting up the heads of the bays, and simultaneously marine erosion is attacking the seaward faces of the promontories. The island fringe, a natural result

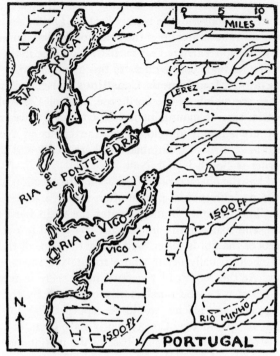

114. The Rias of Galicia, north-western Spain
Land above 1,500 feet is shaded

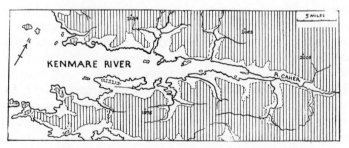

115. Kenmare River, a ria in south-western Ireland
Land above 250 feet is shaded

of the submergence of the folds, is being gradually worn away and the tips of the promontories are being blunted. The waves and tides, however, assist in the silting up of the heads of the rias, and in Dingle Bay the two sandspit barriers that partially close the inner six miles of Castlemaine harbour bear witness to this aggradation.

(iii) **Glaciated Mountainous Coastlines.** Glaciated mountainous coastlines are mainly of the fiord type, examples of which occur on the western side of mountainous countries in the westerly-wind belts. They are strongly developed on the coasts of Norway, of western Scotland, British Columbia, southern Chile and the south-western corner of South Island, New Zealand.

The fiord has the characteristic U-shaped cross-section of a glacial valley, while its longitudinal profile often shows a series of humps and basins. The tributary branches enter the main fiord almost at right angles, and many fiords are roughly Y-shaped in outline. The deepest parts of the fiords (often well below the floor of the adjacent continental shelf) usually occur where tributary glaciers joined the main valley. At the mouth of the fiord itself there is frequently a submarine sill that can be satisfactorily explained by the fact that here the Quaternary glacier would be approaching lowland and would begin to spread out. Consequently its powers of downward erosion, so strong at the immediate base of the mountains, would be greatly reduced, and here the base of the glacier would be less degraded. It is probable, too, that in some cases deposition occurred near the mouth of the fiord. The typical fiord is long and narrow; Loch Etive in Scotland is 15 miles long and about three-quarters of a mile wide. The average length of the Norwegian fiords is 50 to 60 miles. Sogne Fiord, one of the largest, is nearly 140 miles in length, 4 to 5 miles in width, and up to 670 fathoms in depth (Fig. 116). There is, however, no progressive narrowing inland, and frequently a fiord has a number of basins formed by constrictions of the containing walls. The walls of some fiords rise abruptly to 2,000 and even 3,000 feet above the water, and from them waterfalls (hanging valleys) plunge into the waters beneath.

The origin of fiords has given rise to much controversy. In the case of Norway and Scotland the fiords usually follow a south-west to north-east trend, but occasionally branch acutely into an elbow running from north-west to south-east. It has been shown that these directions often correspond to fault lines, but there is little evidence to show that fracturing controls more than the direction of the fiord.

Nor, by any means, are all fiords connected with fractures. The ice, however, would tend to follow river-valleys which existed prior to the Ice Age, and these valleys no doubt often corresponded with fractures. The glaciers would work more quickly along lines of weakness, and, consequently, the larger fiords often partly coincide with lines of fracture.

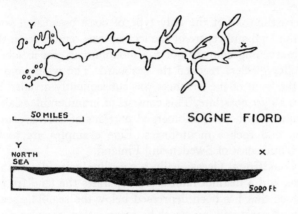

116. Sogne Fiord and Hardanger Fiord, Norway
(After Ahlmann)
In lower figure, dotted line represents the probable bed of the pre-glacial river-valley

It is perhaps unnecessary to point out that fiords occur in what, if world climates were chilled, would be precisely the snowiest parts of the globe. Some of their deepening may well have been aided by river erosion during the warm inter-glacial periods, the deep, narrow incisions being widened and made U-shaped by subsequent ice-advances. But even when this is admitted, the fiord remains the supreme expression of the erosive power of moving ice concentrated along pre-determined lines beneath a thick ice-cover.

The amount of subsidence of fiord coastlines is difficult to assess, since ice can work below sea-level. The weight of an ice-cap would cause an isostatic depression of the land, but during the Ice Age sea-level would be greatly lowered by loss of water. In other words, the Quaternary glaciers were working to a much lower sea-level than that of the present day. With the melting of the great ice-caps sea-level rose again and invaded the glacial valleys. The land recovered from its slight isostatic depression very slowly, and the recovery, which is still in progress, may be marked by the strandflats of the fiords (see page 187).

The fiord coastline is not the only type of coast associated with glaciation. If the highland centre of the ice-cap was remote from the sea and the intervening country not too mountainous, an ice-sheet rather than valley-glaciers reached the seaboard. The wide zone of deposition at the front of the ice-sheet was subsequently invaded by the sea to form a *skjer* coastline. This consists of innumerable islands and small peninsulas, the half-submerged portions of eskers, drumlins, moraines, and roches moutonnées. Fine examples are to be found on the Baltic coast of Sweden and Finland.

(iv) **Fault Coastlines.** Occasionally a coastline coincides with a line of fracture, and a distinct fault scarp separates the land from a lower-lying block that has been depressed below the sea. The scarp of the young fault coast will, on attack by wave action, soon be worn

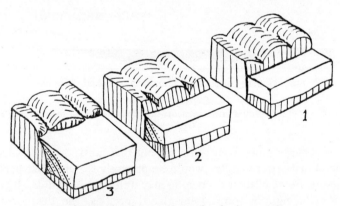

117. Stages in the erosion of a fault coastline
(*After Cotton*)
1. Initial stage   2. Beginning of marine benching and submarine deposition
3. Formation of a narrow coastal shelf

back from the fault line. Eventually wave-cut cliffs will be formed and a narrow bench of débris built up at their base (Fig. 117). These cliffs will usually be dissected by river valleys, as the streams present would be rejuvenated at their seaward ends. Consequently, the coast often consists of an alternation of narrow valleys and of spurs terminating in wave-cut cliffs. In time, provided that longshore currents do not carry away the débris, the coastal slope laid down at the base of the cliffs will grow into a narrow beach. The east coast of the province of Marlborough, in South Island, New Zealand, is a mature, resurrected fault coastline broken by the delta of the Clarence River. The sea cliffs are now bordered by a beach, but they are remarkably continuous. On most other types of shoreline the resistant strata form headlands, but here the cliff is straight irrespective of the varying hardness of the rocks. Thus, the soft mudstones do not form bays, but stand in cliffs which are remarkably well preserved. This stretch of coast, like most fault coast lines, is straight and is bordered close by relatively deep sea, the outer edge of the continental shelf being only about 12 miles from the shore. Since the subsidence of the seaward block along the initial fault was considerable, there is a general absence of islands.

Coasts such as that of the western Deccan in India or of South Africa are also probably due to fracturing, but, beyond the general shape of the shoreline, it is difficult to obtain more detailed evidence of fracture.

(v) **Volcanic Coastlines.** The distinguishing feature of a volcanic coastline is its circular or semicircular shape. The various islands of the Canaries and of numerous other volcanic groups are typically circular in outline, while stretches of the shore of Hawaii closely resemble the waterline of a submerged cone. Even volcanoes situated near the shores of a land mass cause a circular bulge in the coast, as is seen near Mount Egmont in North Island, New Zealand. The face of a volcanic coast is usually cliffed, especially in situations exposed to ocean waves.

Some geographers distinguish a coral type of coastline of which the chief peculiarities are the circular shape of the atoll and the general lowness of the shore. Details of coral growths have, however, already been given.

The Human Aspect

The relations between the type of coastline and man's activities are varied and intricate. The erosive action of waves and currents is usually a distinct nuisance to man, and causes, for example, much expenditure on protective works. Yet the constructive action of waves is also more of a hindrance than a help, as mankind mostly frequents calm waters, where sandbanks and spits are detrimental to shipping. On flat coastlines, where access inland is easy, good natural harbours are few. On mountainous coastlines the indentations often form excellent anchorages, but the mountainous topography obstructs progress inland. In a general sense, a *ria* hinterland is easier to traverse than that of the Dalmatian type of coast; in the latter the rivers follow a circuitous route to the sea and cut deep narrow gorges across the mountain ridges, whereas the ria valley usually opens out seawards. Access inland is most difficult in fiord coastlines, as the fiord usually terminates landward in a gigantic, steep-walled cirque. The fault coastline is, however, often the poorest in harbours owing to its lack of indentations. In all these mountainous and 'submerged' coastlines, human settlement is restricted by the relative lack of flat land near the seaboard.

LIST OF BOOKS

Johnson, D. W. *Shore Processes and Shore-Line Development*, 1919. *The New England-Acadian Shore Line*, 1925.
Ward, E. M. *English Coastal Evolution*, 1922
Bourcart, J. *Les Frontières de l'Ocean*, 1952
Gresswell, R. K. *Sandy Shores in South Lancashire*, 1953
Steers, J. A. *The Sea Coast*, 1953; *The Coastline of England and Wales*, 1964
Guilcher, A. *Coastal and Submarine Morphology*, 1958
King, C. A. M. *Beaches and Coasts*, 1959

Among recent articles are
Steers, J. A. 'The East Anglian Coast,' *G. J.*, Jan., 1927
Gregory, J. W. 'The Fiords of the Hebrides,' *G. J.*, March 1927
Wentworth, C. K. 'Marine Bench-forming processes: Water-level Weathering, *Journal of Geomorphology*, I, 1938, pp. 6–32
Lewis, W. V. 'The formation of Dungeness Foreland,' *G. J.*, Oct. 1932
Shepard, F. P. 'Revised classification of marine shorelines,' *Journ. of Geol.*, 1937, pp. 602–24
Challinor, J. 'A Principle in Coastal Geomorphology,' *Geog.*, 1949, pp. 212–15
King, C. A. M. and Williams, W. W. 'The Formation and Movement of Sand Bars by Wave Action,' *G.J.*, June 1949, pp. 70–85
Arber, M. A. 'Cliff Profiles of Devon and Cornwall,' *G.J.*, Dec. 1949, pp. 191-7
Edwards, A. B. 'Wave action in shore platform formation,' *Geol. Mag.*, 1951, pp. 41–9
Kidson, C. & Carr, A. P. 'The movement of shingle over the sea bed close inshore,' *G.J.*, Dec. 1959, pp. 380–89
Cotton, C. A. 'Accidents and interruptions in the cycle of marine erosion,' *G.J.*, 1951, pp. 343–9; 'Tests of a German classification of coasts,' *G.J.*, 1954, pp. 353–61
Kidson, C. 'The growth of sand and shingle spits across estuaries', *Zeit. für Geom* 1963, pp. 1–22
Dietz, R. S. 'Wave-base . . .', *Geol. Soc. Am.*, 1963, pp. 971–90

CHAPTER TWENTY-ONE

WATER ON THE LAND SURFACES: RIVER RÉGIMES

Drainage of the Land Surfaces

THE nature of ground-water and of springs has been discussed in Chapter Sixteen but it will be realized that vast areas of the land surface of the globe lack rainfall sufficient to feed permanent springs. Regions devoid of surface flow are called areic and occur mainly in deserts, especially in trade-wind latitudes. Regions with permanent or regular seasonal drainage may drain either to the sea, or inland, in which event they are called exoreic and endoreic respectively. Of the world's land surface, about 15 per cent. is areic, just over 10 per cent. endoreic and nearly 75 per cent. exoreic.

Areic and endoreic areas commonly abut and overlap as inland-sloping basins are peculiarly associated with deserts and semi-deserts. The prime causes of inland drainage are rift-faulting, as happens in East Africa, and folding, which alone or aided by faults, may enclose drainage basins, as in high central Asia and the Caspian Sea district. These centripetal drainage basins at high altitudes are liable in time to be captured by external rivers and drained to the sea but those near or below sea-level suffer extinction only through great deposition.

Permanent and semi-permanent rivers drain about 85 per cent. of the world's land surface outside the ice-caps. The geographical interest of these streams lies not only in their work of erosion and deposition but also in their variations in volume. The monthly or seasonal variations in the volume or water-level of a river comprises its régime.

Factors of River Discharge

The main factors which affect the actual volume of a river are twofold: *those causing increase*, such as precipitation; size of drainage basin; and impervious surfaces; and *those causing decrease*, such as evaporation from soil, water and vegetation; and porosity of surface.

The difference between these positive and negative influences results in the river-discharge or run-off which may be measured as a precise volume or as a height of water-level above a fixed mark. Where measurements are precise it is possible to compute the volume of water discharged per unit area of a drainage basin. As the mean depth of rainfall per unit area may be computed from a rainfall map, it is also possible to find the difference between precipitation and run-off, as well as their relationship, the so-called co-efficient of discharge.

The greatest complications of run-off occur where much of the precipitation comes as snow and where extremely porous rocks outcrop. Water stored in the form of snow and ice may stay for long periods on mountains and the precipitation of one year need not enter rivers in the same year as it falls. Porous outcrops, such as Carboniferous Limestone, usually lack surface drainage and their run-off is not easy to measure.

Numerous other local factors influence the nature rather than the amount of river-discharge. Steep slopes drain more rapidly than gentle and steep impervious valley-sides tend to cause sudden spates. Narrow valleys cause greater rises in water-level than do wide, shallow valleys where the water spreads out over the alluvial flats and the peak of the flood is lessened. Lakes similarly diminish the rapidity and height of floods, as is demonstrated in Figure 131. The shape of a drainage basin is also of some importance. A main river which unites a number of tributaries of about the same length within a circular or spatula-shaped basin receives the high water from each at almost the same time and itself floods enormously. On the other hand, in long, narrow river basins the influx of floods of the lower tributaries will often have passed before the effect of floods upstream has reached the lower river.

Vegetation appreciably affects run-off which is usually most rapid where plant cover is least and in the leafless season. Woods and forests not only break the impact of rain and drip it gradually from their surfaces but they also bind soils and supply them with humus and so often increase their porosity. In some temperate areas hardwoods intercept 20 per cent. of the rainfall when in leaf and about 10 per cent. when dormant. Vegetation may sometimes absorb moisture direct by means of its leaf surfaces but by far the bulk of its water needs comes through its roots and passes out through its leaves. A tree of average dimensions exhales into the air about $2\frac{1}{2}$ gallons of water daily and 'to attain a weight of 1,000 lb. an elm tree must absorb from the soil and evaporate into the air 35,000 gallons of water' (Sir Napier Shaw).

Some of the above influences, including vegetation, are largely dependent directly or indirectly on climate. Others, such as mountain structures and porous outcrops are non-climatic and may occur in any areas but, generally speaking, the major control of river régimes is always climatic.

Types of River Régime

River régimes may be classified as single, with one marked seasonal high-water; double, with two periods of high and of low water; and multiple or complex, with several periods of high or low water. It is, however, equally informative to correlate régimes with major climatic types although the drainage basins of many large rivers lie within more than one climatic region.

Equatorial Régimes. In tropical rain forests, as temperatures are uniformly high, rainfall controls river-flow. Considerable areas experience two rainier periods but towards the margins of the forest only one rainy season occurs. The rhythm of two wet and two less wet periods is reflected in the flow of the Congo. Yet this river is so long and its basin so vast that the régime is less simple than it appears at first sight. The equatorial part of the river-basin has heaviest rainfall in spring and autumn but the larger tributaries have a summer rainfall maximum, which means July–September in the north and January–March in the south. Under this balancing influence, the main stream becomes truly equatorial with an abundance of water all the year round.

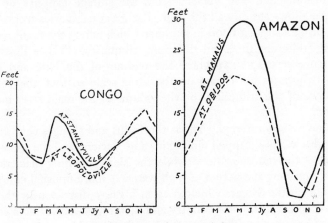

118. Equatorial river régimes
Rise and fall above fixed datum

The basin of the Amazon, by far the world's greatest river, lies mainly south of the equator. It experiences an equatorial river-régime with two high-water periods in its upper course only. Elsewhere so much water comes in from the Brazilian campos that there is one major flood in May or June (Fig. 118).

Some equatorial rivers that rise on wet mountain slopes, especially in lofty islands such as New Guinea, show in addition to their seasonal rhythm a marked diurnal variation. The almost daily rains and rapid run-off cause a rise in river-level during the later part of the day.

Savanna and Monsoon Régimes. In areas such as the Sudan and Monsoon Asia, which experience a damp tropical air-mass in summer and a dry anticyclonic air-mass in winter, the local rivers rise rapidly during the rains and shrink rapidly in the ensuing drought. In the hot season of spring many of the smaller rivers dwindle into pools of water as commonly happens on the streams of the Indian Deccan and even on the Atbara in the Sudan. The Hwang Ho, Yangtze, Irrawaddy, Ganges and Indus are typical of the monsoon régime which differs little from that of the savanna as exemplified by the Orinoco, Zambezi and the great tributaries of the Amazon.

Complications of this simple régime occur where savanna rivers also receive drainage from equatorial regions. The Nile depends on the equatorial climate for its persistency and on monsoon and savanna climates for its floods. The White Nile, which is fed by rains with two maxima and also benefits from the storage capacity of Lake Victoria, flows steadily all the year. The great tributaries from Abyssinia bring a sudden flood in summer which enters the main river at slightly different times on different tributaries. This is illustrated in Figures 119 and 123 where the régimes of the Nile and of its main tributaries are drawn on the same scale so that the relative influence of the tributaries on the volume of the main river may be judged.

Some monsoon rivers, such as the Indus and Ganges (but *not* the Nile) rise in snow-capped mountains and glaciers, and the snow-melt increases the monsoon flood while the slowly-melting glaciers may slightly prolong the period of high water. A more unusual complication affects the Niger, which rises in a rainy district near the sea, flows inland to semi-arid country and then turns seaward to a coastal zone of heavy rains. The great summer flood on the upper river takes five months to reach the lower course where, although lessened, it causes a flood in winter. In the meanwhile the rains of the lower basin have already caused a great flood there in summer.

Mediterranean Régimes. The main peculiarity of Mediterranean climates is the strong summer drought, during which many of

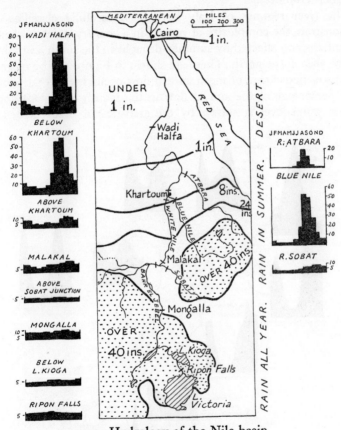

119. Hydrology of the Nile basin

(After Hurst and Phillips)
Graph scales are in 10 million cubic metres per day

the smaller streams dry up. The rivers are fullest in spring and winter, after the heaviest rains, but floods may occur in any of the cooler months. In and poleward of Mediterranean latitudes snow and ice on highlands become increasingly important in river-flow. On the Euphrates and Tigris 'the beginning of the rainy season in late autumn brings down freshets, which continue throughout the winter

and spring, with a general swelling in the flow, till the rising temperature, accompanied often by warm rainfall, melts the snow on the
mountains in April or May. From June onwards the rivers shrink
gradually, until they reach their lower levels in September or
October' (Ionides).

The river régimes of Italy have been carefully classified and
demonstrate the complexity of run-off in a country much broken by
mountains and affected by summer-drought in the south and rainfall
all the year in the north. These are shown in Figure 120 by means of
graphs of percentage of mean annual flow occurring in each month.
The percentage of the annual flow in the month of highest and of
lowest water-level is shown by the statistics in brackets beneath

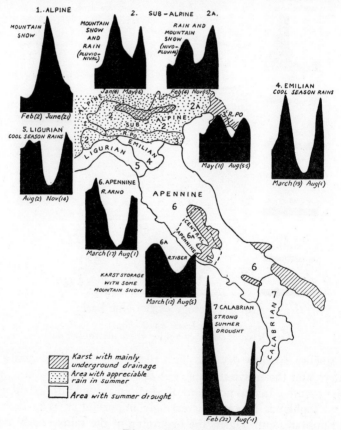

120. River régimes of Italy

each graph. On the Alps and snowy Alpine foothills, spring rain and melting snow or ice cause a high flood between May and July. Upon the eastern sub-Alpine zone and northern Apennines (Emilia), where snowfall is much less, the rivers flood in March and in November. The Po under the influence of Alpine and Apennine tributaries has floods in May and November. In all peninsular Italy, except Liguria, the spring flood exceeds the winter and autumn and increases in relative importance southward as the summer drought lengthens. Where karst or limestone country occurs on the mountains the porous rocks imbibe the rainfall and give it off slowly in the form of strong springs lower down the hillsides.

Rainy Temperate Régimes. In temperate areas where rainfall occurs fairly regularly throughout the year, the season of low-water is summer when evaporation is most active. The Seine, Thames, Severn and Shannon are, on an average, fullest in January or February and lowest in July or August (Fig. 121). The rivers lack

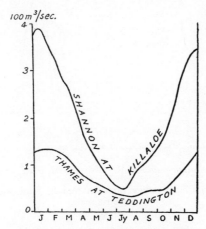

121. Rainy temperate river régimes
Graphs based on mean daily flow for each month

any period of truly low water and heavy rains or snow-melts may cause spates at any time between November and April.

Upon the higher parts of these wet temperate regions and in those with slightly colder, snowier winters, there may be two high-water periods, a greater in spring after snow-melt, and a smaller in late autumn after heavy rains. The Rhine is a complex example of an

Alpine river fed by tributaries draining regions with rain all the year and with, in parts, a little snow-melt in spring.

Cool and Cold Continental Régimes. Continental climates with mean winter temperatures ranging from just below freezing point to well below zero develop peculiar river régimes. The soils freeze solid for several months and vast areas have in addition permanently frozen sub-soils. During winter, precipitation comes mainly as snow; rivers are frozen over and drainage almost ceases. In spring, the snow melts and the soil thaws downward and tremendous floods ensue until the excess of surface water has been removed by the rivers and by evaporation. The thaw in less cold areas comes in April, as on the Dnieper, but in the colder territories is delayed

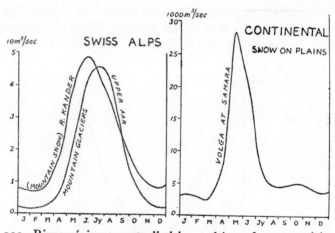

122. River régimes controlled by melting of snow and ice

Graphs based on mean daily flow for each month. Notice great difference in volumetric scales

until May or June (Fig. 122). The vastness and flatness of the drainage basins and, in some, a poleward direction which causes the mouth to thaw last add to the severity of the floods. Yet even the southward-flowing Mississippi and Missouri, partake equally of this régime.

Mountain Régimes. Once an area is high enough to experience regularly a cold-season snowfall, its streams begin to show certain characteristics. Their winter flow is very small and is followed, in late spring or early summer, by a sudden flood, which persists as long as the melting proceeds upwards, maybe to the permanent snow-line. By late autumn the returning cold prevents further melting and the run-off soon becomes remarkably low.

When glaciers and ice-caps occur in river basins, as in the Alps above 10,000 feet, the melt begins later and the great flood comes in summer. The ice continues to melt until autumn when the streams begin to dwindle into insignificance (Fig. 122).

A feature of the warm-season flow of ice-fed streams is their daily rise in water-level. The stream decreases appreciably after sunset and then, soon after sunrise, swells steadily until in the later after-noon it becomes a turbid torrent with perhaps thrice its early morning volume.

Human Significance

The volume and régime of rivers largely control their possible usefulness to man. The greatest potential supplies of hydro-electricity lie on rivers with waterfalls and an abundant flow all the year. Thus, the tremendous unused potential of the Congo far exceeds the output of all the numerous schemes completed on relatively small rivers in Western Europe. On rivers with a marked low water, as in much of southern Europe, either river-flow must be maintained by means of large storage dams or else the hydro-electric power must be supplemented by that from thermal stations.

The value of rivers for irrigation purposes depends mainly on their floods in the warm season. The annual rise of the Nile at Aswan averages 25 feet, while that of the Indus and Ganges is nearer 30 feet. This rise ensures that wide riverine tracts can be flooded by means of simple canals. The incidence of high-water in the warm season is ideal for crop-growth and it is not surprising that enormous quantities of water are taken from rivers for irrigation. Along the Indus about 2,300 cubic metres per second are abstracted for that purpose or an amount exceeding the mean flow of the Rhône. The Indus has—through irrigation, evaporation and seepage—already lost one-third of its sub-montane volume by the time it reaches the Sukkur Dam and this loss increases to one-half before it reaches the sea. For the same reasons the Tigris and Euphrates, although their régimes are less suited for irrigation purposes, dwindle to one-quarter their foot-hill volume by the time they enter the sea. The Nile, which compared with other long rivers has a remarkably small volume, provides a yearly average of about 1,000 cubic metres per second of

w

water for irrigation purposes or twelve times the average flow of the
Thames at Teddington (Fig. 123).

The mineral content of river water is often an advantage for irrigation by inundation canals and direct flooding but it may be a nuisance
in silting up barrages designed for perennial irrigation or for hydro-
electric generation. The Nile, although by no means a heavy carrier
of débris, carries annually at Wadi Halfa about 30 million tons of

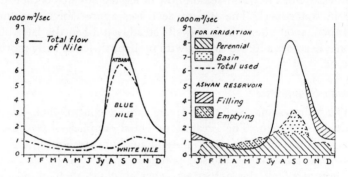

123. Sources and use of Nile water at Aswan

(After Hurst and Phillips)

fine sand, 40 million tons of silt and 30 million tons of clay. In August
and September the amount of solid matter borne by it may exceed
2 million tons daily. To avoid the silting up of reservoirs, as that at
Aswan, the water is not collected until some time after the turbid
flood-peak has passed (Fig. 123).

The navigability of rivers increases with volume and evenness of
régime. On many rivers the great summer flood allows boats to ascend
highest at that season. On the Amazon ocean liners ascend regularly
1,000 miles to Manaus, while large river steamers proceed another
1,000 miles to Iquitos in Peru, where the river is still nearly two miles
wide.

The rise of seasonal floods averages 25 feet on the Nile at Aswan,
25 feet to 35 feet on the great Siberian rivers and up to 50 feet on
the Amazon. These averages, however, hide the danger of very much
higher spates. If two flood-causing factors coincide—such as rapid
snow-melt and heavy rain—or if one flood-forming factor becomes
excessively strong, disastrous floods may occur. The Volga has risen
over 50 feet at Saratov, the Ohio nearly 80 feet at Cincinnati and the

Yangtze in its gorge section has risen nearly 100 feet. It is these unusual floods and the sudden but rare spates of 10 to 25 feet or more in a day that cause so much destruction to life and property.

LIST OF BOOKS

Humphreys, A. A. & Abbot, H. L. *Report on . . . Mississippi River*, 1876
Meinzer, O. E. (ed.) *Hydrology*, 1942
de Martonne, E. *Géographie Physique*, I, 1948
Foster, E. E. *Rainfall and Runoff*, 1949
Pardé, M. *Fleuves et Rivières*, 1955. (By far the best summary of the subject.)
Hurst, H. E. and Phillips, P. *The Nile Basin*, I, 1931; V, 1938
Hurst, H. E. *The Nile*, 1952
Ionides, M. G. *The Régime of the Rivers Euphrates and Tigris*, 1937
Toniolo, A. R. *Atlante Fisico Economico d'Italia*, 1940. Map 9. Idrografia Terrestre
Barrows, H. K. *Floods, their Hydrology and Control*, 1948
Darnault, P. *Régime de quelques cours d'eau d'Afrique Équatoriale*, 1947
H.M.S.O. *The Surface Water Year-Book of Great Britain*
UNESCO. *Arid Zone Hydrology*, 1953
Richards, B. D. *Flood Estimation and Control*, 1944
Pardé, M. *Sur la Puissance des Crues en diverses Parties du Monde*, 1961
Ledger, D. C. 'Some hydrological characteristics of West African rivers', *Inst. Brit. Geog.*, 1964, pp. 73–90
Kolupaila, S. *Bibliography of Hydrometry*, 1961 (Notre Dame, Indiana)

CHAPTER TWENTY-TWO

WATER ON THE LAND SURFACES: LAKES

Nature of Lakes

A LAKE is nothing more than a hollow in the surface of the continents, filled permanently or seasonally with water. The accumulation of water will be permanent if the precipitation exceeds evaporation locally and seepage through the lake floor or if the supply of water from springs or streams entering the lake makes good any natural deficiency of precipitation over evaporation. The lake basin may overflow and, in fact, it is common for lakes to have streams flowing from them. In wet climates lake-basins soon create for themselves an outlet to the sea but in semi-arid districts lakes without overflows or outflows are common. The water in the former, as a rule, remains fresh while in the latter or enclosed type it becomes increasingly saline. The Great Salt Lake in Utah has a salinity of about 220 parts per thousand, while that of the Dead Sea is 238 parts per thousand. To-day, owing to a general desiccation since the last Ice Age, many lake-basins in desert areas contain more salts than water and thousands have become mere salt-pans. A further characteristic of lakes in semi-arid climates is their great periodic variations in size. Between June 1949 and December 1950 the whole of the dried-up floor of Lake Eyre in South Australia was flooded up to a maximum depth of about 12 feet but by January 1953 excessive evaporation and drought had re-converted the 3,000 square miles of lake-bed into dry salt-pans. Similarly, the shallow waters of Lake Chad on the borders of French Equatorial Africa change in area from 6,000 up to 30,000 square miles.

The size of permanent lakes varies from small ponds to Lake Superior, 32,160 square miles and the Caspian Sea, 169,100 square miles. In the world there are 75 lakes exceeding 400 square miles in size and these together cover nearly 500,000 square miles. The sixteen largest lakes are given in the table on p. 325, which also shows their greatest depth below their own water-level and the greatest depth of their beds in relation to sea-level.

It will be seen that many of these big lakes are comparatively shallow and that even the deeper ones occupy relatively insignificant

Lake	Approx. area (sq. miles)	Maximum Depth feet	Relation of Max. Depth to Sea-Level
Caspian . . .	169,100	3,104	−3,189
Superior	32,160	1,007	−407
Victoria	26,800	259	+3,461
Aral	24,000	223	−59
Huron . . .	23,010	732	−151
Michigan . . .	22,400	869	−289
Tanganyika . .	13,000	4,708	−2,149
Baikal	12,200	5,712	−4,196
Nyasa . . .	12,000	2,316	−794
Great Slave . .	11,600	2,014	−1,522
Great Bear . .	11,400	450	−59
Erie	9,940	210	+362
Winnipeg . . .	9,398	62	+634
Ontario . . .	7,250	738	−495
Ladoga	7,230	820	−804
Balkhash . . .	6,800	87	+830

hollows. All told there are only 38 lakes deeper than 1,000 feet and consequently it is not surprising that they are considered as temporary features of the landscape. Many of them receive the sedimentation of sizeable streams and are being reduced rapidly by deltas (*see* p. 248). So great is the silt brought by the Rhône into Lake Geneva that the life of the lake is estimated at less than 40,000 years. Many lakes are also in danger of diminution from the deepening of their outlets. It more or less follows, therefore, that as lakes are so liable to obliteration the majority of those existing to-day must be of fairly recent formation. This contention seems borne out by the general distribution of lakes. Probably 90 per cent. of the world's lakes lie in areas that were glaciated during the last great Ice Age. Most of the remainder occur in rift valleys and folded zones, in volcanic regions and upon riverine flats and coastal plains. Landscapes in many parts of the continents contain the remnants of old lake basins long since either containing a shrunken body of water or turned into seasonal swamps and quite silted up. The extent of marginal lakes during the melting of the great ice-sheets was enormous; glacial Lake Agassiz, the precursor of the present lakes near Winnipeg, alone covered about 100,000 square miles.

To-day lakes form a minor yet interesting part of many landscapes but they predominate the landscape only in certain glaciated regions, as in Finland. Canada has 16 lakes exceeding 1,000 square miles in area, and bodies of fresh water cover no less than 220,000 square miles of its surface. The Laurentian Shield and its margins are in parts an intricate maze of lakes and overflows. There are 3,000 lakes in 6,000 square miles of territory just south and east of Lake Winnipeg, and in a smaller area south-west of Reindeer Lake in Saskatchewan 7,500 lakes have been counted (Plate 23).

Formation of Lakes

Lakes are usually classified according to whether the hollow in which they have accumulated has been formed mainly by deposition or erosion or by earth movements. The difficulty in this or any other classification is that many lakes lie in erosion hollows and yet have accumulated in part behind a deposited barrage. The test, however, seems to be that a basin formed by erosion would retain a lake even if the barrage, where present, were removed, whereas a lake formed by deposition would disappear with the removal of the deposited dam. A few lakes due to meteor impact are in a separate class.

Lakes in Hollows formed mainly by a Deposited Barrage

Unevenly deposited Sheets of Glacial Débris. When a large ice-lobe or ice-sheet is no longer being fed from its source it may become stagnant and melt *in situ*. On melting, it deposits upon the ground beneath an irregular sheet of boulder clay or till or ground moraine. The hollows in this uneven surface often cause a multitude of lakes which are characteristically shallow and very irregular in outline. Such lakes survive in great numbers in the coastlands of the Gulf of Finland and in Minnesota, where, as is common, active steps are being taken to drain them.

Moraines. Where glaciers and ice-sheets move towards their melting edges they tend to deposit ridges and mounds of débris. On the retreat or disappearance of the ice these morainic hills may hold up lakes, especially if the débris contains much clay. Most of these lakes are shallow as the barrier is seldom compact enough to be watertight except in its thicker lower parts. In southern Finland many thousands of lakes have accumulated behind the Salpausselka, a vast series of crescentic morainic ridges. Throughout glaciated regions many of the lakes are in hollows formed partly by ice-erosion and partly by morainic dams. Llanberis lake in North Wales is caused largely by a glacial moraine.

Ice. Where the melting front of an ice-sheet approaches an uphill gradient lakes may accumulate against the ice-front in summer. This occurs commonly in parts of Greenland to-day (Plate 21) and was widespread during the Ice Age. The Great Lakes of the St. Lawrence basin are remnants of vast water-bodies that accumulated at the edge of the Laurentian ice-fronts. The former Lake Pickering in the Vale of Derwent in Yorkshire has completely disappeared (*see* pp. 290–291).

To-day glaciers may also form barriers by crossing and blocking a valley free of ice at least in summer. In mountainous regions a glacier may come from a shaded basin situated well above an adjacent valley and may dam periodically the stream flowing in that valley during the warmest months. Or it may happen that a glacier filling a main valley may cross the mouth of a side valley that is in the sun and contains a stream in summer. In either instance the glacier may when its ice happens to be exceptionally compact cause a lake to be formed. This lake may grow each summer until it overflows or until the ice-dam weakens and breaks. Such lakes, although relatively rare, occur more frequently than was formerly supposed and are of some geographical importance.

The Marjelen See in the Swiss Alps accumulates in a lateral valley against the side of the Aletsch glacier about 150 feet to 250 feet high in the main valley. At times in some summers the ice wall becomes deeply crevassed and the lake may suddenly empty itself causing a wall of water several feet high to sweep down the River Massa into the Rhône. Because of the danger of sudden floods a large overflow tunnel was dug into a neighbouring valley in 1889 and the spates of the Marjelen See have thereby been greatly reduced. In recent years much larger glacial lakes subject to periodic breaks and flash-floods have been found in the Himalayas and the Andes. In the Karakoram range of the Himalayas three glaciers descend steeply into the west side of the deep gorge of the Shyok river (Fig. 124). Occasionally the snout of one of these glaciers crosses and completely blocks the gorge by a wall of ice several hundred feet high. This impounds a lake which, in the warm season, may become 10 miles long and have an average depth of 150 feet. When the dam weakens or bursts a great flood sweeps down the Shyok and Indus. After the burst in August 1929 the peak flood-rise in a few hours was 85 feet just below the ice-dam, 53 feet at 500 miles downstream and about 26 feet at Attock bridge nearly 750 miles downstream.

It seems also that lakes may accumulate in summer upon and within the ice itself and the sudden release of these pockets of water can cause dangerous floods.

Landslips and Screes. Landslips are common in areas much affected by earthquakes as in the Brahmaputra basin in Assam. They are as frequent but less spectacular in areas where heavy rainfall causes a mass-movement downhill of surface soils due either to partial liquefaction or to the lubrication of a slippery sub-surface. When landslips dam a river valley they give rise to lakes which although often short-lived may occasionally endure for a long period. In September 1893, a headwater of the Ganges was blocked by a

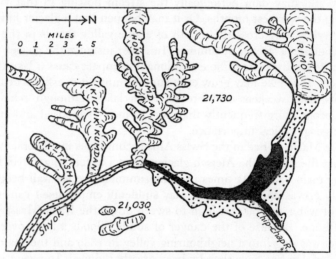

124. Glacier dam across Shyok valley in 1928
(After Mason)
Dammed-back lake is shown in black and valley-floor in stipple. Heights are in feet

landslide near Gohna in Garhwal. The lake formed by it rose steadily until the following August when the dam broke and caused the river-level just below it to rise suddenly by 260 feet. Even at 72 miles from the landslide the river rose 42 feet in less than two hours. During the great Assam earthquake of 15 August 1950, landslips blocked temporarily scores of deep ravines in the mountains. The breaking of the dams caused flash-floods of which that, for example, in the Tidding valley formed a wall of water 60 feet high which stripped the lower parts of the valley of its jungle and buildings and left it twice its former width and littered with vast mounds of stones.

In the Jura mountains in France several long, narrow lakes held up by landslip barriers occur along the River Doubs. Thus the winding Lac de Chaillexon near Brenets is a landslip-obstructed and partially drowned river-gorge over two miles long (Plate 27).

Even a small daily fall of débris from a steep slope in the form of screes may act eventually in the same way as a sudden landslip, and Hard Tarn on the flanks of Helvellyn seems dammed by screes.

Water-Borne Débris. In flat flood-plains streams often create lakes by cutting across meander necks, so forming ox-bows, and by abandoning one arm of a bifurcation to form ribbon-shaped lakes. In deltas, such as that of the Mississippi, lakes are often caused by changes of channel and by deposition across former directions of flow.

Occasionally the excess of sedimentation in a main valley may actually dam tributary valleys that arise locally. In Western Florida, Dead Lake on the Chipola River just before it joins the Apalachicola River is an elongated, lanceolate body of water typical of the many lakes in this area formed in that way (Fig. 125).

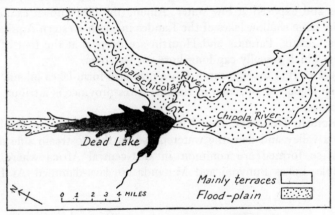

125. Lake dammed by river-alluvium
X Marks cut-off that has built levée across Chipola River

More usual is the diversion and even the damming of main streams by the alluvial fans and cones of tributaries. Mountain torrents commonly build alluvial cones out into the main valley and Styhead Tarn has been dammed in this way by the alluvial fan of a torrent from the Gables. But few tributaries can complete the barrage except in semi-arid climates where sheet-erosion and the mass-movement of waste after heavy rain may cause semi-liquid débris to creep downhill like a lava-flow. Lake Tulare in California owes its basin to a great alluvial fan made by the King River from the Sierra Nevada.

A rarer form of river-created barrier occurs in parts of the Yugoslavian karst where beds of calcareous tufa—a hard deposit derived from lime in solution—may accumulate across stream beds and hold up small lakes such as the Plitvic lakes in Croatia.

Marine Deposits. Sand and shingle spits and bars frequently enclose pockets of sea-water and fresh-water lagoons. Such lakes are not uncommon but they are seldom large or permanent as water tends to percolate through the barrage. Longshore drift of silt and shingle across a bay or river-estuary often creates lakes as already described (pp. 299–303). The haffs of the Baltic and étangs of the Mediterranean coast of France were formed in this way.

Moving sand-dunes on sea-coasts occasionally divert water-courses and may enclose, with the aid of longshore drift, a lagoon. The main action here is that of wind acting on sand driven inland by strong waves and exposed at low water. Among the best-known examples are the large shallow lakes of the Landes in south-western Aquitaine. Lakes Cazaux, Parentis and Hourtin are deepest at the foot of the dunes, which actually cap longshore- or bay-bars.

It seems that moving sand-dunes may also form lakes inland, and the salt lagoon of Epecuen in Buenos Aires province is attributed to this cause.

Volcanic Deposits. In volcanic districts lava-flows are liable to cross a valley and dam the watercourse on the upstream side of it. Lakes so formed are common in east-central Africa where, for example, Lakes Bunyoni and Mutanda are lava-dammed. At Lake

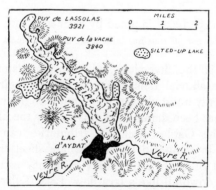

126. Lava flow (La Cheire) damming a river
Present lake shown in black. Heights are in feet.

Kivu the rift-valley was dammed by a line of volcanoes which created a lake and turned the drainage away from the Nile to the Congo. In the Central Massif of France Lake d'Aydat near Clermont Ferrand was caused by a basaltic dyke about 80 feet thick; a number of smaller lakes nearby formed in the same way are now meadowland or marsh (Fig. 126).

The cup-like basins of volcanic craters may collect rainfall as lakes especially if the volcano is dormant or dead. These crater-lakes abound in volcanic regions with a heavy rainfall and numerous examples occur in the East Indies and the Azores (Fig. 127). Most of

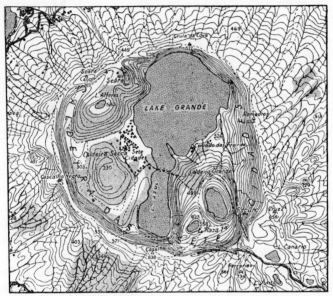

127. Crater lakes in the Caldeira das Sete Cidades, San Miguel island, Azores.

Dotted areas away from lakes are wooded. Heights are in metres

them are quite shallow but some reach a remarkable depth because apparently of a great subsidence of the crater-floor. The large Crater Lake in Oregon, which has a depth of 2,000 feet, and other volcanic lakes are described on p. 207 and depicted in Plate 28.

Organic Deposits. Probably the most ephemeral of all lakes are those caused by organic deposits. Floating vegetation may cause river expansions or lakes as happens during excessive accumulation of 'sudd' on the Bahr el Jebel (Fig. 128). Large log dams on the Red River in Louisiana formed numerous lakes in tributary valleys and when the dams were finally removed the river in one area degraded its bed by as much as 15 feet. Among animals the beaver is the only notable dam-builder and on the headstreams of the Colorado and other rivers in North America numerous lakes and a much greater number of silted-up reservoirs, or beaver-meadows, bear witness to its skill.

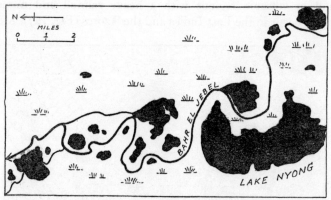

128. Sudd-dammed lakes and river expansions in the Sudan

Lakes in Hollows Formed Mainly by Erosion

Wind Action. In semi-arid and arid districts wind deflation may form hollows below either the local surface-level or even sea-level. The aridity of the climate usually prevents these basins from containing permanent lakes but it is possible for the wind to deflate down to the water-table during a long dry spell and for the hollow to fill with water, at least occasionally, when the water-table rises after rain. Lakes of this nature are common on the low plateau of Patagonia, where most are brackish and known as salitrales.

Ice Action. Although the action of moving ice is to widen rather than to deepen pre-existing valleys and hollows the movement is such that certain parts of the ice-bed are liable to be overdeepened. This overdeepening occurs where the downward pressure of the ice was greatest. There are three main locations at which basins or hollows are likely to be formed.

In the gathering ground of a mountain glacier the formation of a saucer-shaped cirque or corrie was often assisted by a slight rotational movement of the ice which abraded the floor of the cirque. These lakes or tarns are, unless enlarged by barrages of glacial débris, usually small, shallow and circular (Plate 24).

Lakes commonly occur at intervals along the floor of a heavily-glaciated valley. They are typically ribbon-shaped and relatively deep (Fig. 129). Many are true rock-basins and the larger have smaller basins and rock sills in their floors and where near the sea are nothing more than undrowned fiords. Of the many Scottish lochs of this nature, Morar is the deepest. It descends to 1,017 feet and since its surface is only about 30 feet above the sea (from which it is separated by a narrow sill) its bed is in part 987 feet below sea-level.

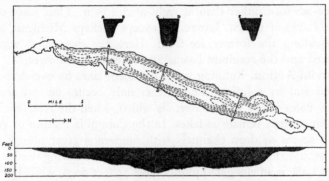

129. A typical glacial trough lake: Coniston Water in the English Lake District

(*After Mill*)

Depths are in feet. Longitudinal and cross-sections shown in black.

The world's largest glacial lakes lie along the former outer edge of the Pleistocene ice-sheets and glaciers. Thus the north Italian lakes are situated where the main Alpine ice-lobes met the flat peripheral lowland. It is presumably at this sudden break of slope that the downward thrust of the ice was strong and consequently many of these lakes, although far inland, have floors well below sea-level. The table on p. 334 of the eight deepest lakes in Europe contains four lakes in Norway, three in Italy and one in Scotland. All are in glaciated valleys either near the sea or at the junction of glaciated mountains and plains.

Many marginal glacial lakes have morainic bands around their extremities farthest from the ice-advance and a few—such as Lake

Name	Maximum depth (feet)	Depth of Maximum below sea-level (feet)
Hornindalsvatn .	1,686	1,512
Mjösa . .	1,473	1,076
Salsvatn . .	1,460	1,417
Tinnsjö . .	1,437	837
Como . .	1,342	689
Maggiore . .	1,220	576
Garda . . .	1,135	922
Morar . . .	1,017	987

Garda—change from a typical ribbon-shape on the highlands to a more circular or spoon-shape on the submontane plain. It is noticeable, however, that where the pressure came from ice-sheets as distinct from ice-lobes and glaciers the lakes are usually alined along the ice-advance rather than at right-angles to it. Thus most of the Great Lakes of the St. Lawrence, except perhaps Michigan, are in basins along the former ice-front. Here the deepening was less localized and the resultant basins are wider and shallower.

Fluvial Action. Running water can form lakes by overdeepening its bed and by solution. The former only occurs on any scale at plunge-holes below large waterfalls which, when the river-course is abandoned, may remain as lakes. In the Columbia Plateau of Washington scores of deep channels with numerous great plunge holes (now occupied by lakes) were cut by flood-waters during the Ice Age. Moses Lake is an elongated plunge-hole 75 feet deep, while the great plunge-hole basin of another spillway contains Rock Lake, which is seven miles long and up to 250 feet deep. At Jamesville in central New York State an old overflow plunge-hole of the Great Lakes is occupied by a lake still 60 feet deep.

Lakes caused by the solvent action of water are not uncommon in some limestone districts. In the Yugoslavian karst some of the larger dolines and poljes have been eroded down to the underlying clay and carry permanent lakes. In the Swiss Alps the Muttensee has been formed by the uniting of three dolines. A rather different effect operates along the River Shannon where solution has widened the Carboniferous Limestone of the river valley to form Loughs Ree and Derg. In some rainy low-lying areas the solvent action of rain-water may riddle a surface with lakes as happens in the Lake District of Florida.

Lakes in Basins formed by Earth Movements and Volcanic Explosions

Folding. Mountain building, especially of a folding nature, must often give rise to lakes most of which disappear as the stream network integrates and deepens the valleys. Yet lakes due directly or indirectly to folding persist to-day, as for example in Lake Fahlen in the Säntis Massif of Switzerland; and in Lake Joux in the Jura where a typical enclosed drainage basin has been formed largely by a mountain fold running obliquely across the valley between two parallel ridges (Plate 29). Large high-level lakes such as Titicaca are obviously connected in part with a major orogenesis.

Warping. Where land surfaces undergoing surface tension or isostatic pressure neither fold nor break warping may occur. Isostatic depression at the edges of great continental ice-sheets is often postulated. Thus the great terminal basins or marginal lakes of the St. Lawrence system and even of north Italy may be due partly to isostatic depression during the Ice Age. But surface warpings occur also in non-glaciated areas. In western Uganda, streams which once flowed to the Congo have been reversed by upwarping and tilting into Lake Victoria, which, with its outflow Lake Kioga, occupies the down-warped basin. It appears that the axis of the arching is about 23 miles east of the rift valley. Eastward of this line the streams, reversed in direction, flow sluggishly to Lakes Victoria and Kioga; westward they flow vigorously into the rift valley; and the divides between them are swampy segments of the old continuous valleys (Fig. 130).

Fracturing. As has already been described (p. 198) rift-valleys encourage the formation of long, narrow lakes, often of a considerable depth and with floors well below sea-level. Lakes Tanganyika and Nyasa, and the Dead Sea possess these qualities as also does Baikal, the deepest lake in the world and the one that descends farthest below sea-level. These rift-lakes are among the more enduring bodies of inland water as often they are immune from the lowering of level by the erosion of water exits, although in rainy climates they overflow to form gorge-like discharge channels (Plate 30).

Faulting also causes lakes when fault-scarps or tilted fault-blocks hold up the drainage of the area sloping towards them. Examples abound in the western cordillera of the U.S.A. where Lakes Abert and Warner in Oregon and the Klamath Lakes are impounded against high fault-scarps (Fig. 59).

In areas of large-scale uplift and wide-scale faulting as in the Great Basin of the western U.S.A. lakes may accumulate in rather shallow hollows and need not assume an elongated trough-like nature. Thus

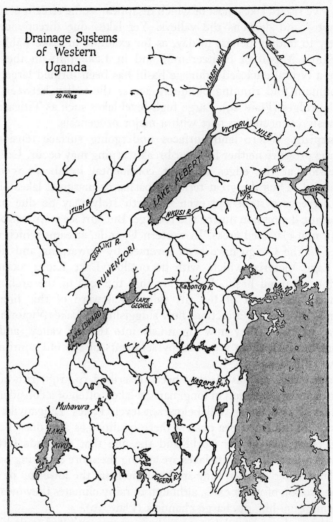

130. Drainage systems of western Uganda, to show how warping has severed formerly continuous streams and backtilted their headwaters with the formation of Lake Victoria.

(*King, 'South African Scenery.' Oliver & Boyd*)

Arrows show direction of stream-flow

the Great Salt Lake of Utah, the shrunken remnant of the vast
Ice-Age Lake Bonneville, is formed by the combined action of
orogenesis and fracturing.

A rather different form of fracturing may occur during earthquakes when the sudden sinking of small areas may cause lakes.
Reelfoot Lake, 20 miles long and up to 20 feet deep, in the Mississippi valley on the borders of Tennessee and Kentucky, was formed
largely in this way after severe earthquakes in 1811 and 1813.

General Uplift of Coastal Areas. The uplift of coastal areas may
well be associated with warping or regional tilting but it may in
some areas be connected with a general uplift of the coastlands.
Shorelines newly emerged from beneath the sea often contain a maze
of lakes as in parts of Florida where Lake Okeechobee and others
near it occupy depressions in the original sea-floor.

Volcanic Explosions. Hollows formed by volcanic explosions
and not by the volcanic deposition of a crater are in a sense minor
earth-movements. Such explosive hollows usually lack a containing
rim and are presumably formed by explosive action. The many
circular lakes known as *maare* in the Eifel upland of western Germany
are of this type.

Lakes and Human Affairs

To a minor degree lakes affect mankind in much the same way as
do oceans. They act as political boundaries and lines of defence;
they provide sources of salt and soda in dry climates; and frequently
give rise to considerable fisheries. Where of any size they become an
important means of communication and the Sault-Ste-Marie ('Soo')
canals on the Great Lakes have in recent years carried 115 million
tons of shipping annually as against 110 million tons on the Suez
Canal. They form landing-bases for seaplanes and in cold countries,
when frozen, are used as landing grounds for ordinary planes. Their
influence on the climates of their shores is discussed in Chapter
Twenty-three.

But many lakes have a quality not to be found in oceans: the latter
are essentially terminals whereas many lakes are intermediaries. This
intermediate position gives some lakes a great importance as filterers
and regulators of river-water. Rivers leaving lakes have, except
perhaps in time of high flood, been well filtered so that industries
such as textiles which need clean water find it best to be on the exit
from the lake.

x

The expanse of water in a lake greatly lessens the flood-peaks of rivers entering it; a lake too, by giving off water slowly and evenly, tends to prevent excessive low-water in time of drought (Fig. 131).

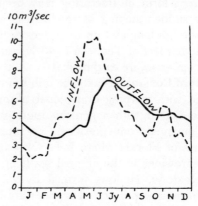

131. Regulating effect of Lake Garda on flow of Mincio river
(After De Marchi)

Consequently many lakes form natural storage reservoirs and regulators of much importance for water-supply and hydro-electric generation.

It is obvious, however, as has already been noticed in the discussion on glaciation, that where multitudes of lakes dominate a landscape they may be more of a hindrance than a help to man. In Finland they hinder agriculture and land-communications, but, above all, here and elsewhere they cover land that might otherwise be suitable for agriculture.

LIST OF BOOKS

General
Collet, L. W. *Les Lacs*, 1925
Halbfass, W. *Die Seen der Erde*, 1922 (Heft 185 of *Petermanns Mitteilungen*)
Hutchinson, G. E. *A Treatise on Limnology*, I, 1957 (1,015 pp.; with full bibliography)

Regional
Russell, I. *Lakes of North America*, 1895
Mill, H. R. 'Bathymetrical Survey of the English Lakes,' *G. J.*, 1895, pp. 46–72; 135–66
Murray, J., and Pullar, L. *Bathymetrical Survey of Scottish fresh-water Lochs*, 1919
Lucas, K. 'Bathymetrical survey of lakes of New Zealand,' *G. J.*, 1904, pp. 645–60; 744–60
Collet, A. 'Alpine Lakes,' *S. G. M.* 1922, pp. 73–101

Delebecque, A. *Les Lacs Français*, 1898
Davis, W. M. 'Lakes of California,' *California Journ. of Mines and Geology*, 1933, pp. 175–236
Thienemann, A. *Die Binnengewässer Mitteleuropas*, 1925
Horie, S. 'Morphometry of Japanese lakes,' *Jap. Jour. Limnol.*, 1956, pp. 1–28
Zumberge, J. H. *The Lakes of Minnesota*, 1952
Keller, R. 'Die grossen seen Nordamerikas,' *Erdkunde*, 1959, pp. 319–43
Langbein, W. B. Salinity and Hydrology of Closed Lakes, *U.S. Geol. Surv. Prof. Paper*, 412, 1961, pp. 1–20
Jennings, J. N. 'Floodplain lakes in Ka Valley, New Guinea', *G.J.*, 1963, pp. 187–90
Charlesworth, J. K. 'The bathymetry and origin of the larger lakes of Ireland', *Proc. Roy. Irish Acad.*, 63B, 1963, pp. 61–69

Particular
For ice-dams and floods of R. Shyok:
Mason, K. and others. *Himalayan Journal*, 1929, 1930, 1932, 1940. 'The study of threatening glaciers,' *G. J.*, Jan. 1935, pp. 24–49
Marcus, M. G. 'Periodic drainage of glacier-dammed Tulsequah Lake, British Columbia,' *Geog. Rev.*, 1960, pp. 89–106
Stone, K. H. 'The annual emptying of Lake George, Alaska', *Arctic*, 1963, pp. 26–40; 'Alaskan ice-dammed lakes', *Ann. Ass. Am. Geog.*, 1963, pp. 322–49
Landslides:
Sharpe, C. F. S. *Landslides and Related Phenomena*, 1938
Kingdon-Ward, F. 'The Assam Earthquake of 1950,' *G.J.*, June 1953, pp. 169–82
Alluvium:
Vernon, R. O. 'Tributary valley lakes of western Florida,' *Journ. of Geomorphology*, 1942, pp. 302–11
Volcanic rim:
Williams, H. *Geology of Crater Lake National Park, Oregon*, 1942
Organic deposits, log dams and vegetation:
Veatch, A. C. *Geology . . . of northern Louisiana*, U.S. Geol. Surv. Prof. Paper 46, 1906
Hurst, H. E., and Phillips, P.F. *The Nile*, 1931–38
Beavers:
Ives, R. L. 'The Beaver-meadow complex,' *Journ. of Geomorphology*, 1942, pp. 191–203
Plunge-hole erosion:
Bretz, J. H. 'The channeled scabland of eastern Washington,' *Geog. Rev.*, 1928, pp. 446–77; *The Grand Coulee*, 1932
Rift; warping; and volcanic barrages:
King, L. C. *South Africa Scenery*, 1963
Worthington, E. B. 'The Lakes of Kenya and Uganda,' *G.J.*, April 1932, pp. 275–97. *Inland Waters of Africa*, 1933
Biological aspects:
Hutchinson, G. E. *op. cit*; and for British Isles, Macan, T. T. and Worthington, E. B. *Life in Lakes and Rivers*, 1951

Section Five

CLIMATE, SOILS AND VEGETATION

CHAPTER TWENTY-THREE

THE INFLUENCE OF SURFACE COVER
ON LOCAL WEATHER

The Need for Surface Recordings

AS already shown the main factors which control climate are
latitude, altitude, distance from water bodies, the character of
prevailing winds or air masses and local factors. The last named are
always present in land-climates but many of them are of little
importance except in calm, clear weather. To-day these local factors
are attracting increasing attention because most of the temperatures
used in climatic statistics are taken within the Stevenson screen in
which the thermometers are situated at 4 feet above ground-level
and are shaded from the sun.

It will be remembered that the atmosphere is heated (and cooled)
mainly from below by contact with the earth's surface and that this
heat is transferred upward especially by convection and turbulence.
Consequently, the air nearest the earth's surface tends to acquire
the temperature of that surface much more closely than air at several
feet above it. Just as the soil heats relatively quickly by day and cools
relatively quickly by night so the air in contact with it assumes a
relatively great diurnal range (Fig. 132).

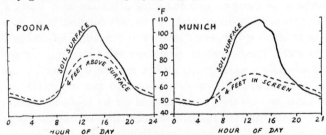

132. Soil and air temperatures: Poona and Munich

(*After Ramdas and Geiger*)

Temperatures refer to clear weather in January at Poona and April-May at Munich

The screen temperature is presumably about the same as that
experienced by a person standing in the shade but it hardly applies
either to people in the sun or to all those plants which grow and
mature in the open and never attain a height of four feet. To give

the weather conditions actually experienced by animals and plants is the chief aim of the study of climate near the ground. Unfortunately no suitable technical term has yet been coined to designate this study. The term micro-climatology is often used to mean the study of climatic factors over a small area but the term is inadequate and is better replaced by eco-climatology or the study of local climates. Yet it is obvious that any locality has firstly a regional setting or macro-climate, secondly a local setting or eco-climate, and thirdly within itself, it has a host of minute climatic differences (such as between leaf and tree-trunk, or moss and marsh) which are essentially micro-climates.

It should be noticed that temperature differences between soils and different vegetation and between hill-tops and hollows will be greatest when clear, calm weather prevails. Yet the study of eco-climates involves much more than temperatures and some important weather elements such as surface winds increase in importance with their strength. Without entering into a discussion of elaborate micro-climatological investigations such as the climate enjoyed by beetles in hollow tree-trunks, we may assume that the study of climate near the ground includes the influence of the shape and aspect of land-forms and of the nature of surface cover, whether bare soil, or vegetation, or water-bodies, or built-up areas.

General Influences

Influence of Shape. The shape of a surface may exert a considerable influence on its climate. As already discussed (pp. 100–105) the main contrast is between convex and concave land-forms during spells of calm, clear weather when inversion of temperature may occur. Generally speaking the convex feature is cooler by day and warmer at night than the concave shape. In non-mountainous areas the katabatic or downhill creep of cold air by night into hollows is more important than the anabatic or uphill flow during the heat of the day. But in mountainous areas, especially outside the Tropics, both winds (here called mountain- and valley-winds) are of great importance. This arises because the local factors in mountain belts are augmented by a regional factor whereby the whole mountain zone is heated more by day and cooled more by night than are the adjacent plains so causing a general drift of air towards the mountain tops by day and outwards from them by night (Fig. 22).

In moderate and undulating topographies the downhill creep of

29. Lakes dammed by up-fold: Lac du Joux (to left) and smaller lake. In middle distance is an up-fold that runs obliquely across the synclinal valley
(*By courtesy of Swiss National Tourist Office*)

30. A rift-valley lake: Lake Manyara in the great East African rift-valley, Tanzania. Scarp of rift-fault is seen on left
(*By courtesy of Public Relations Department, Tanzania*)

[*Rev. T. L. Jackson*

31. Inversion fog in the Coxwold–Gilling gap, North Riding. The Howardian Hills rise (in middle distance) above the shallow blanket of fog

[*Fairchild Aerial Surveys, Inc.*

32. Smoke pall over Manhattan, New York, on a calm, sunny day in winter

cold air into 'frost-hollows' occasionally gives remarkably low temperatures. At the Gstettneralm doline, a large solution hollow in the limestone of the Austrian Alps, the floor supports only tundra and the upper slopes stunted pines whereas the plateau above is clothed with fine coniferous forest. The floor of this hollow has had temperatures of −60° F. whilst a nearby mountain top, the Sonnblick at 10,200 feet, rarely experiences below −2° F. In a frost-hollow at Rickmansworth in the Chilterns ground frost may be expected at least once or twice in any month of the year. Here on 29th August 1936 the night temperature even in the screen dropped to 34° F. and the day maximum rose to 84·9° F. giving a diurnal range of nearly 51° F. During 1935 this hollow had fog on 50 mornings compared with 9 at Rothamsted a few miles away on flattish ground (Plate 31).

Another minor climatic influence of concavities of some importance in cool and cold regions is their tendency to fill with snow in the winter half-year. When sheltered from sun and wind, snow patches may linger long into the summer on uplands in Britain.

Influence of Aspect or Orientation. Just as the influence of concavities and convexities applies to all solid objects as well as landforms, so the influence of orientation applies also to any features— hills or trees or houses—projecting above the earth's surface. Orientation has, for our purposes, two main qualities; first, in respect to the sun's rays, and second, in respect to prevailing winds.

Insolation. During the apparent course of the sun in the heavens from its rising in the east to its setting in the west, when the daytime sky is not overcast, sunward-facing slopes will be warmer than those nearby in the shadow (see pp. 99–100). This effect of surface slope on insolation varies considerably with latitude. Where the noon sun is almost or quite overhead few slopes can directly face the sun and flat surfaces will be the hotter. But outside the Tropics sunward-facing slopes receive most insolation. In Britain, temperatures at the surface of soils have been 15° F. higher on sunny aspects, and it is usually assumed that a slope of 1 in 10 to the south will increase surface temperatures by about 15 per cent. on clear days.

What might be called a 'riviera' effect occurs where steep slopes directly face the sun and receive a very high insolation for the latitude. On calm, bright days the reflection and radiation of heat from scarps and cliffs may give great heat to the lowland strip beneath them.

The problem of aspect or exposure is, however, complicated by the fact that any slope receives solar radiation both direct from the sun and indirect from the diffuse radiation of heat rays from the atmosphere. On overcast days diffuse radiation prevails and as this comes from all directions there is not much difference in the heat received by different slopes in the same locality. On bright sunny days direct solar radiation comes into full play and differences of exposure become of real importance on lowlands as well as on mountains (p. 109). The term 'climate' is in fact derived from a Greek word meaning 'to slope' and to the Greeks climate was the hill-side exposure to the sun. Figure 133 shows the average altitude of the noon

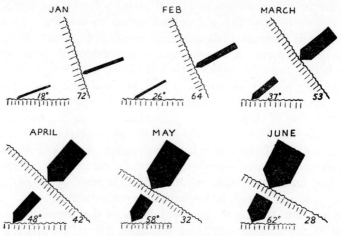

133. Insolation on horizontal surface and on slope of maximum sunniness at Kew, 51° N.

(After J. M. Stagg)

Angular measurements give respectively mean altitude of noon sun and angle of slope of surface of maximum insolation

sun at Richmond (Kew Observatory) in Surrey for the first six months of the year. The gradient of the southward-facing slope which would receive the greatest amount of direct insolation is also shown and the amount of heat received is denoted by the width of the black band. Irrespective of the units of measurement (which are gm. cal./cm.² per day) it will be seen that in Britain the higher the sun the lower the gradient of the sunniest southern slopes. But it will be noticed that at, for example, Brighton (51° N.) slopes of much less than 28° (about 1 in 2) cannot receive the maximum direct radiation for their locality whereas at Nice (44° N.) that would be possible on slopes of 21°. Figure 133 also shows the greater insolation in Britain on slopes as opposed to horizontal surfaces.

Much the same conclusions are represented graphically in Figure 134 which relates to Potsdam (52° N.). Again the favourable radiation

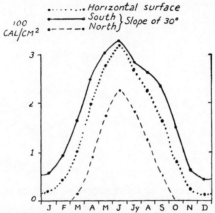

134. Mean daily totals of insolation on differently oriented surfaces at Potsdam

(*After Geiger*)

upon south slopes is emphasized and the misfortune of a poleward exposure. In this city, east and west slopes of 30° were found to experience a mean daily solar radiation slightly less than that on an horizontal surface.

Exposure to Winds. Orientation in respect to prevailing winds or air-movement may be of great climatic importance locally. It cannot be emphasized too strongly that a relief feature of any strength tends to deflect surface winds into a direction parallel to itself. Two coastlines or ranges of hills will canalize or 'steer' winds blowing almost in their direction. East-west surface winds predominate in the Strait of Gibraltar and in the Firth of Forth; Belfast Lough changes north and east winds to north-easterly; in north-south valleys winds of those directions prevail (Fig. 135). Where a marked topographical gap occurs the canalization of an air-movement may become excessive locally. In the Rhône Valley the mistral is a 'ravine' wind of great strength and of exceptional gustiness where it breaks out of the local defiles. Similar ravine winds occur in many parts of the world where, under a regional barometric influence, cold air rushes down-hill or is drawn up-hill. Even streets may canalize winds and some towns on rough days are very gusty.

On the other hand obstacles, such as hill-ranges and belts of trees, at right angles to a prevailing wind tend to break its force. The high

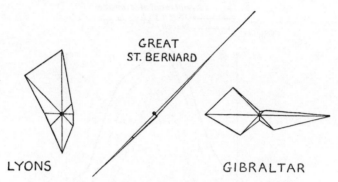

135. Influence of relief trend on surface winds

Wind-roses show percentage total winds blowing from each main direction

speeds of coastal gales from off sea surfaces soon diminish inland because of surface obstructions and surface friction. With winds of about 30 m.p.h. a dense wind-break often causes a reduction in speed of about 15 m.p.h. even at a distance six times the height of the wind-break. A less dense obstruction, such as a belt of trees, usually restricts the wind-speed less but affects a wider zone. Thus an obstruction of moderate density will cause a reduction in wind-velocity of at least 20 per cent. over a width equal to 15 shelter-belt heights on the leeward side and to 2 shelter-belt heights on the windward. Hence it is not surprising that wind flow diminishes rapidly towards the surface of the ground and in forests and cities.

These obstacles, however, have a turbulent effect on wind move-ment. As a rule lee sides of hills are relatively sheltered but whether the down-slope wind will be relatively warm or relatively cold is by no means certain. The föhn and chinook (pp. 105–107) are warm: the mistral and bora are cold, as also is the helm wind of the North Pennines, a violent cold easterly blast rushing down the steep western slope of the Crossfell range in Cumberland. It is usual, however, for the lee-sides of cold-air approaches and exposures to warm-air approaches to have some temperature advantages. In addition the leeward side is often a rain-shadow.

The climatic effect of gustiness beyond the shelter zone provided, for example, by a belt of trees has its meteorological counterpart on mountains in the formation of eddies or standing-waves in the atmosphere above the lee side of the mountain range. These standing-waves and turbulent eddies, which are of great interest to gliders and aviators, may also occur under certain atmospheric conditions to leeward of mountainous islands.

Influence of Colour and Composition. A third condition that is of universal application in eco-climatology is colour which affects the heating qualities of all objects receiving solar-radiation in any form. Generally speaking the darker the object the more heat it absorbs, a black asphalt path being in this respect almost the opposite of a smooth snow-cover. In hot countries black objects exposed to the full glare of the sun often become 50° F. up to 70° F. hotter than those in the shade. Throughout low latitudes dark-coloured objects in the sun may reach temperatures of 150° F. up to 170° F. and even 180° F. in the very sunny regions athwart the Tropics. Black cloth exposed at Khartoum ($15\frac{1}{2}$° N.) acquired a heat of 184° F. (temperature in shade 118° F.); bare dry sand has risen in temperature to 183° F. at Loango (5° N.) and 175° F. in the northern part of the Lower Peninsula of Michigan. At Poona in May, black soil rose in temperature to 148° F., brown soil to 136° F. and soil covered artificially with a white powder to 118° F. In higher latitudes, outside the Arctic and Antarctic circles, the highest temperatures of dark, dry objects fully exposed to the sun are likely to be about 130° up to 145° F. C. E. P. Brooks has summarized the *relative* effect of colours roughly as follows:

Black	100
Dark blue, brown and green	85–90
Grey	75–85
Reds, and mixtures, and lighter shades of dark colours	70–75
Yellow	50–55
Whites	40–50

It happens that colour and composition usually have to be considered together. The water content of a surface is of great importance as moisture uses up heat in evaporation processes; likewise, the compactness and opacity of a surface largely decides the depth to which solar radiation will penetrate. The influence of these factors are dealt with later. Suffice it to say here that moisture greatly reduces the albedo, or reflective power, of a surface. A black soil reflects about 14 per cent. of the incident solar radiation when dry and about 8 per cent. when wet; corresponding statistics for yellow sand are about 20 per cent. and 10 per cent. respectively.

Influence of Surface Cover

Bare Soils. Differences in the nature of soils are sufficient and sufficiently sustained to cause appreciable differences in their climates. The main contrasts are between dry, sandy soils and damp clay soils, the former having relatively great extremes of temperature and the latter relatively small. The following screen temperatures for clear weather on 8-11 May 1941 at Lynford on the sands of Breckland and at Cambridge demonstrate this.

	Max.	Min.	Range
Lynford	54° F.	17	37
Cambridge	52	29	23

As a rule, the diurnal range of temperature over bare ground increases with the air content of the soil and is greatest over dry sands, and decreases through light and heavy loams to damp clay. Figure 136

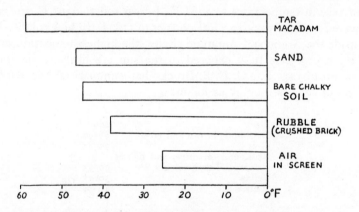

136. Mean monthly range of temperature just below surface and in screen on Salisbury Plain in June, 1925

(After Johnson and Davies)

shows the mean monthly range of soil temperatures just below certain surfaces on Salisbury Plain in June 1925. The *mean* monthly maximum temperatures recorded here in June were: tarmac, 109° F.; bare chalky soil, 96°; sand, 95°; rubble, 88°; and the atmosphere in the screen, 71° F. The main significance of these and other similar findings is that improvement in the moisture and humus content of poor dry soils tends to raise the minima temperatures just above them and so to diminish the risk of night frost in the spring growing season.

Vegetation: Grass. The minima temperatures of air resting on grass surfaces are nearly always appreciably lower than those of the air at 4 feet in the Stevenson screen (Fig. 137). When the night is clear and calm the differences may amount to 8° or 10° F. but they seldom exceed a few degrees on windy, cloudy nights.

It is common also for minima air temperatures over grass to be 1 or 2 degrees lower than those over adjacent cultivated ground and in calm clear weather this difference may increase to 5° F. or even 6·5° F. The lowering of temperature arises largely from the evaporation of moisture from

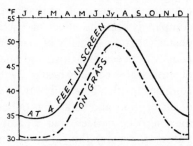

137. Mean minima temperatures at Oxford for 55 years

the blades of grass and the check on air-movement imposed by the vegetation. It will be appreciated that, as air over bare soil is warmer than that over grass at night, tillage tends to lessen the risk of damage by frost to visible plant-growth in the growing season. Confusion sometimes arises in explanations of the effect of grass on air temperatures as a grass sod insulates the soil *beneath* and minimum temperatures *beneath* it are often higher than those *beneath* a bare soil nearby but the maximum temperatures are lower, especially in the warmer months.

The effect of grass surfaces on air temperatures depends partly on the height of the grass which decides whether the air rests *upon* or largely *within* the grass cover. Grass minima temperatures relate to air just above a surface covered by grass only 1 or 2 inches high. The general relative effect of various grass-crops and of other surfaces is well illustrated by the following minimum air temperatures recorded by Cornford at 3 feet above flat ground on the morning of 30 May 1937, at East Peckham in Kent.

Bare soil	49·5° F
In wood	49·0
Among raspberry canes	48·2
Meadow	45·7
Grass and clover	43·5
Tall mowing grass	43·0

Other Crops. The effect of most other crops resembles that of grass in so far as they check air movement, tend to increase the surface area to be heated and to create surfaces with different rates of absorption of incoming radiation. The check on air movement is often remarkable. Bigger surface obstacles diminish wind-speeds quite effectively and at a height of 5 feet the wind-velocity is usually only about 75 per cent. of that at 33 feet. But this diminution is small compared with the effect of frictional resistance of smaller surface obstacles, and especially of vegetation. The wind speed at 2 inches is usually between one-third and one-half of that at 5 feet. The increase of surface area actually heated by direct radiation in a vegetation cover compared with a bare soil may be small nor need the variable powers of absorption and reflection (albedo) of plants have much effect. It must also be noticed that a plant-covered soil will, for all practical purposes, receive the same amount of incoming radiation and lose the same amount of outgoing radiation as will the same area of bare soil. Yet plant-cover, once it is close and continuous, has a distinct effect on the diurnal fluctuations of temperature. Vegetation as a rule tends to create relatively humid and equable conditions compared with those over bare earths. This is because most plant communities and many crops grow sufficiently high to form an upper surface and a soil surface with a vertical air-space in between. The upper surface absorbs the bulk of the insolation and shelters beneath it an air-space that near the ground is usually more humid, and a few degrees cooler by day and several degrees warmer by night than the atmosphere above the plant-cover and above adjacent bare soils.

Within a ripening wheat-crop in Perthshire during August the air temperatures at ground-level were usually 2° or 3° F. warmer by night and 2° to 4° F. cooler by day than the air at the top of the crop. At the same time the air at the top of the wheat experienced temperatures slightly cooler by night and warmer by day than that at the same height (3 ft. 6 ins.) over bare soil. But the importance and complexity of the air-space between the ground and the wheat-ears may be grasped from the following table (for a sunny day in August) which shows that half-way up the wheat the maxima temperatures rose occasionally 5° F. above those at the ear-level and 8° F. above those in the stagnant air nearest the ground.

8th Aug. 1947	At 42 ins. over bare soil in open	At 42 ins. amid ears of wheat	At 18 ins. within wheat	At ground-level in wheat
Min.°F.	43·5	42·5	42·5	47·5
Max.	70·5	72·0	77·0	69·0
Range	27·0	29·5	34·5	21·5

Woods and Forests. The problem of the absorption and reflection of incoming radiation becomes truly complicated in woods and forests where the tree-growth allows some sunlight to reach the ground direct. There are then two main surfaces absorbing and radiating heat, the crowns—or primary radiator—and the ground which may itself be covered with a variety of plants. It appears that the diurnal range of temperature is greatest at crown level and that minima temperatures are least severe half-way between the ground and the crowns.

Equally important from a geographical point of view is the relative effect of woods on wind-speeds and on mean surface temperatures. Forests greatly check wind-movement and within a few hundred yards of a forest edge wind-velocities decrease by one-sixth or more (Fig. 138). In dense evergreen coniferous forest this check causes relative calm to prevail in the inner parts of a woodland.

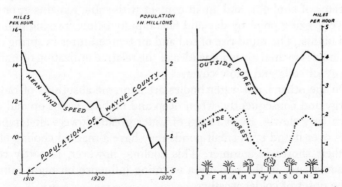

138. Influence of built-up areas and forests on wind-speed

Growth of Detroit or Wayne County is represented by increase of population. Notice effect of forests when in full leaf, (*After Zon*)

Y

The amount of light which penetrates to the ground in a forest is reduced by the foliage by 75–95 per cent. in thick fir forest and by 60–80 per cent. in a pine forest. When in leaf ash woods have been known to reduce the insolation at ground-level by 70 per cent. and oak woods by 80 per cent. As a result woods have the effect of lowering the temperature of the air beneath them by day. It happens that this daytime cooling is much more effective than the raising of the night minima temperatures by the shelter afforded by the tree growth against out-going radiation. In Sweden maxima temperatures in July and August averaged 6° F. lower near the ground in a pine-wood than over a nearby grass field while minima temperatures were only 1° F. higher. In Germany the daily temperature range proved to be 4° F. less in a forest than in the open, the forest maxima being 3° F. lower and the mean minima 1° F. higher. Much, of course, depends on the nature of the tree-growth but on an average forests compared with open ground reduce the daily mean temperatures near the ground by about 4° to 9° F. in the warm season and raise them by 1° or 2° F. in the cold season. Forest soils are much less liable to frost than non-wooded soils in countries with cold winters.

In addition to ameliorating temperatures of the air-space within them, forests also increase its relative humidity by as much as 3 to 10 per cent. Extensive forests *may* also slightly increase rainfall but this is not certain except perhaps in forest clearings which seem to experience somewhat more rainfall than open areas outside the forest edge.

Water Bodies and Snow. Smooth fresh snow is the supreme reflector of sunlight, and air in contact with snow remains at or just above freezing point by day and falls rapidly below freezing point on cold nights. The rapid rise of soil and air temperatures in spring once the snow is melted gives some idea of the relative utilization of radiant energy of soils and snow-cover.

On the other hand water bodies are supreme absorbers of radiant energy and consequently, when deep and mobile, warm up and cool down very slowly. Evaporation of water greatly lowers air-temperatures over it and even small ponds may have a marked cooling effect on their shores in summer. This cooling, however, is partly compensated later in the year by a warming influence on night minima at the onset of colder weather.

The fact that water-bodies occupy topographical concavities is important as owing to their warmth by night and in the cold season—

as long as they are unfrozen—they convert what might otherwise be frost-hollows into 'heat-islands.'

The main climatic influence of water bodies is seen in the qualities of a marine coastal strip which enjoys to the full land- and sea-breezes. Yet many lakes create powerful land- and lake-breezes while the Caspian Sea actually causes a 'monsoon' type of seasonal wind-reversal—outflowing in summer and inflowing during winter.

The climatic effect of prevailing winds crossing lakes is well seen in the fruit belt on the eastern shore of Lake Michigan which does not freeze completely over (p. 48); and in the heavy rainfall of the north-western shores of Lake Victoria. Even the Sea of Galilee (66 sq. miles in area) gives rise to a cool, humid, refreshing lake-wind in the morning and forenoon of summer. The Dead Sea, which is ten times bigger, consists of a large northern part up to 1,312 feet deep and a small shallow part (up to 26 feet deep) joined to each other by a narrow strait. The deep northern basin heats slowly by day and creates a cool, vigorous lake-breeze which in the early afternoon suddenly lowers the temperature of the northern shores by several degrees. On the arid lands around the lake the temperatures of the surface soils often rise to over 130° F. and the relative coolness of the pleasant lake-breeze comes as a great boon which may be felt up to about 15 miles from the northern, 9 miles from the western, and 6 miles from the southern shore. The small shallow southern basin of the lake heats much more rapidly by day and its temperature may rise to 109° F. in July. Consequently the temperature differences between the water and its shores are not great and here the lake-breeze is weak and does not cause a distinct drop in the furnace heat of the early afternoon. In the dry heat of the Jordan valley any slight increase in humidity is welcome but it will be obvious that in less arid countries a small shallow lake will not always be beneficial climatically during a hot summer as it may become very warm and active evaporation from it may make the lowest layers of the air very humid.

Built-up Areas. Man's alteration of the earth's land-surface reaches its climax where he covers large areas with a continuous coating of houses, pavements and roads. At least outside the Tropics sloping roofs may receive greater insolation than nearby horizontal surfaces and the architectural custom of orientating buildings to face the sun increases this possibility. But the exceptional heat-acquiring qualities of built-up areas depend also on at least four other factors.

Rows and groupings of houses, or urban patterns, tend to hinder wind-movement. The great decrease in the horizontal air-movement within a city is well illustrated by conditions at Detroit where the average wind-speed has dropped almost as rapidly as the city has grown (Fig. 138). Buildings and streets often protect pockets of relatively still air which heat up quickly by day and form 'heat-islands' that may persist long into or even throughout the night.

Many cities are built mainly of darkish materials which, being good absorbers of radiant energy, rise to high temperatures in day-light.

These dark surfaces are exceptionally well-drained and are, from the habit of leading all water as quickly as possible underground, kept remarkably dry. As a result relative humidity is usually lower in cities than in adjacent open country and loss of heat by evaporation is reduced to a minimum.

To these conditions already very favourable to high tempera-tures, man adds his own quota in the form of dust, smoke, warmth and fumes from his mechanical contraptions and heating contriv-ances. In hot, sunny spells in summer the effect of man's activities is

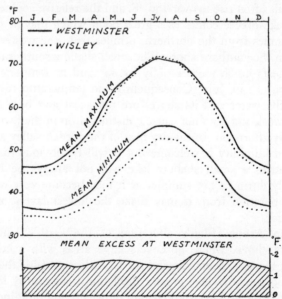

139. Mean monthly maximum and minimum temperatures of urban and rural (Wisley) areas

not sufficient to raise temperatures appreciably except after sunset and the dust of his commotion actually tends to lessen the sunlight hours. On windy days the warmth of cities is often soon dispersed into the general atmosphere and the dust and smoke fall to earth over wide distances. But on gentle and calm days in the cooler seasons, the heat and dust arising from human activities and household and factory chimneys appreciably increase temperatures both at night and in daylight (Fig. 139).

Generally speaking the dust, smoke and warmth of human bustle and combustion may in themselves be said to have three main climatic effects which are seen best in coal-burning cities in temperate latitudes during anticyclonic weather.

First; atmospheric pollution tends to keep out direct sunlight from ground surfaces. In cities, such as Vienna, which are not highly industrialized the loss of insolation compared with that on the adjacent countryside, is 16 per cent. in winter and 13 per cent. in summer. In central London the relative loss—compared with open country—is 56 per cent. in December (when household grates are in full use) and 6 per cent. in June. All told, this city centre is deprived of about 300 hours of sunshine every year, of which 100 hours are lost in the three winter months alone (Fig. 140).

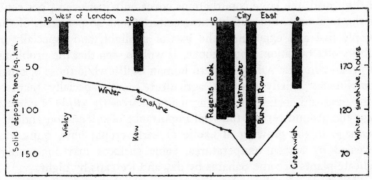

140. Section across London from west to east to show how atmospheric pollution affects winter sunshine.

(*C. E. P. Brooks and National Smoke Abatement Society*)

Second; although atmospheric pollution deprives a city of direct insolation by day, it may, on calm nights, form a lid or pall which blankets in the outgoing radiation. This pall, coupled with the sheltered position of the urban 'heat-island' within its ramparts of houses, keeps a city snug and warm after sunset. At Nashville, Tennessee, the inner city has had temperatures of 35° to 39° F. while the open country suffered killing frost. At Vienna on calm spring nights the city centre has been 14° F. warmer than the adjacent countryside.

Third; visibility in cities is lower than in rural areas. This is especially so in time of fog, the frequency and density of which are much increased by atmospheric pollution. The dense urban fog, popularly known as 'smog,' owes much of its opacity to the fact that dust and smoke particles are settling locally instead of being diffused over a wide area (Plate 32).

The general result of all these factors is that large cities are less windy, less sunny, slightly drier, warmer in winter and at night, and foggier and 'stuffier' all the year than the nearby countryside. Whether the open country ever *feels* hotter than a conurbation—as screen temperatures in summer indicate—is doubtful as in the city reflection on to the skin from hot arid surfaces is strong and trying.

Human Significance

As yet an insignificant part of the land-surface is occupied by built-up areas. Even in England, one of the densest populated parts of the world, the large conurbations cover only just over 4 per cent. of the total area. But within these conurbations 40 per cent. of the people live and experience the loss of sunlight, and especially of ultra-violet radiation, noted above. It will be seen that the problems of urban climates mainly concern human health while those of other eco-climates chiefly concern agriculture, and especially market-gardening and horticulture. In agricultural areas the study of climate near the ground reveals the vital importance of orientation and air-drainage in the growing of exotic or early crops; how, quite un-revealed by screen temperatures, some surfaces may for young, delicate plants become Saharas by day and ice-caps by night.

The study of urban eco-climates reveals that man, in spite of architectural planning and careful arrangements to obtain the most suitable views and weather, is his own worst enemy. Even in almost smokeless conurbations dust itself can be a serious atmospheric pollution. Unfortunately in cities, as in agricultural districts, it is the young and delicate who are affected most. The effects of impure air and loss of sunshine are, however, insidious and difficult to dis-tinguish. Consequently the urban climate has become more notorious for its 'smogs' during which, in the subsiding air of calm anticyclonic weather, smoke, fumes and obnoxious gases (mainly from factory chimneys) may accumulate in pockets of sufficient concentration to affect the throat and lungs of human beings. Very rarely the untimely death of many people in an industrial city of Europe or the United States draws attention to the dangers of this abnormal concentration of substances which in ordinary times diffuse more widely into the atmosphere.

The need for placing industrial establishments to leeward of cities and for curbing or prohibiting the emission of smoke and fumes is more than obvious. Nor is human health the only sufferer. Atmospheric pollution includes gases, such as sulphur dioxide, which, although in very weak solutions, corrode metals, paint and stonework. It is somewhat ironic that the Houses of Parliament in Westminster, built partly of magnesian limestone, are being turned slowly by the London atmosphere into Epsom salts.

LIST OF BOOKS

The standard work on eco-climatology (with a detailed bibliography) is Geiger, R. *Climate near the Ground*, 1950 (4th edition in German, 1961)
There is a great deal of eco-climatology from the human point of view in:—
Franklin, T. B. *Climates in Miniature*, 1955
Brooks, C.E.P. *Climate in Everyday Life*, 1950
Landsberg, H. *Physical Climatology*, 1958
Piéry, M. (ed.) *Le Climat de Lyon et de la Région Lyonnaise*, 1946
Smith, L. P. *Farming Weather*, 1958
Other useful books and references are as follows (*Q.J.R.M.S.* denotes *Quarterly Journal of the Royal Meteorological Society*)

General
Balchin, W. G. V. and Pye, N. 'A micro-climatological investigation of Bath . . . district,' *Q.J.R.M.S.*, 1947, pp. 297–323
Brunt, D. 'Some factors in micro-climatology,' *Q.J.R.M.S.*, 1945, pp. 1–10; 1946, pp. 185–88
Matthews, H. A. 'Reality in Climate,' *Geog.*, 1937, pp. 87–100

Shape, Aspect, Colour, etc.
See references to Heywood, G.S.P.; Cornford, C. E. and Garnett, A., on pp. 109–110
Hawke, E. L. 'Frost-hollows,' *Weather*, 1946, pp. 41–45
Scorer, R. S. 'Theory of waves in lee of mountains,' *Q.J.R.M.S.*, 1949, pp. 41–56; 'Mountain-gap winds,' *ibid.*, 1952, pp. 53–61
Manley, G. *Climate and the British Scene*, 1952
Caborn, J. M. 'The influence of shelter-belts on microclimate,' *Q.J.R.M.S.*, 1955, pp. 112–15; *Shelterbelts and Microclimate*, 1957
Barnes, F. A. 'Shelter and exposure in West Anglesey,' *Weather*, 1949, pp. 110–13; 183–9
Lee, D. H. K. 'Thoughts on housing for the humid Tropics,' *Geog. Rev.*, 1951, pp. 124–47
Atkinson, G. A. 'Architecture in the Trade Winds,' *Weather*, 1953, pp. 313–15

Soil and Vegetation Cover

Salisbury, E. J. 'Ecological aspects of meteorology,' *Q.J.R.M.S.*, 1939, pp. 337–57
Johnson, N. K. *A study of the Vertical Gradient of Temperature in the Atmosphere near the Ground, Met. Off. Geoph. Mem.*, 46, 1929
Johnson, N. K., and Davies, E. L. 'Some measurements of temperatures near the surface in various kinds of soils,' *Q.J.R.M.S.*, 1927, pp. 45–59
Ramdas, L. A., and Dravid, R. K. 'Soil temperatures . . . ,' *Proc. Indian Acad. Sc.*, 2, 1936, pp. 131–43
Ångström, A. 'The albedo of various surfaces of grounds,' *Geografiska Annaler*, 1925, pp. 323–42
Waterhouse, F. L. 'Microclimatological profiles in grass cover . . . ,' *Q.J.R.M.S.*, 1955, pp. 63–71
Paton, J. 'Temperatures and airflow within a wheatfield,' *Weather*, 1948, pp. 22–26
Zon, R. 'Climate and the Nation's Forests,' *Climate and Man*, 1941, pp. 477–98
Day, W. R. 'Local climate and the growth of trees . . . ,' *Q.J.R.M.S.*, 1939, pp. 195–209
Braun-Blanquet, J. *Plant Sociology*, 1932
Richards, P. W. *The Tropical Rain Forest* (Chap. 7 Microclimates), 1952
Bernard, E. *Le climat écologique de la Cuvette Centrale Congolaise*, 1945
UNESCO X. *Climatology*; XI. *Climatology and Microclimatology*, 1958

Water-bodies

Ashbel, D. 'Influence of the Dead Sea on the climate of its neighbourhood,' *Q.J.R.M.S.*, 1939, pp. 185–94
'Conditions of the winds on the . . . shores of the Sea of Galilee,' *Met. Mag.*, 1936, pp. 153–55

Built-up areas

Kraus, E. 'Climate made by man,' *Q.J.R.M.S.*, 1945, pp. 397–412
Parry, M. 'The Climates of towns,' *Weather*, 1950, pp. 351–56
Brooks, C. E. P. *Atmospheric Pollution in Great Britain*, 1947 (a pamphlet of the National Smoke Abatement Society); 'Climate and the deterioration of materials,' *Q.J.R.M.S.*, 1946, pp. 87–97
Marsh, A. *Smoke, the problem of coal in the atmosphere*, 1947
Meetham, A. R. 'Natural removal of pollution from the atmosphere,' *Q.J.R.M.S.*, 1950, pp. 359–71; *Atmospheric Pollution*, 1956
(Beaver Report). *Air Pollution Committee Report*, H.M.S.O., 1954
Hewson, E. W. 'Atmospheric Pollution,' in *Compendium of Meteorology*, 1951, pp. 1139–57
Thring, M. W. (ed.). *Air Pollution*, 1957
Smokeless Air. Quart. Journ. of National Soc. for Clean Air
Pollution of the Atmosphere in the Detroit River Area (International Comm., Washington & Ottawa), 1960
The Air over Louisville, 1956–7; *The Louisville Air Pollution Study*, 1961 (Cincinnati)
Smoke and Air Pollution, New York: New Jersey, 1958 and Ann. Repts. of Interstate Sanitary Commission
Smoke Control (England and Wales) 1962–1966 H.M.S.O.
Chandler, T. J. 'London's urban climate', *G.J.*, 1962, pp. 279–302; *The Climate of London*, 1965

CHAPTER TWENTY-FOUR

SOILS: THEIR FORMATION AND DISTRIBUTION

SOILS attract numerous scientists who study them in many ways and for many reasons. The pedologist deals with the study of soil development and distribution; the farmer is interested primarily in the fertility and amelioration of soils; geologists are more concerned with the formation of soils and the existence of old soils (palaeosols) buried beneath new; geographers are particularly interested in the distribution of soils and their productivity, actual and potential. The modern tendency is for soil scientists to combine the study of natural soils with that of cultivated soils and so to compile a section of knowledge indispensable to agricultural and geographical studies.

Formerly the pedologist usually defined soil as a natural body of mineral and organic constituents, differentiated into horizons of variable depth, which differs from the underlying parent material in appearance, physical arrangement, chemical properties and biological characteristics. To-day most pedologists favour the definition of soil as 'the collection of natural bodies on the earth's surface, containing living matter, and supporting or capable of supporting plants'. This newer definition favours the inclusion in pedology of man-altered soils, as well as of thin surface collections of weathered material without well-developed soil-horizons and of deep superficial coverings of fine weathered material reached only by the longest tap roots.

The Nature of Soils

Mineral Content. The lithosphere, or Earth's crust, from which soils are derived through weathering and erosion, consists by weight of about 95 per cent. igneous rocks and 5 per cent. sedimentary (4 per cent. shales, 0·75 per cent. sandstone, and 0·25 per cent. limestone). The sedimentary layer, which is thin except under orogenic belts and absent over large areas, is derived directly or indirectly from igneous rocks and does not differ greatly from them in average chemical composition. When the average percentage composition by weight of all crustal rocks (down to about 10 miles depth) is computed, the twenty chief elements prove to be those listed in column A in the following table.

		A	B			A	B
O	Oxygen	46·60	70·00	P	Phosphorus	0·12	0·70
Si	Silicon	27·72	0·05	Mn	Manganese	0·10	trace
Al	Aluminium	8·13	0·02	F	Fluorine	0·07	—
Fe	Iron	5·00	0·02	S	Sulphur	0·05	0·20
Ca	Calcium	3·63	0·50	Sr	Strontium	0·04	trace
Na	Sodium	2·83	0·05	Ba	Barium	0·04	—
K	Potassium	2·59	0·20	C	Carbon	0·03	18·0
Mg	Magnesium	2·09	0·07	Cl	Chlorine	0·02	0·1
Ti	Titanium	0·44	—	Cr	Chromium	0·02	—
H	Hydrogen	0·14	8·00	N	Nitrogen	0·02	0·50

There are at least another 80 elements that exist in minute relative proportions in the Earth's crust and some of these, such as nickel, tin and copper, form important metal ores when concentrated locally while others, although remaining dispersed in very minute quantities, are important to agriculture as is the case with zirconium, molybdenum, boron and iodine.

In the soil the above elements exist almost entirely as complex oxygen compounds and particularly as silicates of aluminium, calcium, sodium, magnesium, potassium and iron. When viewed in a more physical sense, the mineral skeleton of the soil may be said to consist of unweathered minerals, insoluble secondary salts, secondary minerals (clay particles), and exchangeable ions, that is atoms or molecules with a positive (*cation*) or negative (*anion*) electric charge. This viewpoint is important because the *texture* of a soil is judged on the relative amounts of coarse and fine inorganic solid material that it contains. A widely-accepted international scale classifies mineral particles over 2 mm. (·04 inch) in diameter as gravel, 2 mm. to ·02 mm. as sand, from about ·02 mm. to ·002 mm. as silt, and below ·002 mm. (or 2 microns) as clay. A typical textural classification of soils is expressed in Figure 141. It is apparent that for a given volume of soil, the finer the particles the greater their total surface area and because important chemical and physical-chemical reactions in soils take place mainly at the surface of particles, the greater also their reactivity. As sands and silts are relatively large they are relatively inert chemically and their effects in the soil are mainly physical. On the other hand, the clay particles are chemically active and besides having important physical effects also have very significant chemical effects. Thus normally the clay fraction controls the chief chemical properties of a soil. But it must be noticed that clay is defined pedologically by *size of particle* and that in fact the clay-size fraction of a soil consists of clay minerals proper, and accessory 'non-clay minerals' such as chlorite, and especially quartz (SiO_2) which is very resistant to chemical attack. The clay minerals proper are for the

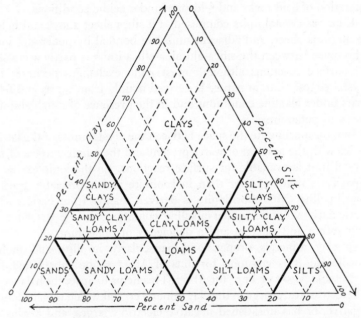

141. A simplified system of soil nomenclature according to the texture or size of the inorganic fraction

most part crystals in which the crystalline or molecular structure is determined by the spatial arrangement of oxygen atoms, which by weight form half the mineral content. The atoms of oxygen are tightly packed around atoms of silica and aluminium usually in the form of tetrahedra and octahedra respectively and these are arranged systematically and bonded together in very thin sheet-like lattices (Fig. 142). Consequently most clay minerals occur as platy particles almost of only two dimensions. Chemically all are hydrous silicates, chiefly of aluminium and magnesium. The layered clay minerals that concern us most are kaolinite, illite and montmorillonite, so named from Montmorillon near Vienne, France. These differ in their ionic exchange properties according to their inter-layer cations and their residual surface charges. As shown in Figure 142, kaolinite consists of a pair of silica-alumina sheets bonded together by hydrogen ions. The space between the pair of crystal units is fixed and is largely inaccessible for surface reactions, so that ionic exchanges occur only at the crystal edges. In nature kaolinite is derived mainly from the alteration of acid rocks and feldspars under acidic conditions.

Illite has crystal units composed of a silica sheet above and below an alumina sheet, and adjacent units are bonded by potassium ions. The space between the silica-alumina-silica units is partly accessible for surface reactions although most ionic exchanges occur at the crystal edges. Illite in nature is derived mainly from micas and feldspars under alkaline conditions and in the presence of much aluminium and potassium.

In montmorillonite the space between the crystal units of the lattice varies with the amount of water present and the whole surface of the crystal unit is available for surface reactions. So the ionic exchange capacity is much greater than in kaolinite and appreciably greater than in illite. Montmorillonite is derived mainly from basic rocks, particularly volcanic, under alkaline conditions in the presence of magnesium and calcium and a deficiency of potassium.

In clay minerals proper, and particularly in illite and montmorillonite, it is common for ions of Al^{+++} to be substituted for Si^{++++} and Mg^{++} or Fe^{++} for Al^{+++}. This helps to upset the balance of the electrostatic charge in the clay mineral crystal which then becomes negative, or has unsatisfied negative electric charges, and begins to attract and hold positive charged ions (cations) such as H^+, Ca^{++}, Mg^{++}, K^+, etc. The attracted ions are held in a state of dynamic equilibrium with similar ions in the soil solution and can be replaced (or exchanged) from the soil particle in response to changes in the chemical nature, or concentration, of the soil solution. This ionic exchange, including the reaction with water molecules, provides the main key to soil fertility and to soil properties such as plasticity and cohesion.

Organic Content. Most soils develop in association with the growth and decay of plants and animals which form humic compounds. This decomposition of organic matter, which is partly due to the action of bacteria and other micro-organisms, reduces the decaying tissues to small fragments, then to colloids (minute crystals

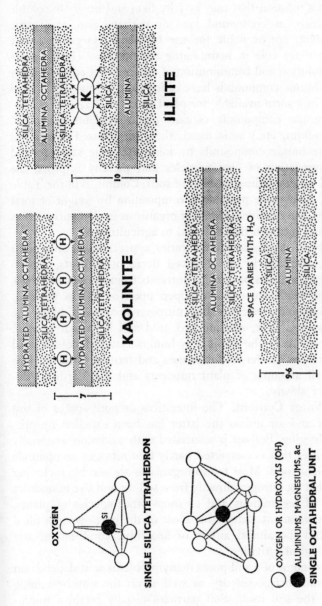

KAOLINITE

HYDRATED ALUMINA OCTAHEDRA

SILICA TETRAHEDRA

HYDRATED ALUMINA OCTAHEDRA

SILICA TETRAHEDRA

⊢— 7 —⊣

Ⓗ Ⓗ Ⓗ Ⓗ

ILLITE

SILICA TETRAHEDRA

ALUMINA OCTAHEDRA

SILICA TETRAHEDRA

SILICA

ALUMINA

SILICA

⊢— 10 —⊣

MONTMORILLONITE

SILICA TETRAHEDRA

ALUMINA OCTAHEDRA

SILICA TETRAHEDRA

SPACE VARIES WITH H_2O

SILICA

ALUMINA

SILICA

⊢— 9·6 —⊣

OXYGEN

Si

SINGLE SILICA TETRAHEDRON

○ OXYGEN OR HYDROXYLS (OH)

● ALUMINIUMS, MAGNESIUMS, &c

SINGLE OCTAHEDRAL UNIT

142. Diagrammatic expressions of the physical-chemical structure of clay minerals

(After Grim and M. B. Russell)

In the molecular diagrams the actual electric charges or bonds are shown by thick lines and the imaginary shape of the arrangement of the ions by dotted lines. In the layer-lattice structures the bases of tetrahedra or octahedra are always in the same plane and so arranged as thin plates. Numbers denote scale in Angstroms or thousand-millionths of a metre suspended in a solution that may be jelly-like) and finally to soluble molecules which, as compound ions or perhaps also as organic molecules direct, are available for use by plants. Organic matter plays an important role in maintaining soil fertility. Humic acids assist in the solution and comminution of mineral elements. In other words, some humic compounds have cations and anions which hold mineral ions in a form available for plant-use.

If these humic compounds contain plant nutrients (nitrogen, potassium, sodium, etc.) as is usual, they are released slowly. In addition some humic compounds by ionic exchange stabilize and cause the aggregation of clay particles and thereby improve the structure and water-holding capacity of soils. Column B in the Table above shows, by average percentage composition by weight of total organic matter, the relative richness of organisms in elements such as phosphorus and nitrogen that are vital to agriculture.

The quality of vegetable humus varies appreciably. Some plant communities, particularly broad-leaved deciduous forests, absorb relatively large quantities of mineral nutrients from the soil and these on leaf-fall or death are either dropped upon or left in the soil. Where its potassium, calcium and nitrogen content is high, the vegetable mould is dark, alkaline and mild and is known as *mull*. The other extreme is the acidic raw humus (*mor*) of cold humid climates where the coniferous forest uses and returns to the surface relatively small amounts of plant nutrients and the leaf-litter decomposes very slowly.

Air and Water Content. The interstices or pore spaces of soil contain water and air unless the latter has been expelled by prolonged waterlogging. Soil-air is associated with oxidation, especially of organic matter that is converted partly into nitrogen compounds valuable as plant food. Most micro-organisms are aerobic including the bacteria that fix nitrogen directly from soil-air and live in nodules on the roots of leguminous and of some other species of plants. Beneath the permanent water-table most of the pore-space is filled with water and the almost airless or anaerobic condition greatly decreases biological activities.

The water content of a soil poses many problems as it depends on climate and sub-surface seepage as well as on the water-retaining properties of the soil itself. Soil particles usually retain a microscopic film of moisture that adheres tightly to them through a variety of forces (ion adsorption, hygroscopic absorption, etc.). This very

thin film is not easily removed even by evaporation and is not available
for plant-use.

Normally, except in top soils exposed to hot dry weather, a thicker
outer film of moisture also adheres to the soil particle and its microscopic moisture film. This outer film is held less tightly (mainly by
surface tension) and the bulk of it can be used by plants.

As is shown in Plate 33, the two associated films usually occupy
only the peripheries of the pore spaces leaving the remainder of the
pore space filled with air. But in time of rain, snow-melt and flood
the accumulation of water on the surface of the ground soon acquires
sufficient weight or pressure to overcome the surface tension of the
moisture-films around the soil particles. Then gravitational or free
water percolates down the pore spaces towards or to the water-table.
An idealized arrangement of these three kinds of soil moisture is
shown in Figure 143.

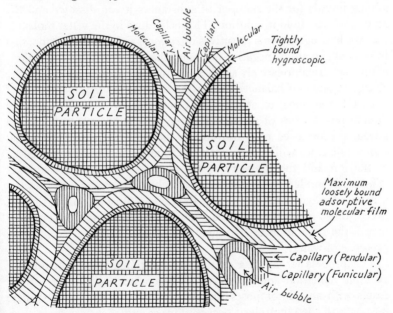

143. Idealized forms of water in the soil at some distance from the
water-table

(*After Plyusnin*)

In nature the hygroscopic and adsorptive molecular films form first and persist
longest. They pass imperceptibly into the capillary or free-water where and when it
exists. Near and at the water-table the pendular or angular capillary water
becomes funicular and gradually fills all the pore space

This over-simplified scheme of downward-moving water is often reversed in time of strong evaporation or excessive transpiration at and near the top of the soil. At such times the soil particles nearest the surface lose almost all their moisture film and may draw up towards them, by surface attraction in the capillaries or pore spaces, moisture from the damper soil below. This capillary action in nature only becomes really significant where the water-table is fairly shallow and the water-filled pore-spaces in contact with it are large enough to allow a rapid upward movement. In fact the effective capillary rise may be more significant in sands than in clays. Sometimes if the top-soil experiences marked heating by day, moisture may rise from the subsoil in the form of vapour also.

The great importance of the presence and movement of water in a soil cannot be appreciated fully without further reference to the processes of ion exchange and adsorption. Water is unequalled among liquids for the number of substances it can dissolve and the amounts it can hold in solution. This is because in the water molecule the two hydrogen atoms (positives or *cations*) are so deeply imbedded in the oxygen atom (negative or *anion*) that the resulting molecule is almost spherical, relatively small and strongly cohesive as the electric charge at one end balances that at the other (Fig. 144a).

In the soil, ions, or atoms with electric charges, have as it were double possibilities of attracting water dipoles because cations will attract the negative pole and anions the positive pole. How many water molecules will become attached to any ion depends on the size of the ion and the intensity of the electric charge on its surface. When surrounded by a number of attached water molecules an ion is said to be hydrated (Fig. 144b). In reality almost all ions in the soil except those in silicate minerals are hydrated and hydration virtually means the separation of hydrated ions from silicate surfaces in the presence of water.

The process is one of a number of reactions, which include for example oxidation and hydrolysis. When water (as $H+$ or $OH-$) reacts with and decomposes the complex silicates that predominate in soils, so-called hydrolysis occurs. This reaction, as with hydration, depends largely on the size of the ion and the intensity of the electric charge on its surfaces. Three-dimensional crystals like feldspar have broken chemical bonds around their edges, where positively charged sodium, potassium, calcium and barium occur in the interstices of the negatively-charged framework. These crystals are unstable and vulnerable to hydrolysis, particularly with decreasing size of particle. Most of them, except possibly quartz, are unstable in sizes smaller than silt. On the other hand, minerals with crystals with a very thin two-layer lattice structure, like most clays and micas, have broken chemical bonds on their thin edges only, and therefore chemically are stable especially in small sizes.

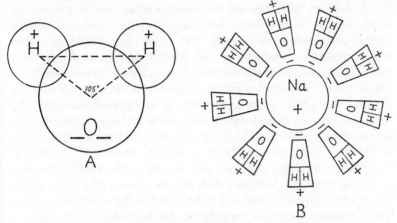

144. Water molecule (A) showing its dipole nature; and (B) hydrated sodium ion

As already noticed, once the clay-size has been reached the particles have virtually acquired their greatest potential for ion exchange and adsorption. The minute crystals are colloids intermediate in physical properties between true solutions and solid particles suspended in solution. When resembling liquids and moving freely, colloidal solutions are called *sols*; when like gelatine and showing signs of slight rigidity they form *gels*. Their reactions vary with their electric properties and their chemical environment.

Clay colloids tend to gel in soil solutions (sometimes called basic and neutral), in which their edges have positive charges and in which reaction with cations, other than Na, can take place. Thus hydroxides of Al, Fe, and Ti tend to have positive charges (or to be basic and alkaline solutions) and these tend to stabilize the aggregation of ions. In contrast, clay colloids can be dispersed or become sols in acidic solutions and if their positive charges are blocked by chemical re-agents.

z

For the sake of simplicity it is usual to regard soil colloids as a clay-humus complex which is negatively charged and which reacts with adjacent cations (K+; Ca++; Mg++; etc.) so as to aggregate or stabilize the molecule. But in reality although the humus charge is usually either positive or neutral, occasionally it is negative and unfavourable to aggregation. Normally the cations are adsorbed or attached mainly to the surface of the adsorbent atom and are fairly tightly held so that they are not easily washed out by percolating water. But, as farmers know well, salts (cations linked with mobile anions) are washed out of soils. Yet cations not linked to mobile anions cannot be washed out and, in fact, soil losses by leaching become proportional to the fraction of total cations not balanced by or linked with immobile clay anions. Losses from the soil will increase if anions are added either as fertilizer or as carbonic acid (H_2CO_3) from microbes or in soil with a low cation exchange capacity. The action of carbonic acid in leaching nutrient ions from a soil is shown in Fig. 145. For a given addition of H_2CO_3, the extent to which H ions displace exchangeable ions and thereby render them vulnerable to loss by leaching, increases with the abundance of a given cation and the intensity with which it is held on the clay molecule surface. This intensity increases roughly in the following order: sodium (Na+); potassium (K+); magnesium (Mg++); calcium (Ca++). It will be obvious that ions which are vulnerable to leaching are normally also those most readily available for plant-use.

Inevitably in wet climates the basic cations are gradually replaced by H ions especially of carbonic and humic acids and the soil solution becomes increasingly acid unless rich in natural supplies of bases or supplied them artificially by liming. Consequently the H ion concentration in a soil solution forms a measure of the extent to which basic cations have already been displaced. This is measured in the laboratory from the proportion by weight of free H ions in the soil solution and is conveniently expressed inversely on a so-called *pH scale* by the negative logarithm of the concentration. Thus a concentration of $\frac{1}{10,000}$ gram per litre or of 10^{-4} is pH4; a concentration of $\frac{1}{100,000}$ gram per litre or of 10^{-5} is pH5. In simple language, acidity becomes ten times less intense for each rise of one unit on the pH scale.

A soil in which the H ion concentration is similar to that in pure water at 20° C. is neutral (pH7) and soils with negligible exchangeable hydrogen ions commonly show this value. Solutions with a concentration greater than that of pure water at the same temperature are said to be acid and those with lesser concentration are termed alkaline. Extremely acid solutions (under pH4) usually contain strong acids either organic or sulphuric. Soils of pH4 to 5·5 have a substantial content of exchangeable H or of ions such as Al capable

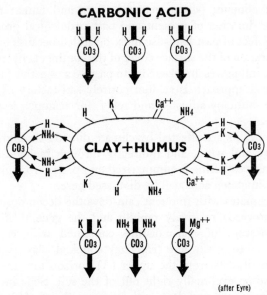

(after Eyre)

145. Diagrammatic representation of leaching
action of carbonic acid in soils

of neutralizing OH. At pH6·5 to 7·5 nearly all the negative charges have been balanced by basic cations. At pH7·5 to 8·5 calcium carbonate is present and at still greater alkalinity probably soluble carbonates, other than $CaCO_3$ are present.

For farmers the pH value affords a useful measure of the rate of turnover of nutrients held in organic combination and of the availability of inorganic ions. Most cultivated crops seem to prefer slight acidity (pH6 to 6·5). For pedologists and especially for geologists interested in weathering and sedimentation, acidity is equally important as it largely controls the mobility of colloids and the solution of and precipitation of hydroxides. In acidic solutions (pH4 and less) alumina and silica, other than quartz, dissolve readily whereas in neutral and alkaline solutions of pH5·5 to 8·5, alumina scarcely dissolves at all and silica, including quartz, quite appreciably. The implications of changes of pH value and of changes of temperature will be referred to again later.

Factors in Soil Formation

Soil results from the interactions of five main local or regional influences: climate; parent material; flora and fauna; relief; and the relative duration of present and past pedological processes. To these, the effect of past land-use must be added because of the widespread alteration of the surface cover of the continents by man. Sometimes these influences are classed into passive and active factors but a purely causal approach is seldom entirely satisfactory as the interactions are mutually adjusted and relief, for example, is both cause and effect.

Climate. Soils are affected by climate directly through moisture, temperature and wind, and indirectly through biological agencies. However, as weathering is summarized on pp. 215–16, only the main effects of climate on soils will be discussed here.

In wet climates with frequent rain-days the decomposition of soil particles proceeds relatively rapidly and the general effect of the frequent descent of soil-moisture combined with complex ion exchanges is to wash out (*eluviate*) colloidal clay particles and soluble constituents from the upper (A) horizon into the lower (B) horizon and so eventually right out of the soil. Soils in which the lower horizons are enriched with compounds of *Al*uminium and iron (Fe) are called *pedalfers*. The development of pedalfer soil-profiles differs appreciably with climate. In continuously wet climates, cold, cool or warm, acidic organic compounds combine mainly with Al and Fe compounds which are washed downward and accumulate in a specific lower horizon leaving a high proportion of whitish-grey quartz silica in the overlying *podzol* layer. In warm humid climates with a marked dry season, because of the intermittent drier conditions, the whole soil profile is rich in local enrichments (spots, mottles, concretions, etc.) of Al (especially kaolinite) and of reddish iron oxides. In other words, tropical ferruginous and ferralitic (lateritic) soils differ from tropical and extra-tropical podzols in the fate of the mobilized Al and Fe rather than in the nature of their upper horizon which usually is rich in very insoluble quartz.

In drier climates where for long periods evaporation and transpiration predominate over rainfall, an accumulation of calcium carbonate or of some other alkali occurs somewhere in the soil, which is termed a *pedocal*. In very arid climates, and elsewhere if the water-table is within 6 feet or less of the surface and so encourages appreciable upward movement of capillary soil-moisture, salts may accumulate on or near the surface. In semi-arid tracts the accumulation, particularly of lime, may be at a depth of a few feet, this being the point most commonly reached by descending soil-moisture. In areas with a relatively long period of moisture surplus, as in the humid eastern border of mid-latitude steppes, the calcium carbonate enrichment may form scattered nodules only. It is interesting to notice that in most pedocals the highly soluble salts are found at greater depth than moderately soluble salts which could only be so if the upward capillary movement of soil moisture was negligible.

Air and ground temperatures, which largely control evaporation and transpiration, also, as noticed above, greatly influence chemical and organic processes in the soil. In humid climates above freezing point the average rate of organic decay and of solution of mineral particles, particularly silica, doubles with a 10° C. (18° F.) rise in temperature. In tropical rain-forest the natural annual production of vegetable matter may reach 80 tons or more per acre but humus rarely accumulates where the mean temperature exceeds 25° C. (77° F.), unless the litter is too poor in mineral nutrients to support microbial activity.

Wind is also a significant factor in soil-formation as it increases evaporation and may remove fine, dry, loose surface particles. Exposed dried-out soils are liable to wind-erosion which may truncate some soil-profiles. The redistribution of salts from saline surfaces such as playas may become important in and near deserts. The *loess* which covers large areas in Eurasia and the Americas is considered by many authors to consist of fine, wind-borne dust, particularly because it occurs at various altitudes on the down-wind side of deserts, glacial outwash plains, and wide alluvial-floored valleys. The loamy loess of western Europe (*limon* in France) and of North America may have been slightly altered by moisture or weathering during deposition but, like the loess in China, it has escaped the extremes of podzolization and is usually homogeneous, friable, rich in lime and of an excellent physical structure.

Influence of Parent Material. The influence of the material from which soils develop, whether residual rock or relatively newly-transported drift, usually weakens with time, and in mature deep soils, especially in the tropics, there may be little similarity between soil and bed-rock. Yet in a more restricted time-sense geology is important as it influences the proportion of chemical elements available and the rate of soil-formation.

We have already noticed (p. 361) that the average composition of the sedimentary part of the lithosphere differs little from that of the igneous part. This fact is re-affirmed in the following table of the average chemical composition of rocks in which minor elements are omitted and water which forms on an average about 5 per cent. of shales, 1·6 per cent. of sandstones, 0·33 per cent. of limestones and 1·15 per cent. of igneous rocks. Column A refers to the crystalline surface of the world's continental shields (about 40 million square miles); Column B to the folded, sedimentary areas (about 16 million square miles); Columns C, D, E refer to shales, sandstones and limestones respectively.

	A	B	C	D	E
SiO_2	66·4	51·9	58·1	78·3	5·2
TiO_2	·6	·5	·7	·3	·1
Al_2O_3	15·5	11·4	15·4	4·8	·8
Fe_2O_3	1·8	2·6	4·0	1·1	·5
FeO	2·8	2·0	2·5	·3	—
MgO	2·0	3·8	2·4	1·2	7·9
CaO	3·8	12·6	3·1	5·5	42·6
Na_2O	3·5	1·3	1·3	·5	·1
K_2O	3·3	2·4	3·2	1·3	·3
P_2O_5	·2	·1	·2	·1	—
CO_2	—	11·4	2·6	5·0	41·5

The remarkable similarity in the chemical composition of igneous rocks and shales is even more striking when it is realized that shales form over 75 per cent. of sedimentary rock. It seems obvious that if major differences in soils due to geological influences do exist they will occur mainly on the relatively small tracts of sandstone and limestone. Both are usually very porous but limestones are of especial interest as some are so pure and so alkaline that their insignificant clay residue develops into a soil extremely slowly.

On soft, loamy limestones where weathering frees lime almost as fast as leaching removes it, the soil forms more quickly but remains highly calcareous and dark-coloured. These *rendzinas*, which seem to favour tall grass rather than thick forest, occur on some chalk uplands in Europe. The patches of dark grey to reddish clayey soils (*terra rossa*) that have accumulated in depressions, especially in the limestone karsts and causses of southern Europe, have a more complicated origin. They probably developed under a forest cover on various rocks as well as limestone and were originally *brown forest*

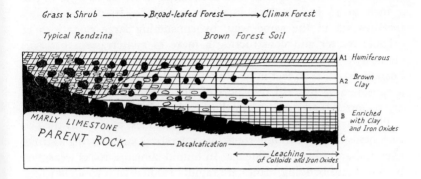

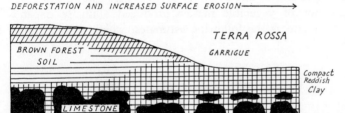

146. The development of rendzinas into brown forest soil and of deforested brown forest soils into terra rossa

(*After Duchaufour*)

Oblique shading denotes humiferous horizon; horizontal shading denotes clays, and vertical shading iron oxides. White particles in and near rendzina are lime mottles. Vertical arrows show main direction of leaching process

soils. Deforestation led to the removal of brown forest soil except in hollows, which happen to abound in limestone tracts, and to the degradation of the surviving pockets into terra rossa which now acts as a parent material for the development of a new soil (Fig. 146).

Influence of Flora and Fauna. Soil biota or biological soil-formers are often grouped into macroflora (plants) and microflora (fungi and bacteria) but to these must be added a larger fauna ranging in size from earthworms to burrowing mammals. Soil biology has evoked a large literature on the amazing variety and abundance of the visible and invisible inhabitants of the soil. Among the more obvious participants are earthworms in cool climates and termites in warm climates. Darwin estimated that earthworms brought to the surface in Britain material equivalent to a layer of soil 1 inch to $1\frac{1}{2}$ inches thick in about ten years. In warm climates termites act in the same way, by consuming large quantities of organic matter and constructing a maze of subterranean passages connected with a similar labyrinth above ground. These termite heaps may reach 20 feet in height on well-drained sites and represent a considerable upward transport of fine particles which eventually are re-sorted, and partly washed away, by rain.

Plants usually attack the parent material and less commonly the bed rock beneath it by sending roots into interstices and exerting a strong chemical and physical influence. In humid climates, rocks weakened by joints and bedding planes are rapidly riddled with rootlets which gradually prise open the lines of weakness. Even massive rocks if relatively soft and soluble will be pierced deeply by roots which occupy passages largely dissolved out by weak acid solutions.

The amount of vegetable matter produced annually depends mainly on climate as the following estimates, which exclude roots, will show.

Type of Vegetation	Vegetable Matter Tons per acre
Alpine meadow	under $\frac{1}{2}$
Short grass prairie	under $\frac{3}{4}$
Tall grass prairie	$2\frac{1}{2}$
Tropical savanna	13
Tropical monsoon forest	22
Tropical rain forest	40 to 90

The plant cover is important pedologically for at least five main reasons.

1. Vegetable matter supplies food for the soil fauna whose action decomposes it into humus particles and colloids.

2. Humus, particulate and colloidal, tends to improve the drainage and aeration of fine soils and to increase the water holding qualities of coarse soils.

3. Humus colloids link up with clay colloids and at the same time humic acids derived from the oxidation of organic matter assist in

the breaking down of mineral particles and so further increase ion or base exchange.

4. Roots and especially root hairs (Plate 34) are relatively powerful adsorbers of certain important cations. For example with regard to potassium, the lattice structure of clay minerals often enables them to hold potassium ions strongly enough to prevent washing out by rain water but not strongly enough to prevent adsorption by root cells.

5. Mineral bases that roots raise from the lower soil horizons are eventually for the most part returned to the ground surface and so help to maintain soil-fertility. This nutrient cycle is largely in the shape of leaves and woody matter and partly as potassium, phosphorus and nitrogen excreted and dripped from the leaves. It is illustrated in Fig. 154 and referred to again on p. 391.

Influence of Time-Scale and of Past Land-Use. On a geological time-scale soils form rapidly. Sand-dunes at Southport, Lancashire, developed a crude soil-profile in about four centuries. Newly-exposed surfaces on Alpine moraines and on the floor of drained Lake Ragunda in northern Sweden showed signs of podzolization after 100 years and will probably have a podzolic profile in less than 1,500 years. However, in all these examples the parent material was already comminuted.

In the humid tropics soil formation proceeds rapidly and a small East Indian island covered by pumice during the great eruption of Krakatoa in 1883 had nearly 14 inches of so-called soil within fifty years. Some tropical soils are over 40 feet thick and locally overlie several hundred feet of fine weathered parent material that must be of considerable antiquity whereas upon large areas of the northern land-masses the present soils probably began to form after the last major ice-retreat between about 20,000 and 10,000 years ago. The thinness of soil on some of the resistant rocks of Laurentia and Scandinavia demonstrates that in some climates resistant strata break down very slowly and that on all slopes and convex landforms soil accumulation depends largely on the relative rates of weathering and of denudation or erosion.

The time-scale in soil-formation also often involves the activities of animals, particularly man. During the last 10,000 years or more man has altered the vegetation of vast areas by burning, tree-felling and cultivation. In addition, he surrounds himself with numerous herbivores, such as horses, goats, sheep, oxen and camels, which devour seedlings, saplings, twigs and bark. Everywhere cultivation and grazing, particularly if too intensive, leads to deforestation and often to the destruction of grass swards. The overall effect on surface vegetation is increased by the fact that in many rather dry areas most of the taller plant-growth probably developed on a Pleistocene climate that was slightly cooler and wetter than at present. For example the early post-glacial vegetation of the Mediterranean coastlands of North Africa included tall forest and shrub that had grown up when the annual evaporation was lower and the water-table higher than to-day. Within the last few thousand years man has felled the tree-growth, increased surface run-off, and in many areas further lowered the water-table by deep wells so accelerating the natural process of desiccation in a slightly warmer, drier climate.

In some regions, long cultivation has produced new soils, notably the deep plaggen soils of the Netherlands and the paddy soils of south-east Asia. Similarly, the clearing of forests and their replacement by heath in parts of western Europe created humus podzols with a new B horizon.

Influence of Relief. Universally, relief exerts a strong influence on the depth and the drainage of soils. As a rule soil-depths increase with flatness and decrease with declivity. On steep slopes, particularly in mountains, denudation is rapid and the bare rock is often exposed except at the foot of the slopes where a scree of debris accumulates usually at too fast a rate to allow a soil-profile to develop. These fans and screes of coarse rock-debris are called *lithosols*.

In hilly country that has developed a more or less continuous cover of fine materials, the soil, although fairly similar over wide areas, varies slightly with variations of slope, especially when the slope itself is due to variations in bed-rock. These local topographic–lithologic correlations are often depicted by means of cross-sections or soil catenas (Fig. 147).

Where the relief is flat or almost flat over wide areas the relative absence of erosion favours the retention of senile soils with thick soil-horizons (*planosols*) in humid climates and with a strong hardpan in sub-humid districts. Relief influence, however, is equally striking where concave landforms, like basins and hollows, encourage the accumulation of soil, rock-waste, organic matter and water. Seasonal or spasmodic floods may deposit layers of silt and cause at least temporary waterlogging. At such times, oxidation is prevented and various iron compounds gradually accumulate in the lower soil

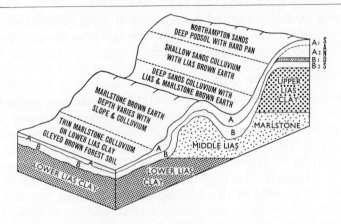

GEOLOGICAL CATENARY SEQUENCE (about 1 mile long; average slope 1 in 10)
COMBINED WITH BLOCK DIAGRAM AND
GEOLOGICAL CROSS-SECTION

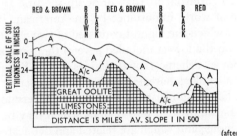

(after G.R. CLARKE)

**TRUE CATENARY SEQUENCE ON SAME ROCK
FORMATION ON COTSWOLD DIP SLOPE**
SOILS DIFFER THROUGH SLOPE, DRAINAGE AND ASPECT.
THE WHOLE SLOPE IS CULTIVATED.

147. Types of soil catenas

Notice the great exaggeration of the vertical scale, especially in the
lower diagram

horizons giving them a distinct bluish tinge. The process is known as
gleization and the zone of iron-eluviation as the glei or G horizon.

The influence of great local differences in relief or altitude on
climate and vegetation has already been discussed (pp. 98–110; 432–
436). On mountains the strong altitudinal differences in vegetation

leads to a marked altitudinal arrangement of zonal soils but often their distribution is so disrupted by lithosols that in small-scale maps they are usually grouped as mountain soils.

World Classifications of Soils

Probably the earliest notable classification of soils was made in China about 2,300 B.C. when nine grades were recognized according to productivity, the best being the soft yellow soils of Shensi and Kansu. The first elaborate modern classification was evolved after 1882 when V. V. Dokuchaiev, author of *The Russian Chernozem*, was officially appointed to map, analyse and classify the nation's soils and their crop-yields. With a few decades the Russians had established the concept that a *soil* consisted of superimposed horizons, unified by a common genesis in which climate and vegetation predominated. The three main categories suggested were *normal* (zonal); *developed* but not as normal (intrazonal); and *undeveloped* (azonal). These were subdivided into a total of 13 sub-classes (7 in normal, 3 in developed abnormally and 3 in undeveloped) and given names such as tundra, light-grey podzol and chernozem. Against such a background, C. F. Marbut in 1927 drew up the first truly comprehensive scheme of world importance by an American. It comprised six categories, of which the highest consisted of two classes (pedalfers and pedocals) and the lowest included several thousand soil types.

The latest Soviet scheme (1962) divides the U.S.S.R. into 'soil bioclimatic belts' (polar; boreal; sub-boreal; temperate-warm subtropical) which have a similarity of radiation and thermic conditions. Each belt may be subdivided into regions which may be either lowland or mountainous, and which are related and delineated by similar conditions of moistening and of temperature régime. The regions are subdivided into zones and these into provinces (and successively into areas and districts) according to the effects of variations in the zonal soil-forming process, including the indirect effect of landforms and the direct effect of land use. In such a complex classification, particularly if extended to the world, it is virtually impossible not to be either contradictory or in practice quite unmanageable.

The latest American classification (1960) attempts to provide a scientific and universal system of soil description and is based on criteria that can be measured and observed rather than on the genesis of soil horizons about which much has yet to be learnt. The smallest unit, a *pedon*, is a block of soil, up to about 10 square metres in surface area and adequate for the investigation of soil horizons and their inter-relationships. Contiguous similar pedons form a *polypedon* and contiguous similar polypedons a *soil series*. The soil series forms the lowest category of a major classification which in ascending order of magnitude is:

Soil Series: Families: Subgroups: Great Groups: Suborders: Orders.

The scheme recognizes 10 orders, about 40 suborders and 120 great groups and these are considered of universal application. But for the United States alone the subgroups number about 400, the families about 1,500 and the series at least 7,000.

The 10 orders are based on gross composition, number (or absence) and quality of soil horizons and the degree, actual and potential, of weathering of the mineral constituents. Where possible statistical limits are given to the quantities involved. Thus histosols (peats) contain 20 per cent. or more of organic matter.

Order	Approximate Descriptive Equivalent
ENTISOLS	Azonal; lithosols, regosols, alluvial.
VERTISOLS	Grumusols.
INCEPTISOLS	Brown acidic soils; some Brown forest and Humic gley soils.
ARIDISOLS	Desert; reddish desert; sierozem, solonchak, some reddish brown soils.
MOLLISOLS	Chestnut; Chernozem; Brunizem (Prairie: Brown Earth); Rendzinas; Some Brown forest and associated solonetz and Humic gley soils.
SPODOSOLS	Podzols; Brown podzolic; Ground-water podzols.
ALFISOLS	Grey-brown podzolic; grey-wooded; noncalcic brown; de-graded Chernozem and associated planosols and some Half-bog soils.
ULTISOLS	Red-yellow podzolic; reddish-brown lateritic and associated planosols and Half-bog soils.
OXISOLS	Lateritic (ferralitic). Latosols (ferruginous)
HISTOSOLS	Bog soils (peats; organic; etc.).

The orders are divided into Suborders mainly on differences in the chemical or physical properties that reflect the presence or absence of waterlogging and on genetic differences due to climate and vegetation. The Great Groups are defined largely on the presence (or absence) and arrangement of soil horizons. Each Great Group has a most distinct or central type surrounded by intergrades (i.e. subgroups) with a weaker or different combinations of properties. Thus a characteristic Spodosol (*Podzol*) has a bleached A2 horizon and a B horizon that has accumulated free sesquioxides and humus but not silicate clays, while a characteristic Alfisol (Grey-Brown podzolic soil) has a browner eluvial A2 horizon and a B horizon enriched with illuviated silicate clay. In the terminology used, the former has a spodic B horizon and the latter an argillic B horizon. At intergrades between these two Great Groups, the soils (Subgroups) often contain two illuvial B horizons separated by an eluvial horizon. The upper B horizon is like that of a podzol or Spodosol and the bottom like that of a Grey-Brown Podzolic or Alfisol. Within the Subgroups the Soil Families are differentiated primarily on properties important to plant growth, such as soil texture, thickness of horizons, supply of plant nutrients, etc.

The best that can be done in a short summary is to outline the Suborders and Great Groups for one of the Orders. In most instances the technical terms used are self-explanatory.

ORDER 6. SPODOSOLS

Soils with either a spodic horizon of illuvial accumulation of mainly humus (organic colloids) in conjunction with sesquioxides of iron and aluminium or both, or a thin iron-cemented horizon that overlies a fragipan and acts in the same way as a spodic horizon. The spodic horizon is far more common and is noted for its high cation exchange capacity.

SUBORDER 6.1 AQUODS

Spodosols saturated with water at some season and possessing one or more of the following characteristics associated with soil wetness: a histic epipedon (thin organic horizon with 30 per cent. or more organic matter); mottling in an albic horizon or in the top of the spodic horizon; and a duripan (layer cemented mainly by silica) in the albic horizon.

GREAT GROUPS

6.11 *Cryaquods*. Arctic and Alpine. Mean annual soil temperature below 47° F. and usually below 32° F. Often permafrost at 30 inches or less. Continuous spodic horizon.

6.12 *Normaquods*. Typical groundwater podzols. Spodic horizon has dispersed humus and aluminium but in no part with over 5 per cent. by volume of nodules cemented by iron.

6.13 *Sideraquods*. The upper part of the spodic horizon contains more free iron than do the overlying horizons; mottles are present but strong cementation in albic horizon and thin ironpan are both absent.

6.14 *Placaquods*. Aquods with a thin, wavy ironpan resting either on the spodic horizon or on a fragipan and causing saturation by water above it at some season. Usually the albic horizon is mottled.

6.15 *Thermaquods*. Aquods that occur in the warm tropics and subtropics where mean annual soil temperature exceeds about 60°F, and especially on quartz sands. Upper 3 inches or more of spodic horizon is enriched with dispersed humus or humus and kaolinite. The albic or leached horizon may exceed 7 feet.

6.16 *Duraquods*. Having a strongly cemented or indurated albic horizon and usually lacking a histic epipedon. The soil temperatures are warmer than those of Cryaquods.

6.17 *Fragiaquods*. Aquods that have a fragipan below the spodic horizon but lack a thin, wavy continuous ironpan above the fragipan. Also lacking a histic epipedon.

SUBORDER 6.2 HUMODS

Spodosols that lack the effects of waterlogging apparent in Aquods but have a spodic horizon enriched mainly with dispersed humus or humus and aluminium. The spodic horizon normally has in its upper ·4 inch (1 cm.) either ·7 per cent. or more free iron (Fe) and ·5 per cent. or more organic matter, or over 1 per cent. organic matter and ·35 to ·7 per cent. free iron. Humods occur commonly under heath in western Europe but their classification is still incomplete.

SUBORDER 6.3 ORTHODS

These are spodosols of the core or characteristic areas, popularly known as true Podzols and Brown Podzolics. Their spodic horizon is enriched with both humus and iron and contains, in its upper 1 cm., over 2·1 per cent. free iron (3 per cent. Fe_2O_3) and less than ·5 per cent. organic matter.

GREAT GROUPS

6.31 *Cryorthods*. With mean annual soil temperatures below 47° F. and typically below 32° F., with a cemented or indurated spodic horizon (*ortstein*) or 2 to 10 per cent. organic matter in the upper 4 inches of the spodic horizon.

6.32 *Placorthods*. Orthod with ironpan resting on spodic horizon.

6.33 *Normorthods*. Characteristic Orthods, without ironpan and with a continuous spodic horizon that is very firm when moist (*ortstein*) and contains 2 to 10 per cent. organic matter in its upper 4 inches.

6.34 *Fragiorthod*. Orthods with a fragipan below the spodic horizon and without a thin wavy ironpan above the spodic horizon.

SUBORDER 6.4 FERRODS

Spodosols in which iron has accumulated to a considerable concentration without a comparable accumulation of humus. The upper 1 cm. of the spodic horizon contains more than 2·1 per cent. free iron (3 per cent. Fe_2O_3) and less than ·5 per cent. organic matter. Silt-sized pellets of humus and iron do not occur nor mottles in any albic horizon that may overlie the spodic horizon. These iron-rich soils are common in the humid tropics and need much further classification.

This excellent soil classification is not yet complete and probably for decades will not replace the existing Russian–American schemes based mainly on soil processes and soil profiles and grouped under three Orders, zonal, intrazonal and azonal. Zonal soils are mature and have well-marked horizons developed on average parent material and average relief under the dominant control of regional climate and vegetation; intrazonal soils also have observable horizons but these have developed under the predominant influence of exceptional local factors; azonal soils consist of surface mantles that are immature or skeletal pedologically such as mountain screes, marine mudflats, newly exposed river alluvium, new morainic debris, active sand-dunes and recent volcanic deposits. These three Orders are usually subdivided successively into Suborders, Great Soil Groups, Soil Families, Soil Series and finally Soil Types (equivalent to pedons). The smaller categories are needed for the mapping and detailed study of small areas when pedology becomes an elaborate assemblage of local names and physical–chemical–agricultural soil properties. For these local studies the student should consult detailed soil memoirs and attempt also to construct soil catenas. For most global and continental purposes the greatly simplified classification given below will probably suffice provided it is always realized that the typical soil described in a Great Group is midway in nature between a fairly wide range of Soil Families that merge gradually into other Suborders.

SUBORDERS OF ZONAL SOILS	GREAT SOIL GROUPS
1. Soils of Cold Zone	Tundra Soils
2. Light-coloured Podzolized Soils of timbered regions with rainfall all the year	Podzol Grey-Brown Podzolic Red-Yellow Podzolic
3. Ferruginous (latosol) and Ferralitic (lateritic) Soils of warm humid forests and humid tropics with dry season	Ferruginous Soils Ferralitic (lateritic) Soils
4. Dark-coloured Soils of semi-arid, sub-humid and humid grasslands.	Brunizem (Prairie Soil: Brown Earth) Chernozem Chestnut
5. Light-coloured Soils of warm and hot arid regions	Sierozem Red Desertic
SUBORDERS OF INTRAZONAL SOILS	
1. Halomorphic (saline and alkali) of ill-drained arid regions and of littorals	Solonchak Solonetz
2. Hydromorphic (marshes, swamps, etc.)	Bog Meadow (Ground-water Podzol Ground-water Laterite)
3. Calcimorphic	Rendzina Tropical Black Clays
AZONAL SOILS	Lithosols (mountain scree, etc.) Regosols (sand-dunes, etc.) Alluvial.

Tundra Soils. The true zonal Arctic soil occurs only in better-drained tracts amid large areas that are waterlogged and so intra-zonal. But in this climate where permafrost prevents downward drainage, waterlogging becomes so common that intrazonal soils may almost be considered zonal. Indeed, the Russian term tundra denotes a marshy plain. Here, except on better drained hummocks and sands, soil temperatures remain low and permafrost (permanently frozen subsoil) seldom retreats more than a foot or two from the surface. During the brief growing season, the decay of humus and any bacteriological activities are hindered by waterlogging. As the soil is frozen for long periods the chemical breakdown of minerals is also slow and mechanically-broken angular fragments form most of the parent material. Over large areas the layer affected by thaw is barely deep enough to provide plants with sufficient mineral elements whereas mosses and lichen develop normally and sphagnum turf-peats may become a few feet thick. Soil horizons are thin and not very distinctive. Typically an inch or two of raw humus, largely lichen and moss, overlies a thin brown-grey clayey or muddy horizon (mainly mica and/or montmorillonite), which downwards rapidly becomes bluish due to gleization. The best soils occur on well-drained slopes and river flood-plains where the permafrost may thaw in summer down to 6 or 7 feet and the humus horizon may be nearly a foot thick. The effect of freeze-thaw and of solifluxion on tundra soils in the warmer season is to sort out the coarser material from the finer. The assorted stones and gravel are rearranged into elongated strips (stone stripes) on slopes and into polygonal or circular outlines around patches of fine particles on flat areas. The main methods of soil improvement are better drainage and increase of nitrification by liming, manuring and sowing leguminous plants.

Podzols. Podzols may form under forest in any climate with rainfall all the year, including tropical woodland and coniferous woods in temperate western seaboards neither of which has cold winters nor permafrost. But in climates with warm winters podzols are intrazonal. Thus in the tropics, soils on free-draining coarse sandstones and quartzites, in for example British Guiana and Borneo, are often so poor in mineral bases that they become very acid and allow acidic peat to accumulate at the surface and typical podzol horizons to develop beneath it.

The prime location of podzols is a wide belt, south of the tundra, in the northern continents where the colder months bring severe freezing and the relatively rapid snow-melt and soil-thaw in spring cause waterlogging. Over vast areas in the U.S.S.R. the thaw proceeds down to a permafrost layer at several feet depth. The summer warmth is greater than in the tundra and favours the growth of trees, particularly conifers. The characteristic podzol of northerly latitudes develops under predominantly coniferous forest (taïga) in areas greatly affected directly or indirectly by Pleistocene glaciation. Parent material is often glacial drift and intrazonal tracts of stony moraine, swamps and lakes abound. The typical podzol is heavily leached largely because of spring melt-water (or of heavy rainfall in warmer areas) and of high acidity. Its fertility is low because of slow organic decay and poor physical structure. A characteristic soil-profile under mainly conifers consists of A_0, a litter layer (2 to 3 inches) composed largely of conifer needles; A_1, a shallow (1 to 2 inches) grey to reddish-brown acidic layer, with some humus; A_2, a well-marked, leached, ash-grey or whitish-grey horizon, a few inches up to two feet thick, loose, structureless or horizontally stratified (laminated on clays) and relatively rich in quartz silica. The term *podzol* (Russian: ashes beneath) was derived from this layer. Beneath is an equally characteristic B horizon packed with organic colloids, bases and, to a lesser extent with iron and clay minerals (dominantly illite) washed down from above. It is commonly reddish-brown, nutty in structure and enriched partly with oxides of iron and aluminium that sometimes cement the bottom of the horizon (B_2) or accumulate sufficiently to form nodules and ochrish-brown hardpans (Fig. 148 and Plate 35).

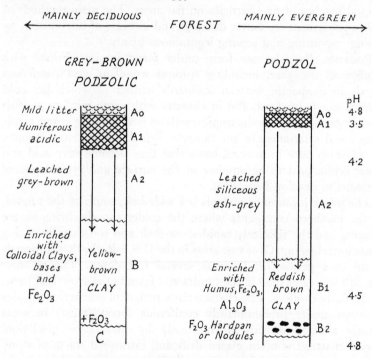

148. Profiles of podzol and grey-brown podzolic soils

Under natural conditions podzols vary greatly in fertility but when strongly podzolized they suffer from excessive acidity and unfavourable physical properties. On the other hand they have a positive water balance and experience frequent moistening. The northerly podzols occur where long warm summer days encourage the growing of crops such as vegetables, especially potatoes, and small grains and grass. After clearing of conifers, the liberal application of lime and fertilizers, particularly phosphate and potash, and the sowing of grasses and legumes will weaken the hardpan. If the depth of ploughing is increased gradually and organic matter added, the agricultural productivity of the soil benefits rapidly from the intermixing of the upper horizons and the slow breaking up of the hardpan layer. The Dutch have perfected a system of podzol reclamation in which the horizons are excavated and then replaced with the B horizon above the A_1 and A_2 horizons.

Grey-Brown Podzolic Soils. On the southern fringes of the main podzol belt although the climate is warmer the continental interiors remain very cold in winter and the change in vegetation is slight. Here *grey podzols* or *grey wooded* podzols develop which in a simple classification may be grouped with typical podzols. But towards the coast, where the increase of winter warmth is more pronounced and the annual rainfall greater, conifers give way to mixed and deciduous forest and the soil horizons become thicker and less distinct. The soils are less leached, less acidic and more retentive in their upper horizons of hydroxides of iron which, with the organic matter, give a grey-brown colour. They do not show true podzolization but experience clay eluviation. The characteristic profile now becomes: A_0 a thin cover of mainly broad-leaved litter; A_1, about 3 to 6 inches of dark grey to brown silty loam, rich in humus; A_2 about 5 to 12 inches of brownish-grey, leached, platey-textured, silty loam; B from 10 to 36 inches of yellow-brown to brown silty clay loam, acidic, enriched by illuviated colloidal clays and bases, and often arranged in coarse, angular blocky aggregates which tend to increase in size downwards; C, a light yellowish-brown clay parent rock, often glacial drift (Fig. 148 and Plate 35).

These *grey-brown podzolic soils* cover much of central and western Europe, most of the north-eastern United States and large areas in north China and Japan. Although weakly acid and only moderately fertile in a natural state, they respond rapidly to good management. During the last 4,000 years man has replaced most of the broad-leaved forest with a cultivated landscape and today produces here a wide variety of cereals, fodder grasses, legumes, roots, fruits and vegetables.

Red-Yellow Podzolic Soils. Where rainfall is abundant (often over 40 inches annually) and falls in all seasons and temperatures are warm in winter and hot in summer, bacterial decay proceeds rapidly during most months and in at least the hot weather, silica is dissolved relatively rapidly. The resultant processes are characterized by clay movement and loss of bases but podzolization is weaker than in podzols and laterization is virtually absent. These intermediate soils occur in large tracts of the sub-tropics, especially those with eastern marginal climates, as well as in the tropics. Outside the eastern margin climates, red-yellow podzolic soils often seem to be end stages in the development of Grey-Brown Podzolic and of mature tropical soils as the following diagram illustrates.

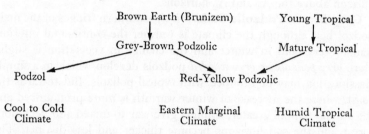

Profiles of Red-Yellow Podzolics vary considerably but a common succession consists of a thin upper layer of organic matter, grading into a friable pale-yellow to yellowish-green A_2 horizon and so to a red or yellow B horizon enriched mainly by downwashed alumina oxides. After clearing, these soils are soon exhausted of their organic content and need heavy fertilizing to maintain high crop-yields, particularly as their calcium content directly available for plant-use is low. Yet, as a rule, they are fine in texture, well-drained and moderately rich in reserves of unweathered minerals.

Ferruginous and Ferralitic Tropical Soils. Soils of the humid tropics are characterized by their relatively great thickness and their relatively high content of sesquioxides of iron and alumina (Fe_2O_3; Al_2O_3) and their relative poverty in bases, alumino-silicates, and organic matter. These characteristics are caused mainly by the rapidity of the dissolution of bases and alumino-silicates and by great bacterial activity in hot damp conditions. But it is essential to notice that the acceleration of soil-forming processes, including the weathering of parent material, is necessarily associated with rapid surface erosion and that soils will be greatly affected by their age, or state of survival, as well as by lateral drainage, or position in the relief. Equally accelerated will be the effects of man's interference with the vegetation cover. Nor can climatic change be ignored as many of the older hardpans seem to date back to long before Pleistocene times.

Tropical soils are classified mainly by differences in their relative content of oxides of iron and of alumina, especially kaolinite, and in the distribution of these oxides in the various soil-horizons.

In regions with a marked dry season, as under poor bush-growth in the Sudan, *ferruginous soils* often occur. These contain a high proportion of iron oxides, often partially dehydrated and crystalline in structure and sometimes creating small stable concretions about the size of sand particles. Such soils usually lack free alumina and in hollows with a shallow water-table will form a compact iron-oxide enriched horizon at the surface.

Where the climate is wetter, and especially under rain-forest, *ferralitic* (lateritic) *soils* predominate. These contain free aluminium oxide (Al_2O_3; gibbsite), the amount varying partly with the parent material. On acid parent material, rich in silica, the soils are feebly ferralitic, and contain small amounts of free alumina and large amounts of kaolinite. On non-acidic parent materials (poor in silica) free aluminium oxide is common and the soils become typically ferralitic. A characteristic Red Ferralitic Soil under rainforest has a very thin veneer of leaf-litter, poor in exchangeable bases. Beneath are the following horizons: A_1 Up to 7 inches, greyish with humus content of over 5 per cent., pH5; A_1 Leached horizon, 6 inches up to 4 feet thick, with quartz particles; B Very compact but not indurated, brick-red clay, 4 to 7 feet thick, with ochreous blotches and small concretions, and usually containing over 50 per cent. kaolinite and only 10 to 20 per cent. of free sesquioxides; C Clay, with yellow and red blotches, 6 to 10 feet thick, grading into 10 to 20 feet of ochre-coloured parent material. The typical profiles and evolution of ferralitic soils are shown in Fig. 149.

Generally, the process of laterization or desilication is rapid and old soils may contain less than 3 per cent. silica in contrast to over 20 per cent. in young, but the process is probably never completed and stops just before the clay minerals have lost all their silica.

In humid climates with a dry season and in forest clearings and areas exposed to severe erosion in less wet parts of the equatorial zone, the surface desiccation and erosion may expose an indurated B horizon. The desiccation may partially dehydrate the colloidal iron and aluminium oxides which become partly crystalline and help to cement together concretions already in the soil. In addition lateral downhill seepage at the lower end of a slope and the upward movement of iron-enriched water in hollows with a high water table often form strong cemented hardpans. The latter situation probably also

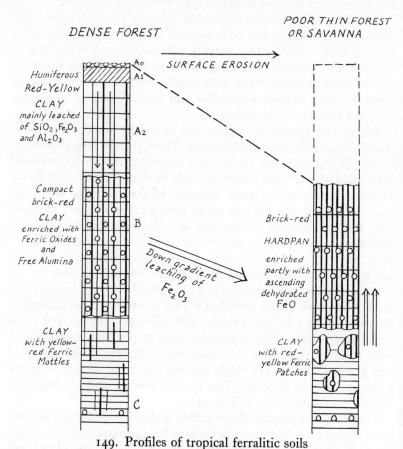

149. Profiles of tropical ferralitic soils

(*After Duchaufour*)

For key *see* Fig. 146. Open circles denote free alumina

includes incrustation by iron-oxides that have been partly dehydrated and then carried upward by capillary water (Fig. 149). The term laterite is often applied to-day to these hard brick-red ironpans that cannot be broken up in the hand whereas the name was first used for compact red and mottled clays (B horizon) that were plastic when dug and shrank and hardened on exposure so as to be suitable for building (Latin, *later*, a brick).

From an agricultural point of view, ferralitic hardpans form very poor soils and tend to repel plant roots. On the other hand, parent materials rich in exchangeable bases (apart from the coral-reefs) provide the best soils in the tropics, as, for example, the black 'cotton soils' on the basaltic lavas of the Indian Deccan. The typical ferralitic soil in the rainforest or the wooded savanna soon becomes deficient in humus and lacking in fertility once its plant cover is removed. Often even more disastrous is the rapid surface erosion of the A horizon. Consequently successful agriculture depends on retaining as much plant-cover as possible and on growing crops, especially trees, that will root deeply and widely in search of nutrients. Thus rubber, cacao, coffee and other crops that will tolerate some shade can be interplanted in forest and given more light by forest thinning as they mature. For ground crops, such as corn and cassava, the unsatis-factory native method of clearing by burning a circular patch and then abandoning it after the crop-yields dwindle, can be made scientific by cutting strips or corridors and allowing them to revert to forest after four or five years. Within ten to twenty years the strips have regained most of their former fertility. The widespread practice of burning is disastrous as it destroys the organic matter and in the drier savanna prevents natural re-afforestation.

In localities where the parent material contains very little weather-able silica and most of the soluble silica has been removed by leaching, the proportion of alumina may rise sufficiently to form bauxite. In many areas, bauxite (as in the Guianas), oxides of iron (limonite), and manganese have been concentrated locally by water action in proportions worth exploiting for their mineral content.

Chernozemic Soils. The transition from forest to grass is best expressed by the degraded chernozems (Fig. 150) that extend between the grey-brown podzolic soils and the true chernozems in, for example, parts of eastern Europe and western Siberia. These ex-panses of slightly podzolized chernozems are thought to have been invaded by forest during the wetter, cooler climate of the last main glacial advance of the Pleistocene period. To-day in subhumid climates with an appreciable period of excessive evaporation, forest usually gives way to grass, the calcium content of horizons increases and pedocals replace pedalfers. Under a vigorous grass sod in which microbial decomposition is hindered both by winter frosts and summer droughts, the surface layer becomes rich in organic matter[1]

---

[1] Forests return as much organic matter to the surface but in forest micro-climates some humus decomposition continues virtually all the year.

and darkens into a black-grey earth (*chernozem*) while the depth of the humus-enrichment is increased to 3 or 4 feet by the penetration and decay, often annual or biennial, of roots. As the horizons are rich in lime, the humus is mild and the whole soil-profile has a cloddy-granular or nutty-prismatic structure excellent for agriculture.

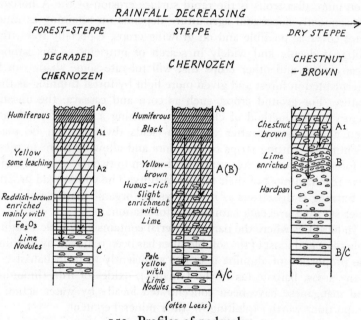

150. Profiles of pedocals

For key *see* Fig. 146. White blotches denote lime mottles and nodules

Within the framework of these general characteristics, Chernozemic soils vary mainly according to parent material and the length and frequency of spells when surface evaporation exceeds precipitation.

In wetter parts, especially those adjoining the grey-brown podzolic soils, the annual precipitation of 30 to 40 inches appears adequate for tree-growth yet in large areas, as in the middle United States, an open tall grass prairie prevailed when first discovered by Europeans. The chief natural influence favouring grass rather than trees may be the frequent drying out of the upper horizons to a depth of a foot or so between summer rains. The characteristic *prairie soil* or *brunizem* is dark brown or dark grey-brown in colour, and although less rich in calcium than are true chernozems and experiencing clay eluviation lacks distinct podzolization (Plate 36). The dark A horizon is up to 2 feet thick and very slightly acid; the B horizon is often a yellow-brown, neutral or calcareous loam up to about 2 feet total thickness and composed largely of illite and montmorillonite. The soil has an excellent structure, a high humus content and good moisture storage. Its agricultural value may be judged from the United States corn-belt which lies mainly on Prairie Soil developed from fine fluvio-glacial and aeolian deposits rich in lime. Here a rotation of corn, grasses and legumes can keep the humus content at about 70 per cent. the natural proportion.

Where the annual rainfall decreases to between about 15 and 30 inches a true *chernozem* develops under prairie and steppe grasses (Fig. 150 and Plate 36). Typically, under a dense grass-sod, a black or dark-grey A horizon rich in humus and often 1 or 2 feet thick grades slowly into a yellowish-brown B horizon, dominantly of illite and montmorillonite, commonly 2 to 3 feet or more thick, and containing tongues of humus and a slight accumulation of colloids and bases. The C horizon is pale yellow and normally contains nodules of lime. Deep chernozems are regarded by the Russians as the best soils in the world not least because their abundant nutrients, such as phosphorus, potassium and calcium, are present in a form readily available to plants. They are particularly ideal for the large-scale cultivation of small-grained cereals and form the world's chief 'bread baskets'. Like ferralitic soils they have the quality of being most productive naturally when first ploughed whereas podzols improve in fertility with destruction or removal.

In the driest grasslands, particularly short-grass prairie and semi-arid steppe, the shallower penetration of the rainfall, causes shallower, lighter-coloured soil-horizons, less leached of lime and less rich in organic matter. These *chestnut* soils acquire, at a depth of a foot or two, an accumulation of calcium and other alkaline salts that may concentrate into hardpan (Fig. 150). Yet they retain a prismatic structure favourable to tillage and would be more widely cultivated but for the risk of occasional prolonged drought and excessive soil erosion.

Desertic Soils of Semi-arid and Arid Regions. The semi-arid and desert environment contains large areas of bare-rock, coarse rock-debris and sand but where the finer material becomes stabilized, especially on alluvial fans and wadi-floors, a sparse scattering of shrub may provide an incipient soil-profile. In deserts with a cool or cold season the soils tend to be light-grey above and reddish-brown below. These *sierozems*, or almost-desert soils, usually have within 2 feet or less of the surface, an horizon with salt or lime accumulation which tends to form a hardpan. They are deficient in humus and nitrogen but being almost unaffected by leaching are usually rich in alkaline and saline salts and in soluble mineral constituents. The clay minerals are largely montmorillonite and illite. Where irrigation is possible good crops can often be obtained.

In hot deserts without a cool season the surface mantle, where particulate, is typically reddish grey to deep red. These *red desertic soils* may form hardpans but usually lack distinct horizons and most irrigation on them is connected with river-water which supplies some alluvial clay.

Intrazonal Soils

Halomorphic: Excessive Salinity. In deserts and areas of interior drainage with great evaporation, strong concentrations of salts form at and near the surface. In the playas of the North American deserts the common salts are sodium carbonate (Na_2CO_3; alkaline or a black alkali), borax ($Na_2B_4O_7$), calcium carbonate and various sulphates and chlorides. The salt flats of the Great Salt Lake, Utah, are predominantly saline or of the white alkaline group. The Russian terminology for these so-called soils is as follows: Where the salt content is high and increasing, *solonchaks*; where rainfall and drainage-water transfer the salt accumulation mainly to the B horizon, *solonetz*; which in steppes may be further leached into a *soloth* that shows signs of podzolization (Fig. 151). Solonchaks are typically whitish-grey or grey and structureless whereas solonetz usually have a platey A horizon (with pH of about 7) and a prismatic columnar B horizon (with pH9 or more).

Hydromorphic or Glei Soils: Excessive Water-supply. Where soils are frequently waterlogged, oxidation is replaced by gleization in which humus decomposed under slow anaerobic conditions combines with ferrous oxides (Fe_2O_3) to form a soluble humic-iron complex which can be washed out from the profile altogether or re-deposited as spots, mottles and concretions. The typical profile has an organic A_1 horizon, an intermediate A_2 horizon characterized by the precipitation of blotches and small concretions of ferric oxides, and a true *Glei* or G horizon where reduction dominates and greenish-grey ferrous iron compounds accumulate, sometimes in small con-cretions. The G horizon coincides with the lowest common level of the water table while the A_2 horizon is the zone up which the water-table periodically rises. Thus in the true Glei soil, the movement of ferrous iron is upward in contrast to that of many so called *pseudo-gleys* where the movement is downward.

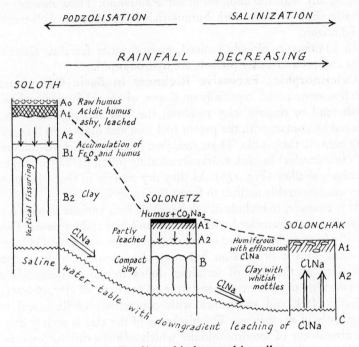

151. Profiles of halomorphic soils

In marshy and shallow, water-filled hollows, especially in cool and cold climates, partly decayed vegetation may form several feet of highly-acidic peat above a glei horizon of grey-blue, sticky, structure-less clay. This *bog-peat* abounds on wet European moors and, being almost contantly waterlogged, is only cultivable after drainage, liming, deep ploughing and admixture with sand or silt.

Where the vegetation happens to be subjected to washing by fresh-water rich in lime or alkaline salts, it gradually decays into a less acidic black, spongy peat which on drainage forms excellent arable land. These *fen peats* are the basis of the agricultural prosperity of the English Fens where drainage and ploughing have so accelerated oxidation and comminution that the ploughed surfaces have been lowered several feet and wind erosion may be serious in hot dry weather.

Soils somewhat resembling fen-peats may develop on river flood-plains where an appreciable growth of thick vegetation is waterlogged periodically with the addition of some alluvium. These *meadow soils* usually show a thick, dark humus-rich layer above a well-marked glei horizon.

In addition, as already noticed, ground-water ferralitic (lateritic) soils and ground-water podzols are hydromorphic.

Calcimorphic: Excessive Richness in Basic Compounds. On limestone rocks, especially on slopes where solifluxion and rain-wash tend to remove clay residues, the soil is continually recal-cinated by contact with the parent rock and also often with limestone fragments in the profile. These *rendzinas* are usually brown-coloured, rich in microbial humus, markedly alkaline (often pH8 or more) and relatively shallow (Fig. 146). As they dry rapidly in time of drought they are favourable neither to forest nor to agriculture.

It is common to include in calcimorphic soils various *tropical black clays* which are saturated with calcium and magnesium bases. These usually form on basalts and on calcareous sedimentary clays in humid climates with a dry season but may also occur in hydromorphic situations at the foot of ferralitic slopes in the rainforest. They contain little organic matter and over 50 per cent. of clay, character-istically black and sometimes with concretions of lime and less frequently of manganese. The blackness of the clay is mainly due to the abundance of montmorillonite which adsorbs cations (especially calcium) and also has the interesting property of swelling when hydrated and so causing a hummocky surface. Although compact and not easy to till with light implements, Black Clays are among the best agricultural soils of the tropics and include the *regur* or black cotton soil on the basaltic lavas of the Indian Deccan. The Tirs of Morocco and Black Waxy soils of Texas probably also belong to the Black Clay group.

Azonal Soils. The immediate and skeletal surface mantles of mountain scree and other new deposits are best studied in connexion with the landform-making agencies already discussed on pp. 185–312.

Man and Soils

The main object of modern farming is the maximum production of food per acre compatible with the minimum deterioration of the soil. In cultivation this is achieved by mixing soil-horizons, by adding manure and fertilizers and by ensuring a favourable soil-structure. The maintenance of fertile pastures also faces many difficulties, especially of over-grazing, but these are small compared with the prevention of soil-impoverishment and soil-erosion during the cultivation of crops that are planted and harvested annually. In fact the main object of harvesting and grazing is to take advantage of soil constituents that have been converted by plants and animals into a concentrated edible form.

Decrease in fertility can be largely offset by crop rotations which include nitrogen-fixing plants, especially legumes; by occasional ploughing in of green crops; by liberal applications of animal and artificial manures; and, where needed, by draining and liming to improve soil-structure. Soil erosion, although inevitable where a fine seed-tilth is prepared and where harvesting exposes much bare soil, can be greatly decreased by good agricultural techniques. Where wind erosion of fine particles is serious, as happened in the 'dust-bowl' of the American mid-West, linear wind-breaks of trees are helpful, but the best remedy is to plough up strips only, leaving most of the surface under grass. Where rainfall in the warm season is torrential, as is common in many countries, exposed cultivated surfaces may suffer severe soil-wash. On gently-sloping surfaces fairly heavy rain may remove a thin sheet of fine surface particles but on steeper surfaces and when the rainfall begins to be excessive (from 1 to over 4 inches an hour or from 4 to over 10 inches a day) the sheet-wash soon forms rills and the rills combine into gullies (Plate 37). Sheet- and gully-erosion is common in areas stripped of tree-growth in the eastern United States, around the Mediterranean, in tropical Africa and monsoon Asia. The Tennessee Valley Scheme in the United States successfully restored a deforested drainage basin (of 42,000 square miles) ruined by gully-erosion back to much of its former fertility. The methods included numerous dams across streams and rivers to control floods and the extensive replanting of trees and grasses.

In any cultivated area liable to sudden heavy surface run-off, ploughing along the contour is necessary (Plate 38), and flow in natural drainage-channels must be checked as often as possible to prevent channel-erosion. The main remedy, however, is well-controlled, carefully-maintained terracing along the contour, which may even include the laborious carrying back to upper terraces of soil washed from them. Elaborate terracing is almost universal in warmer climates and may be seen, for example, in southern Italy, Madeira, Peru, and above all in China where it was understood and perfected when Britain was in the Stone Age.

LIST OF BOOKS

General

Joffe, J. S. *Pedology*, 1949
Brade-Birkes, S. G. *Good Soil*, 1944
Plyusnin, I. I. *Reclamative Soil Science* (Foreign Publishing House, Moscow)
Robinson, G. W. *Soils: Their Origin, Constitution, and Classification*, 1959
Kellogg, C. E. *The Soils that Support Us*, 1951
Duchaufour, P. *Pédologie* (Nancy), 1965
U.S. Department of Agriculture. Yearbook. *Soil*, 1957
Millar, C. E., Turk, L. M. and Foth, H. D. *Fundamentals of Soil Science*, 1958
Rode, A. A. *Soil Science* (Trans; Jerusalem), 1962
Bunting, B. T. *The Geography of Soil* (Hutchinson Univ. Library), 1965
Soil Science (periodical)

Soil Factors

Russell, E. W. *Soil Conditions and Plant Growth*, 1961
Russell, E. J. *The World of the Soil*, 1957
Jenny, H. F. *Factors of Soil Formation*, 1941
Eyre, S. R. *Vegetation and Soils*, 1963

Geochemistry and Soil Chemistry

Mason, B. *Geochemistry*, 1958
Grim, E. *Clay Mineralogy*, 1953
Keller, W. D. *The Principles of Chemical Weathering*, 1955
Bear, F. E. (ed.) *Chemistry of the Soil*, 1965 (advanced)

World Classifications

Simonson, R. W. 'Soil classification in the United States', *Science*, Sept. 1962, pp. 1027–34
U.S. Department of Agriculture. *Soil Classification, 7th Approximation*, 1960; with Supplement, 1964
Muckenhirn, R. J. *et al.* 'Soil classification and the genetic factors in soil formation', *Soil Science*, 1949, pp. 93–106
Thorp, J. and Smith, G. D. 'Higher categories of soil classifications', *ibid*, 1949, pp. 117–26.

Regional Classifications

Gourevitch, A. (trans.) *Soil-Geographical Zoning of the U.S.S.R.*, 1963 (London)
Mohr, E. C. J. and Van Baren, F. A. *Tropical Soils*, 1954
Kubiena, W. L. *The Soils of Europe*, 1953
Tamm, O. *Northern Coniferous Forest Soils*, 1950
Wolfanger, L. A. *The Major Soil Divisions of the United States*, 1930

Local and Practical Soil Surveys

U.S. Department of Agriculture. *Soil Survey Manual, Handbook 18*, 1951
Clarke, G. R. *The Study of Soil in the Field*, 1957
Taylor, J. A. 'Methods of soil study', *Geog.*, 1960, pp. 52–65
Bunting, B. T. *Annotated Bibliography of Memoirs and Papers on the Soils of the British Isles* (London School of Economics), 1964
Mackney, D. and Burnham, C. P. *Soils of the West Midlands*, Soil Survey of Great Britain, 1964
Eden, T. *Elements of Tropical Soil Science*, 1964 (A small non-technical book of considerable practical value)

Man and Soils, and Vegetation

U.S. Department of Agriculture. Yearbook. *Soils and Men*, 1938 (1232 pp.)
Bennett, H. H. *Elements of Soil Conservation*, 1955
Smith, G.–H. (ed.) *Conservation of Natural Resources*, 1950
Wagner, P. L. *Human Use of the Earth*, 1960

CHAPTER TWENTY-FIVE

VEGETATION

Nature of Vegetation

AS plants and animals have evolved simultaneously all vegetation could in a sense be considered natural until the wholesale burning and clearing of forest by man in the last 10,000 years. The animal kingdom depends for food, directly or indirectly, on the vegetable kingdom and animals are as a rule destructive of vegetation. Whereas nearly all animals are mobile and many are warm-blooded because of the generation of internal heat and not a few construct artificial shelters, the natural plant is fixed, exposed and lacking in internal-heat generators. Consequently, vegetation is strongly influenced by the local environment and as most plants grow partly above and partly below the surface of the ground this local environmental control includes soils and drainage. The problem of these geographical interactions is exceedingly complex but the complexity is heightened and taken beyond the bounds of a simple summary by the fact that plants, as well as animals, evolve. Moreover, the gradual evolution of plants has been affected by changes in world climate and in the relative distribution of land-masses and oceans. There are also complications due to the migration or spreading of plant species which is always occurring, although normally the individual plant is fixed.

To-day the plant kingdom is usually considered to consist of four primary divisions or phyla, in each of which the plants appear to have a common ancestry. In three of the phyla, Thallophyta (algae, fungi, lichens); Bryophyta (liverworts, mosses); and Pteridophyta (ferns, horsetails, club mosses, etc.) reproduction is by means of spores. In the fourth phylum, the Spermatophyta, reproduction is by means of seeds and the phylum is subdivided into two main groups, Gymnosperms, with naked seeds, and Angiosperms, with seeds enclosed in an ovary and borne in flowers. A considerable number of gymnosperms are coniferous and needle-leaved but many are broad-leaved including some conifers. Most of the angiosperms are broad-leaved.

The earliest forms were probably algae from which evolved mosses and liverworts as well as the more elaborate spore-bearing forms, horsetails and clubmosses that form the bulk of the coal-measures of Carboniferous times (about 350–275 million years ago). Before the Triassic period (225–180 million years ago), gymnosperms had become of great importance and they remained dominant until the end of the Cretaceous period (*c.* 70 million years ago). About this time, angiosperms, which had been slowly increasing in numbers probably since the end of the Carboniferous, began to spread rapidly at the expense of the gymnosperms and by the Eocene (70 million years ago) or beginning of the Tertiary era had come to dominate the world's land plants. Although some gymnosperms have managed to survive in abundance, to-day angiosperms form the bulk of the world's vegetation and comprise probably over 200,000 species, including most of the plants cultivated and used by man.

The nature of plant evolution can hardly be summarized satisfactorily in a brief discussion. Evolution proceeds continuously but at an uneven rate. The inheritance of characters between parent and offspring is effected by means of chromosomes which form part of all cell nuclei. The characters are resident in the chromosome as *genes* and an alteration in the genes means an alteration in the characters that are transmitted to offspring. Hence changes in the number of chromosomes and changes in the genes (*gene mutation*) may both lead to new species of plants. The interbreeding of relatively unlike parents (or *hybridization*) usually leads to these changes, and so often will strong alterations in the environmental factors. The latter idea involves the concept of natural selection, as stressed by Darwin, in which a variation or adaptation in the plant which enables it to deal more successfully with its environment, naturally increases its powers of survival and of dominance. However, simple modifications in a plant due to environmental conditions are not usually transmitted direct to progeny whereas gene mutations and genetic recombinations resulting from hybridization are always heritable.

These inherited qualities are of great importance as the geographical distribution of a plant is largely limited by its *tolerance*, or its capacity to tolerate or to respond to its environment, which is governed by the 'laws of evolution and genetics' (Cain, p. 10). Some species have a wide tolerance or adaptability to climate, soil and other environmental influences whereas others seem to be highly specialized. When a plant's capacity for adaptation or 'genetic innovation' becomes too small to allow it to meet the challenge of a changing or changed environment it will deteriorate or die and may survive only by migration.

In fact, whether a plant or species is present or not in a region depends partly on migration (whether the plant arrived or not), on natural competition (whether it survived the competition of other plants in the locality) and on its chance survival of catastrophes, such as destruction by animals, fire, and sudden climatic change. The Palm family is restricted to the warmest parts of the tropics and sub-

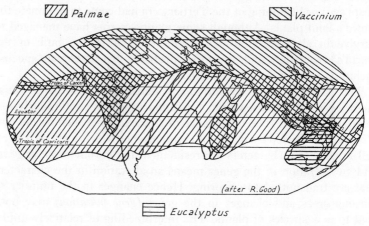

152. General limits of distribution of certain plants

tropics but within this extensive zone it is widely distributed. On the other hand, the genus *Eucalyptus* is restricted naturally to Australia and the nearest East Indian islands while the genus *Vaccinium* (bilberry, etc.) is ubiquitous in northern latitudes and common on mountains in the tropics (Fig. 152).

It is obvious that inherited characters make certain plants unsuitable for certain environments. Thus neither a palm nor a teak tree would be suited to the taïga. Among inherited qualities are the length of life, whether annual or biennial or perennial, and the length of the reproductive season (flowering and fruiting) which strongly influences the survival of a species.

How Plants Spread. Some plants spread vegetatively by extending their roots laterally and then sending up vertical shoots or suckers, as do the banana and aspen. A few, such as the banyan, send down aerial roots from their branches. Most plants migrate by dropping seeds which may grow beneath or near them or, as is common, be borne considerable distances by the wind, by animals, especially birds, and by water. During this extension or outward migration of the plant cover, the species that eventually establish themselves and that flourish most are those best suited genetically to local conditions. Eventually in any large area various species adapted to the regional environment will predominate and will compete with each other for light and moisture. Under such competition, often or usually the biggest or tallest species become dominant and the smaller plants, although often much more numerous, fit in functionally with the requirements of the larger. Such a plant community or climax vegetation utilizes to the full the radiant energy received from the sun. It will tend to repel further invasions by migrant species but can readily be disturbed by interference by man and by change of climate.

Climax Plant Formations. After several hundred million years of plant evolution and migration and of geological and climatic changes the world's vegetation has become extremely complex in distribution and in species. There can be no simple satisfactory classification of vegetation types largely because each region tends to have its own combination of species. One popular method of physiognomic classification considers vegetation as a response to climate and groups together large units having the same dominant type of life form and the same general appearance and layer-arrangement of their component flora. The broadest units, forest, shrubs and grass, are considered to be in dynamic equilibrium with the prevailing climate. Such wide global groupings, in which 'the dominants all belong to the highest type of life-form possible under the prevailing climate' are sometimes known as plant formation-types (Weaver and Clements, p. 478). Thus, throughout the tropical rainforest the dominants are broad-leaved evergreen trees whereas in the boreal forest they are commonly needle-leaved evergreens.

These global or zonal plant formation types are necessarily sub-divided further by differences in the species of plants that exist locally, differences, which, as we have seen, depend on many complex interactions, such as migration and competition as well as directly on climate. The tropical rainforest of Australasia differs markedly in floristic composition from that in Amazonia. Similarly, in middle latitudes the winter-deciduous broad-leaved forests may be dom-inated in different continents by either maple-beech or oak-chestnut or oak-hickory. Thus, each of these regional *climax plant formations* has its own distinctive assemblage of species. However, the natural environment of such large areas cannot be absolutely uniform and most climax formations experience slight regional differences in climate and appreciable local variations in edaphic (soil and relief) conditions. These regional and local variations favour some species more than others and assist differences in migrational influences to produce smaller areas with a distinctive dominant species or combination of species. These smaller units, as, for example, ash woods on limestones and oak woods on clays in England, are called *plant associations*.

Usually there are no abrupt boundaries between the vegetation divisions on any one land-mass. The major climax plant formations merge into each other through zones (*ecotones*) in which the species of each zone intermingle or interdigitate. The boundaries between plant associations may be sharper but even on these small scales some intermingling is often discernible.

Plant Successions. If climax plant formations are destroyed locally by natural catastrophes such as fire, flood, ice-advances and landslips or by biotic influences, particularly by man, once the interference ceases the natural vegetation tends to re-establish itself. In many areas a regular succession of plants (termed a *sere*) can be traced in this re-establishment. On rocky talus slopes in middle latitudes a common succession is crustose lichen; foliose lichen; moss; xerophytic herbs; shrubs; climax forest. In the humid tropics, on the parts of Krakatoa that survived the explosion in 1883 but were buried beneath hot volcanic dust the succession was algae; lichen and mosses; ferns; herbaceous and grass savanna; shrubs; slightly xerophytic trees; tall rainforest. The savanna had appeared by 1900, and the taller treegrowth within the next few decades. In large areas of north-western Europe after the last ice-retreat the ice-freed uplands and their fringes were first colonized by tundra with small birch trees; then successively by conifers; mixed broadleaf and needle-leaved forest; and, where the climate was warmer, finally by broad-leaved oak–elm–beech–lime–alder forest with hazel undergrowth.

The silting up of swampy hollows and lakes that form a common feature of many humid regions and particularly of glaciated areas and flood plains, is usually accompanied by a characteristic succession of communities, known as a hydrosere (Fig. 153). A typical lake in middle latitudes first supports in its shallow parts, especially near its shores, submerged plants including buttercups and bladderworts. These assist sedimentation and as the water shallows, floating plants, including waterlilies and pondweeds, begin to invade the lake shore, mainly by means of rhizomes rooted in the mud. When the water has shallowed to less than a few feet, plants that root on the bottom and grow mainly above water (bulrush, reed, etc.) invade freely and

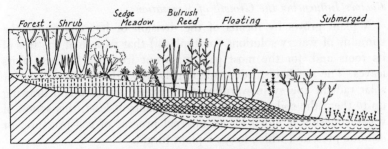

153. A hydrosere in middle latitudes
(*After Firbas*)

increase the rate of infilling. With further shallowing, other species such as sedges, rushes and cotton grass invade also and form a sedge-meadow stage which eventually becomes too compact and dry for hydrophytes. Then species of trees (willow, alder, etc.) find a favourable habitat and as humus begins to accumulate a mixed forest of alder, ash and oak takes over and gradually matures into a forest climax.

Because of the differing timespan during which these natural successions have been operating and of the extent and length of plant-migrations and adaptions, the present plant formations and associations on the land-masses show one outstanding contrast— the antiquity and richness of the floristic composition of the tropical rainforest as opposed to the newness and poverty of the flora in the large areas glaciated during the Pleistocene period. However, two other important aspects of vegetation must be added: vast areas are virtually devoid of plant cover because of aridity and of ice-caps, and tremendous areas have been cleared and altered by man and other animals.

To-day it seems rational to talk of climatic climax vegetation only in areas such as selva and taïga where man and animals have not greatly modified the plant cover. However, the problem is further complicated by the effects of climatic changes during and since the Pleistocene period. In most regions to-day with a marked dry season and a low annual rainfall, as occur widely in the tropics, subtropics and continental interiors elsewhere, the increased climatic warmth of the last 15,000 years has helped man, herbivores and fires to alter the vegetation.

Factors Influencing the Growth of Vegetation

A plant grows by means of the division of its cells under the stimulus of watery solutions (protoplasm) that pass in at the tips of its roots and, for the most part, pass out from the surface of its leaves. The mechanism depends largely upon the action of heat or solar radiant energy in warming the exposed surfaces and drawing up to them the soil-solutions. In some plants at least, this action of water cohesion is aided by an internal process of cell-pressure, or osmotic pressure, which keeps a plant turgid.

Temperature and Light. For survival, terrestrial green plants must obtain carbon dioxide from the atmosphere. This is achieved through small pores (stomata) in the skin or epidermis of aerial leaves and stems, which open in daylight and allow photosynthesis to proceed. During darkness the non-green and the green (chlorophyll-containing) parts of plants release carbon dioxide and consume oxygen. Provided moisture is available, stomatal transpiration increases with temperature at least until the plant begins to wilt. In addition, respiration or a slow general breathing occurs at the surface of all living plant tissues, green or non-green, in light and in darkness. This doubles in amount for about each 10° C. (18° F.) rise in temperature. Stomatal transpiration is of far greater significance than skin (cuticular) respiration. The former is largely controlled by air temperatures whereas the cuticular respiration being partly or mainly from the roots is strongly influenced by soil temperatures, which, as discussed on pp. 343–54, usually differ slightly from temperatures in the free atmosphere. Moreover, soil temperatures are also important as they control directly the germination of many seeds and so influence considerably the beginning of the growing season. Unfortunately most temperature statistics refer only to air temperatures at 4 feet above the surface.

As a rule each plant has a lower and an upper air temperature at which it begins to perish and a range of (optimum) temperatures in which it flourishes. The lower limit is that at which the stomata fail to raise soil solutions or at which the protoplasm begins to congeal or freeze. For many plants this is a degree or two below the freezing point of water (0° C.), because plant solutions are not pure water and the saps are encased in protective tissues. But tolerance to cold varies greatly with the species. Low forms of Arctic vegetation survive hard frosts while humid tropical plants perish long before the freezing-point is reached. In middle latitudes, it is not uncommon for exceptional or unseasonal frosts to damage only the tender terminal buds and fruits leaving the older, more protected parts unscathed.

The natural adaptions to, or methods of overcoming, cold include leaf-fall in deciduous forests; dormancy without leaf-fall in some conifers; and in annual plants a short, rapid life-cycle that is completed in the warm season, leaving only protected seeds to withstand the cold season and reproduce the species in the following spring. The natural methods of overcoming heat are mainly concerned with reducing transpiration (see pp. 412; 422). If the heat overtaxes the plant's powers of acquiring and maintaining a flow of protoplasm, the turgidity of the cells begins to collapse and wilting ensues.

The length of the period between germination (or budding) and fruiting is usually called the growing season in respect to annual plants and the fruiting-cycle for perennials. This cycle depends mainly on the inherited characteristics of the species and on the intensity and length of daylight. In temperate and sub-arctic latitudes the possible growing season for crops extends from germination to the first killing temperature which is usually a frost. But since the fruiting-cycle depends both on total radiant energy received and on length of growing season, a short hot season may bring the same response as a long cool season. This problem of temperature efficiency is often expressed climatically by means of day degrees (or *growing degree-days* in the United States). Each crop is assumed to have a basic temperature at which it begins to grow and a certain total number of thermal or heat units to bring it to maturity. Thus wheat germinates at a temperature of about 42° F., peas at about 40° F. and sweet corn at about 50° F. To take wheat as the example, the effectiveness of the local temperature in promoting its life cycle is judged approximately by the amount of heat above 42° F. in the following way. Each degree by which each *mean* daily temperature of the growing season exceeds 42° F. is called a growing degree-day. Thus mean daily temperatures of 52° F., 56° F. and 62° F. would add

44 growing degree-days to the accumulated total required for the successful maturing of wheat, which is usually about 1,850 to 2,000 units. This total can easily be achieved in 100 to 120 days in the warm summers of the prairies of the United States and southern Canada but needs double that time in eastern England where wheat must be sown in late autumn or winter rather than in spring.

It is obvious that length of the growing period has little meaning in tropical lowlands where moisture supply tends to be more variable and vital than temperature extremes. But a minor factor that influences the spread of some species in all latitudes is the length of daily illumination or *photoperiodism* which has a considerable effect on flowering. Plants such as dahlias and violets which have short-day affinities are hindered or prevented from flowering by longer photoperiods, whereas long-day species, such as spinach, radish and lettuce, tend to flower badly or not at all in short-day climates. However, many plants can flower even in continuous daylight, and the whole problem is further complicated by the fact that in the same photoperiod some plants are tolerant of shade (sciophilous) and some of sun (heliophilous), a quality which plays a significant part in the layer arrangement in plant formations.

Precipitation. Protoplasm, or the weak solution in which mineral colloids are carried to all parts of the plant, normally contains 90 per cent. or more of water. Consequently large amounts of water are concerned in plant-growth, particularly as most of the water absorbed by roots is neither retained in the plant nor used in metabolism (plant processes whereby protoplasm is converted into living tissue by or in the cells) but is transpired into the atmosphere mainly through leaf stomata. This enormous transfer of water from the soil to the air is difficult to measure exactly. During an average growing season a corn (maize) crop transpires the equivalent of 15 inches of rainfall. Deciduous trees commonly transpire a weight of water equivalent to between 600 and 1,000 times the weight of their dry leaf matter formed during a growing season. If an average-sized elm formed about 50 lb. (dry weight) of foliage in a growing season, the tree would have transpired at least 36,000 lb. or 3,600 gallons of water in that period. For an average-sized beech, if the roots were kept moist, the seasonal amount would probably be more than 5,000 gallons. Even in needle-leaved evergreen conifers the ratio of water transpired to dried weight of leaf matter on the tree commonly exceeds 100 and occasionally reaches 240. An important conclusion to be inferred from these high transpiration amounts is that in climates with a warm or hot season when evaporation greatly exceeds precipitation, tall trees, particularly deciduous species, should not be planted around artificial reservoirs. Usually low, evergreen shrubs and grasses of a shallow-rooted nature would be preferable in all climates. However, it should be noticed that in tropical uplands where savanna is common to-day many of the tall bunch-grasses, for example the East African Kikuyu grass which sends roots down to 19 feet depth in the deep red loams, absorb more water even than local tree growth.

The process whereby plants acquire water and nutrient solutions depends largely on roots and root-hairs, thousands of which may form on a single root-tip. The root grows downward and outward in search of water and of mineral colloids in much the same manner as the exposed plant grows upward and outward in search of light and air. In many common grasses the rate of root elongation in the growing season is over $\frac{1}{2}$ inch a day. The main vertical roots of corn will under very favourable conditions penetrate downward at the rate of 2 to $2\frac{1}{2}$ inches daily for several weeks. The main roots of many grasses attain depths of 5 up to 8 feet; the tap roots of plants such as sunflowers and golden rods often reach 8 or 12 feet; the root-systems of many perennial shrubs and trees will penetrate to over 30 feet and some to 50 feet or more. As the root system deepens and widens the roots and, if present the fine rootlet hairs, absorb moisture either from the capillary film of the soil particles (p. 367) or from gravitational water in transit or from the water-table which sinks lower and lower until replenished. Some small plants do not have root-hairs and absorb solutions directly through their roots but many small and most large plants, particularly of a woody nature, develop root-hairs (Plates 33 and 34) at a tremendous rate during the growing period. In these the functioning root-hairs are nearly always near or at the growing tips of the roots and away from the growing tips the hairs are usually replaced by a somewhat harder, hairless protective layer that normally is not absorptive. Thus rainfall in the growing season must reach the root-hairs or the absorbing parts of roots to be directly useful to the plant.

We have already mentioned the daylight process of transpiration which is associated with photosynthesis or metabolism whereby carbon dioxide from the atmosphere and mineral colloids from the soil interact to form plant tissue. The function of fine roots and root-hairs in absorbing soil solutions will readily be appreciated. There is, however, another important aspect of this process. Root cells during respiration generate carbon dioxide, part of which reacts with water to form carbonic acid which in its turn participates in H-ion (cation) exchange with clay particles (Fig. 145). This direct adsorption of mineral colloids, or base exchange, by root systems supplies mineral matter to the plant.

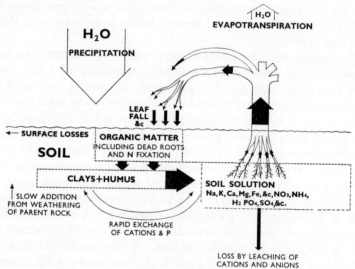

154. The nutrient cycle of vegetation

(After P. H. T. Beckett)

In plant and soil, black denotes movement of organic matter that yields return-nutrients to the soil

It is very significant that fine roots and root-hairs are potent adsorbers of certain plant nutrients, such as phosphorus, nitrogen and sulphur, that are not abundant in the soil. These important nutrients then circulate slowly in the vegetation growth—humus decay cycle (Fig. 154). At any given moment about 98 per cent. of such nutrients present in this circulation is locked up firmly in the vegetation and humus, and less than 2 per cent. is available for immediate re-use by root-cells. If the vegetation and humus are destroyed, particularly by burning, their valuable mineral content is also destroyed. The tropical rainforest will continue to luxuriate if not interfered with but if it is removed the clearings will soon become too infertile for profitable cultivation.

The most recent efforts by climatologists to relate plants to climate concern transpiration and evaporation from exposed surfaces. But whereas evaporation from water surfaces is fairly easily measured, to assess evapotranspiration from plants is difficult and requires elaborate instruments (evapotranspirometers). In these a large tank filled with soil and covered with grass is equipped with devices to supply the grass with all the moisture it needs beyond what falls on

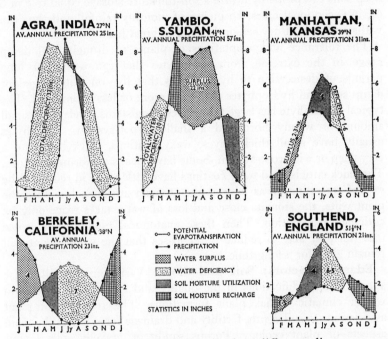

155. Water régime of plants in four different climates

It is assumed that plants can make use of the equivalent of 4 inches of rainfall stored in the soil

it naturally and to collect any surplus water. The total monthly water consumption of the grass can then be measured and related to rainfall and temperature. In compiling graphs of this information, it is usually assumed that grass in extra-tropical climates can in times of drought make make use of moisture stored in the soil equivalent to about 4 inches of monthly rainfall. Fig. 155 shows the resultant graphs for four localities that had a seasonal water deficiency. Such results apply well to large areas where grass, small grains and root crops are grown and often reveal the need for supplementary irrigation in dry seasons in, for example, the Canadian prairies and eastern England. The method seems applicable especially to the main cultivated areas outside the tropics but its weaknesses are many. It is of little value in wet areas without a season of water deficiency and as it applies to short grass it is far less accurate in regard to forests. In the tropics many forest trees and the tall bunch-grasses growing on deep soils can probably utilize a soil-moisture storage equal to 6 or 8 or, in exceptional cases, 10 inches of monthly rainfall and thus can withstand a short dry season in spite of great heat.

The tolerance and adaptations of plants to drought and damp range in the extreme from xerophytes that can withstand long periods of desiccation to hygrophytes that live on waterlogged or damp soils and hydrophytes that exist more or less submerged. The typical hygrophyte has large thin leaves, long stems, a relatively small amount of woody fibre and a shallow root-system. Xerophytes usually have small thick leaves, waxy coatings, thick bark and a very deep or wide root-system. Some have means of storing moisture. The thick cuticles and waxy coatings have little value in reducing the rate of transpiration as long as the stomata are open but once the plant wilts, the stomata close and loss of water takes place only by cuticular evaporation. Then these anatomical modifications come into full play and so reduce water losses that the wilted plant will remain alive for a long time.

Edaphic Factors: Soil and Slope. Variations in soil cause appreciable modifications in the quality and quantity of vegetation within climatic climax plant formations. The relations of plant associations to soil depth, fertility and drainage form one of the main aspects of plant ecology. Porous sandy or gravelly soils favour xerophytic species; a high salt (sodium) content is liked by only a few plants (halophytes), including the coconut. The tolerance or otherwise of particular species to soil reaction or acidity has an important effect on the distribution of plants locally. Tolerance, however, seldom bears a simple relationship to acidity or alkalinity. Acid soils are deficient in calcium and so are sought by calcifuges (lime-haters) and shunned by calcicoles (lime-lovers) but such soils may contain much soluble aluminium and manganese which many plants cannot tolerate. On the other hand, in alkaline soils some minor and trace elements (phosphate, potassium, manganese, boron, etc.) become almost or quite insoluble and certain plants may suffer from a serious shortage of them. Moorland plants such as Calluna, Erica and Vaccinium and flowering shrubs such as azaleas and rhododendrons flourish at pH values below 4·5 whereas many flowering shrubs, excluding the above, and a great many cultivated crops (alfalfa, sugar beet, barley, wheat, etc.) flourish at pH values above 7. Further discussion on the pH scale will be found on p. 370.

Soil, however, as shown on pages 378–80, can seldom be divorced from slope and drainage or landforms. Physiographic factors become especially important when severe or increased erosion leads to shallowing of soil-depth and a more rapid drainage. Such conditions may follow extensive clearing of forests, as on the Brazilian plateau. The natural influence of physiography is, however, greatest in young mountains, many of which are still being steadily uplifted.

Biotic Factors. Although the biotic factor includes all organic or living elements in the environment, in a brief summary it is hardly possible to expand significantly the discussion, on pages 403–406, on the competition and interrelations between local plants and on the great influence that natural selection has upon the distribution of vegetation. The competitive influence is always of prime importance but over large areas has been rendered almost insignificant by the increasing influence of animals, including man.

Anyone acquainted with goats will know too well that herbivores devastate treegrowth by gnawing the bark of trees and by devouring saplings. The millions of bison roaming the Great Plains must have either extended the grassland at the expense of the forest or, at least, prevented the extension of forest. At the same time they induced the selection and dominance of grass species tolerant of trampling and grazing. Multitudes of small rodents can achieve much the same result by destroying saplings and seeds. Once wide grassy glades are formed the danger of fire in hot dry seasons becomes more acute and with each fire saplings and shrubs suffer more than grasses. In this respect, prairie, steppe, pampa, savanna and glades of bamboo and of alang-alang grass in the tropical forest are all alike.

In areas where man became predominantly a herder, he increased his flocks and herds and protected them and himself against carnivores. Thus began a notable decrease of carnivores and a tremendous expansion of semi-domesticated and domesticated beasts that has proceeded ever since. Simultaneously the herder himself discouraged treegrowth by causing fires and by felling trees for his own use. When and where man became a cultivator he deliberately and accidentally increased the burning and clearing of forests at a remarkable rate. To-day over large areas, man, as cultivator or herder, is the dominant factor in the local vegetation cover. In the United States the clearing of forests by European settlers was fantastically rapid, except in the Pacific West. The magnificent stands of White pine (*Pinus strobus*) in Wisconsin and Michigan were cleared so ruthlessly that few survived except in inaccessible spots. The huge *Sequoiadendron giganteum* also survived in deep gorges (in the Sierra Nevada of California) partly because its very thick bark protected it against forest fires. Its maximum age of 3,000 to 4,000 years is surpassed by that of the gnarled Bristlecone pine (*Pinus aristata*) which grows on isolated plateaux in the south-west United States and by that of the fire-resistant African baobab (*Adansonia digitata*). Two of these longest-lived giants are illustrated in Plates 43 and 44.

Animals play an appreciable part in the migration of plants but man has been the only deliberate transplanter, particularly of economically-useful and decorative species. The eucalypt has been widely planted in Mediterranean lands; the Douglas Fir has transformed a small tract of the Scottish Highlands into fine forest; a Russian thistle (*Salsola pestifer*), introduced into North America in 1874 with imported flaxseed and common throughout the Great Plains by 1900, is now a notorious tumbleweed.

Biotic factors also include minor organisms such as insects. The yucca is almost restricted to areas frequented by the *Pronuba* moth which fertilizes it while in many countries boring-beetles have devastated certain species of trees. In the Bermuda islands the fine stands of local 'cedar' (Juniperus) were utterly devastated after 1940 by a beetle introduced in imported timber. The multitude of dead trees has now been largely replaced by shrubs, palms, and species of pine immune to the beetle. A blight that arrived in New York State about 1904 has now devastated the sweet chestnut (*Castanea dentata*) throughout the United States. Similarly the fine chestnut forests (*Castanea sativa*) that flourished at between about 800 feet and 2,800 feet in the southern Alps in Europe have been seriously attacked by blight since 1948.

The Major Plant Formations

To-day the main groups of dominant vegetation are forest, shrub, grass and tundra. The tree is normally over 20 feet high and has a distinct trunk development. The division, however, is necessarily vague in more than a morphological sense. The main plant groupings grade into each other and which of them to-day remain true climatic climaxes and which should be considered plagioclimaxes (formed partly or largely by biotic influences) is not yet clear. If allowance is made for these uncertainties, the following major grouping of plant communities will be found useful in vegetation studies.

FOREST AND SHRUB

1. Tropical rainforest
2. Tropical semi-evergreen and deciduous forest
3. Tropical Thorn forest and semi-desert scrub and desert
4. Middle-latitude evergreen forest
5. Cool-season broad-leaved deciduous forest
6. Intermixed broad-leaved deciduous and coniferous forest
7. Boreal needle-leaved coniferous forest
8. TUNDRA
   GRASSLAND
9. Tropical grassland: Savanna, campo, llano
10. Middle-latitude grasslands: steppe, prairie, pampa
11. VEGETATION ON MOUNTAINS

Tropical Rainforest: Selva. The hot humid tropics although without a dry season has a relatively large amount of sunshine except in cloud-belts on mountains. These areas also enjoy long periods of relative calm apart from the occasional fierce squalls of thunderstorms and on coastal fringes the alternate flow of land- and sea-breezes. Temperatures below 65° F. are rare and growth is unusually rapid, being in exceptional cases up to 2 feet a day.

The plant forms and climax structure represent a competition and intermingling that has gone on for so many millions of years that the species are highly diverse and show a close adaptation to torrential rain and intense transpiration. Most of the trees are evergreen and although a few species shed their leaves for a brief period each tree has its own cycle. The leaves are usually medium sized, 6 to 12 inches being a common length (compare magnolia or laurel) and although often brightly coloured when young, invariably on maturing become leathery and dark-green. Many are glossy and most become highly mobile, turning edgewise to the sunlight. Really large-leaved trees, such as palms, are present but only numerous in open locations particularly riverine swamps and coastal flats and dunes.

The trees have straight, relatively slender trunks that are almost branchless up to their own particular crown-level. All develop an extremely thin, smooth bark, a feature which may be associated with the common habit of cauliflory, or bearing flowers and fruits directly on trunks and branches as in cacao, and with the more unusual habit of many taller species of sending out, at 10 to 25 feet above ground-level, large side-buttresses that join up with lateral roots (Fig. 156).

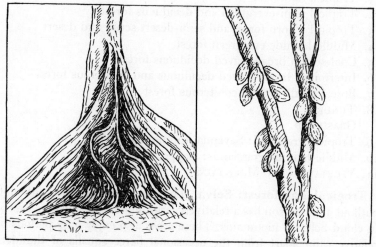

156. Cauliflory (cacao tree) and buttress roots in tropical rainforest

The true rainforest is usually three-tiered. Above, at 100 to 160 feet, stretches the sunlit canopy, uneven in height as it is pierced at irregular intervals with many epiphytes and with individual crowns that are seldom compact. In the gloomy, moist trunk space below flourish shade-loving species that commonly reach heights of 60 to 80 feet and 30 to 50 feet (Fig. 157). The trees of these two lower-tiers often develop umbrella-shaped crowns and leaves with an elongated drip-tip. All three tiers are interlocked by woody lianes that climb from tree to tree, frequently bare and unbranched for 60 or 70 yards (Plates 39 and 40). At least the two upper layers are also greatly thickened by small shrubs and herbs that grow profusely on boughs and twigs. This extra verdure is mainly epiphytic, especially orchids, but some species (mistletoe, etc.) are semi-parasitic, and some semi-epiphytic like certain species of fig (*Ficus*) which germinate

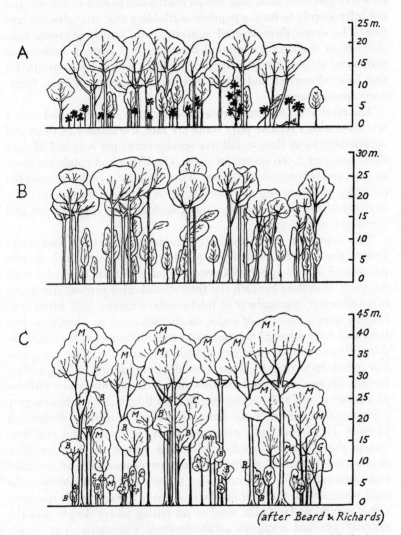

(after Beard & Richards)

157. Profile diagrams of tropical rainforest in Trinidad

A shows montane forest at 800 to 1000 m.; B, rainforest at 300 to 700 m.; and C, rainforest below 300 m. (984 ft.). C represents a forest strip about 210 feet by 25 feet in which *Mora excelsa* is common; where *Mora* is less common the tiers become less definite on a tree and then send long shoots downward to root in the soil and not infrequently to form a pendant scaffolding that strangles the host tree. The dense three-layered superstructure largely prevents sunshine from reaching the ground floor which is poor in herbaceous plants and shrubs, except struggling saplings, and rich in saprophytes that lack chlorophyll and derive their sustenance mainly from fungi-decomposed humus.

The rainforest contains a remarkable richness of species and even a few acres with a relative pure stand are rare. Commonly 20 to 40 and occasionally more than 50 tall tree species occur per acre and Malaya alone has over 2,500 species of trees. The lianes and epiphytes show an even greater number of species. This diversity greatly increases the difficulty of lumbering and harvesting in natural forests, especially as many of the hard, durable woods, such as ebony, greenheart and ironwood sink in water.

Where tropical rainforest is interrupted by clearings and water spaces, the sunlight encourages a dense undergrowth, rich in tree palms and other large-leaved species, that contrasts markedly with the poor, thin flora beneath the three-tiered selva proper. Alongside brackish water, particularly of tidal coasts, a narrow belt, often continuous over hundreds of miles, is characterized by shrubs or low trees that have adapted themselves to reproduction and breathing on diurnally-submerged mudflats outside the surf-line. The mangrove has given its name to these littoral communities although they include species in several families of which the mangrove (*Rhizophoraceae*) is only the commonest. At high tide, masses of bluish-grey evergreen foliage float on the water; at low tide an intricate scaffolding of twisted and bent stilts protrudes from the foetid mud and buttresses the short deformed trunks. The plants develop, on their trunks and branches, shoots which grow downward and outward and take root and in their turn send up new trunks. In addition, the seed when still in the fruit on the tree grows a long radicle that may either extend as far as the mud or on falling pierce deeply into the mudflat or float away to establish itself elsewhere. The whole community exerts a strong influence on the accumulation of fine sediment (Plate 41).

Tropical Semi-evergreen and Deciduous Forest. Where a distinct period of light or unreliable rainfall occurs regularly the selva begins to assume a more seasonal appearance. The change from evergreen to semi-evergreen and so to deciduous is gradual and complicated as wherever soil-moisture is abundant, especially on flood-plains, strips of true rainforest tend to persevere. The transition is also greatly complicated by extensive interference by man.

An important climatological factor, the length of the sun-baking period, is all too frequently ignored in studying these transitional areas. At the Equator, the period of overhead sun, or sun-baking, comprises two separate spells of about 30 days each. Towards either Tropic the period (now single) when the noon sun is almost or quite overhead increases rapidly in length and in strength as the diurnal daylight also lengthens. Near the Tropic the noon sun rises to or almost to the vertical for about 90 consecutive days (Fig. 158) and the sun-baking period becomes intense. This alone would give the areas either side of the Tropic by far the hottest hot-season experienced anywhere on the globe but in addition the dry anti-cyclonic regions athwart the Tropic experience much more sunshine than does the wet equatorial rainforest. Hence Fig. 158, which shows solar radiation outside the lower atmosphere, actually underestimates the torrid nature of summers in latitudes 12° to 35° north and south. Moreover, strong winds and relatively low humidities often further increase the hot-season evaporation and transpiration in these zones.

The increasing strength of the dry season is reflected in the increasing proportion of deciduous trees in the forest climax and in the decreasing amount of vegetation. In South America, the selva is gradually replaced by semi-evergreen forest, often about 70 feet high and two-tiered. In the upper tier about one-quarter of the species are deciduous although all may drop their leaves during abnormal drought. The lower tier consists almost entirely of evergreens, most of which are small-leaved and flower brilliantly in the sunny period. Lianes abound although less prolifically than in the selva but epiphytes are unimportant.

In monsoonal south-east Asia where the annual rainfall averages 80 to 95 inches and the dry season is less than 4 or 5 months, a common form of mainly evergreen climax is dominated by the pyinkado (*Xylia xylocarpa*), a deciduous leguminous tree that grows to 120 feet on deep soils. Where the annual rainfall declines to about 60 to 80 inches, although the pyinkado still predominates, teak may form one-tenth of the stand. The teak (*Tectona grandis*) grows into a good-sized tree of 60 feet in 15 years and when mature rises straight to 100 or 150 feet often from a buttressed base (Plate 42). It has bark

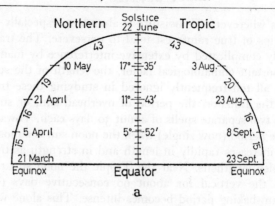

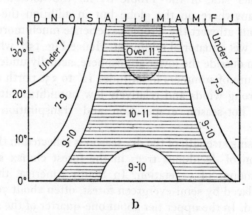

158. Length of sunbaking period in and near
the Tropics

a. On about dates given the sun is overhead at its quarter-
way stages. Numbers on circumference give days taken
for each quarter-way stage. b. represents duration and
strength of sunbaking period. Numbers denote relative
solar energy received per unit of horizontal surface if
there were no atmosphere. In June, if cloudiness and
turbidity are considered, the proportion between actual
insolation at the Equator and 30° N. is about 2 : 3 so
above diagram understates the relative strength of sun-
baking period near the Tropic

about ½-inch thick which generally allows it to survive when thin-
barked evergreen species typical of the rainforest are killed by forest
fires. Its leaves which are usually about 12 to 24 inches long and 6 to
12 inches across, drop a month or two after the drought begins,

presumably when easily available soil-moisture is depleted. The new foliage appears some time before the wet season starts as if in response to increase of spring temperature rather than of rainfall. The ground flora in these wetter teak forests is discontinuous and poor except where bamboos form thickets in abandoned clearings.

Where the dry season exceeds 5 consecutive months and the sun-baking period becomes more excessive, the semi-evergreen forest gives way to deciduous forest. The whole structure tends to be lower, more gnarled, more microphyllous (small-leaved) and more liable to branch nearer the ground yet it retains a two-tiered arrangement, the upper mainly deciduous including teak and acacia and the lower mainly evergreen. The taller trees now tend to develop pronounced umbrella or flat crowns and an abundant ground cover of shrubs and bamboo commonly occurs. Upon laterites and lateritic crusts that dry out rapidly, a gnarled, more open forest with a thick herbaceous ground cover may be present.

These gradations on the periphery of the rainforest are hard to find in Africa where centuries of burning by man have converted what probably was tropical deciduous forest into wooded savanna or open savanna (Plates 51 and 52).

Tropical Thorn Forest and Semi-desert Scrub and Desert. As aridity and the strength of the sun-baking season increase, the deciduous tropical forest gives way over large areas to a more stunted, more xerophytic treegrowth, variously known as thorn forest, thorn scrub and thorn woodland. Such communities abound on the more sterile, more porous soils in the Brazilian caatinga, in Mexico, East Africa and tropical Australasia. Their density varies from thick impenetrable thorn thickets to open stands of low, stunted trees. The thorn thickets grow slowly to a height of only 10 to 30 feet and in parts are interspersed with evergreen shrubs, characteristically deep-rooted and bearing leathery, waxed and resinous leaves. In some regions thin, thorny lianes occur and an abundance of ragged epiphytes resembling lichens and mosses (including Spanish moss). The main growth is sometimes overtopped by taller species such as umbrella-shaped acacias, water-storing candelabra euphorbias, wax-yielding palms (including the valuable carnauba in the Brazilian caatinga) and the bottle tree (*Cavanillesia arborea*) which has a swollen trunk up to 15 feet in diameter containing moisture-filled tissues. After rain, the shrubs bloom brilliantly and may be enhanced by flowering herbs and bulbs. Species of acacia are especially common in Africa and Asia and also compose the mulga and brigalow scrub of Australia.

On its drier fringes the thorn scrub grades into semi-desert with a more open type of highly-xerophytic vegetation. The species are lower; succulents, such as cacti, become more important, and shallow-rooting is not uncommon as the rainfall rarely penetrates deeply. Much of the desert of Australia supports only stunted mallee (*Eucalyptus*), wattle (*Acacia*) and tussocks of the sharp bayonet-like leaves of spinifex or porcupine grass (*Triodia*). In Asia and Africa the strange desert species include thick, low tubers like stunted bottle-trees. Plates 45 and 46 show semi-desert or hot desert scrub in Arizona where many of the scrubs, such as the mesquite, are very deep-rooted.

In hot deserts individual xerophytic species survive only where plant-roots can penetrate deeply to the water-table or where and when occasional rain wets the surface sufficiently to revive shallow roots and to provide moisture for storage in succulent leaves and stems or in underground bulbs and tubers. Many plants and seeds wait long periods for rain and some may be capable of utilizing dew. In the Namib desert of South-west Africa the *Welwitschia mirabilis* has a hollow woody stump, 6 inches high and several yards across, from the rim of which two large ribbon-shaped leaves grow octopus-like for a century or more.

Middle-latitude Evergreen Forest. Outside the tropics the eastern sides of continents tend to receive rainfall at all seasons whereas the western sides pass into semi-desert and desert as described above and eventually into a summer-dry Mediterranean climate. Hence the zonal distribution of climate is complex and it becomes extremely difficult to relate the forests to present-day climates, particularly as the great civilizations of China and the Mediterranean have obliterated vast areas of natural vegetation. Probably during the Pleistocene ice-advances the Mediterranean climate was wetter than to-day and differed little from eastern marginal climates.

To-day eastern humid sub-tropical margins, particularly in southern Japan, central and south China, south-east Australia and the west side of South Island, New Zealand, still retain patches of evergreen broadleaved forest, which abounds in lianes and epiphytes and yet permits a dense undergrowth. In some areas, oaks predominate while in New Zealand tree-ferns are common and in Australia and Tasmania leathery-leaved eucalypts which may exceed 200 feet in height.

Where cold spells occur occasionally and the rainfall is less, the forests commonly contain both evergreen broadleaved and coniferous species. These mixed evergreen forests survive mainly in the southern hemisphere particularly in Chile and southern Brazil where Araucaria, quebracho (*Aspidosperma* and *Schinopsis lorentzii*), evergreen beech (*Nothofagus*) and cedar (*Libocedrus chilensis*) dominate and in northern New Zealand where the kauri pine (*Agathis australis*) was once abundant. Mixed evergreen forest was probably dominant in early post-glacial times throughout areas to-day with a Mediterranean climate. The present landscape around the Mediterranean often contains individual species or clusters of evergreen oak (ilex; kermes; cork), of pine (stone; aleppo; rock) and of cedar (*Cedrus*). These trees, as with the vine and olive, have tremendous root-systems and probably represent the survivors of a climatic climax forest that has been felled by man. The deforested areas are partly cultivated and partly covered with low, dense thickets and more scattered growths of xerophyllous scrub, mainly spiney and evergreen, and variously called maquis, matorral and garrigue.

It is still more difficult to relate vegetation to present climates and environments in western North America where a great belt mainly of tall evergreen conifers stretches from southern Alaska to the Mediterranean climate of central California and in parts reaches inland long distances to the Rockies. The dominant species in the north is Sitka spruce. In the rainy parts of Washington and Oregon, western varieties of cedar (*Thuja plicata*) and of hemlock (*Tsuga heterophylla*) and Douglas fir (*Pseudotsuga taxifolia*) dominate while in the south, especially where coastal fogs drift inland, the dominant is the redwood (*Sequoia sempervirens*), specimens of which live for more than 3,000 years and attain over 300 feet, the tallest being about 366 feet high. On the uplands farther inland, western white pine (*Pinus monticola*) and western larch (*Larix occidentalis*) are often dominant but Douglas fir, various other tall firs (*Abies grandis*; *A. concolor*) and ponderosa pine, lodgepole pine (*Pinus contorta*) and limber pine (*P. flexilis*) are common. Many of these reach an amazing height but none exceeds in bulk the big-tree (*Sequoiadendron giganteum*) of the western slopes of the Sierra Nevada (Plate 44). These stout giants have a very thick bark and many of them show signs of having survived several forest fires. One of them, the General Sherman tree, is 32 feet in diameter at its base, 272 feet high and weighs over 2,000 tons.

Cool-season Broad-leaved Deciduous Forest. Poleward of middle-latitude evergreen forests there stretched in early post-glacial times a wide belt of broad-leaved forest, characterized by relatively slow growth, thin delicate leaves, thick rough bark (a protection against cold and evaporation), and leaf-fall in late autumn. In surviving woodland the crowns often reach 60 to 100 feet and occasionally 120 to 130 feet and according to their shape and density allow a lower tier of smaller trees and shrubs and a rich ground carpet of herbs and grasses. The strong seasonal rhythm is accomplished by means of successive vertical flowering; in early spring when the trees and shrubs are leafless, the ground flora of herbs and bulbs flowers brilliantly; then the shrubs bloom, including the glorious redbud (*Cercis*) and dogwood (*Cornus*); then the smaller trees, including the wild cherry (*Prunus*); and finally the dominant species.

The dominants vary considerably in their distribution. In the warmest parts of western Europe, the pedunculate and sessile oaks with hazel undergrowth predominated, except on calcareous soils where ash (*Fraxinus*) was often dominant or if the limestone was covered with clay-with-flints or some other rather acidic surface material, the shallow-rooted beech (*Fagus*). On shallow, acidic, sandy soils birch or oak-birch, often with an undergrowth of ling (*Calluna vulgaris*), became the chief species. Towards the Mediterranean, elm and lime were common and various other species of deciduous oak as well as sycamore, chestnut and beech. In eastern Europe the beech was usually absent and the oak strongly dominated. In Siberia, especially along the belt now followed by the trans-continental railway, the aspen and birch were dominant except in the far east where many species became important. Over vast areas in the eastern United States the vegetation was also floristically richer than in Europe, hickory being ubiquitous as well as certain species of oak, and basswood, maple and beech were common. Here in the extreme north-east, beech-maple tended to dominate but sweet gums, (*Liquidambar spp.*), chestnut and the tulip tree (*Liriodendron*) were important. The hemlock (*Tsuga canadensis*) increased in frequency southward where chestnut-oak dominated except on sandy and swampy tracts where, after clearance and burning, loblolly, shortleaf, longleaf and slash pines commonly took control. Just west of the Mississippi a more xerophytic association of oak-hickory was

[*Dr. Malcom Drew*

33. Root system and soil particles with attached water-film. Darker spaces are air-filled pores. Water film is shown as lighter rim around soil particles. Actual diameter of main root is ·2 mm.

[*Dr. Malcom Drew*

34. Root and root hairs of Italian rye-grass (*Lolium multiflorum*). Actual diameter of main root is ·2 mm., the upper photograph being double the magnification of the lower

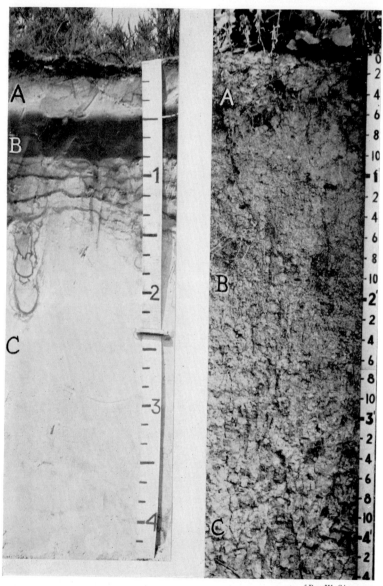

[*Roy W. Simonson*

35. Profiles of podzol and grey-brown podzolic soil. On left, podzol or spodosol
with dark humus-enriched spodic B horizon, formed from sand under heath
near Eindhoven, the Netherlands. On right, grey-brown podzolic or alfisol,
with less leached A horizon and less distinct B horizon, formed from loess
under oak-hickory forest in eastern Iowa

(*By courtesy of U.S. Department of Agriculture*)

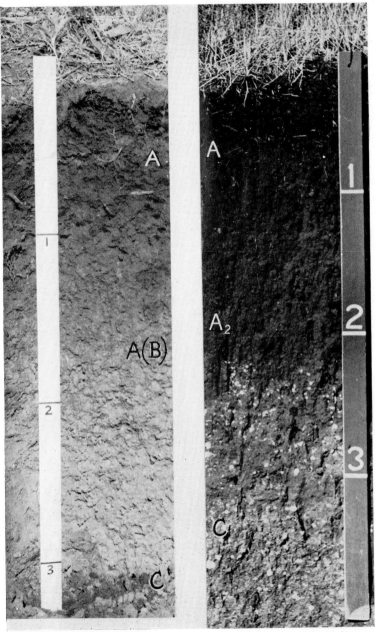

F. W. Cole]   [Roy W. Simonosn

36. Profiles of brunizem and chernozem (mollisols). On left, brunizem or brown earth or prairie soil formed from silty clay under tall grass in Missouri. On right, chernozem formed from glacial till under short prairie in S. Dakota. Scale in feet

(By courtesy of U.S. Department of Agriculture)

[*U.S. Information Service*
37. Gully erosion in Tennessee valley before reclamation

[*E. Cole*
38. Contour strip cropping in Dodge County, Minnesota
(*By courtesy of U.S. Department of Agriculture*)

39. Tropical rainforest, New Britain. Large trunk is *Eucalyptus deglupta* which here forms much of the upper tier

[*Dorien Leigh*

40. Lianes in topical rainforest. Lianes are probably the world's longest plants and may extend for 650 feet

[*Paul Popper Ltd.*

41. Mangroves (*Rhizophora mucronata*) at low tide in Malaya, showing stilt roots, breathing tubes or pneumatores and, in foreground, a few seedlings rooted in the mud

[*Paul Popper Ltd.*

42. Deciduous tropical forest dominated by teak in Burma

[Dr. H. A. Osmaston

43. Trunk of baobab, Tanzania, East Africa. Individual species may survive for about 5,000 years. Notice, at left base of trunk, mound built by termite ants

[*U.S. Information Service*

44. Lower trunks of giant redwood or bigwood (*Sequoiadendron giganteum*) on Sierra Nevada, California. These and the redwood (*Sequoia sempervirens*) are the world's largest trees. Individual species may survive for about 3,500 years

[*Dorien Leigh*

45. Semi-desert and desert scrub near Phoenix, Arizona. Tall whip-like plant is ocotillo (*Fouquieria splendens*); small cacti are *Echinocereus spp.*; tall cacti are saguaros (*Cereus giganteus*) which may reach 70 feet

[*E. O. Hoppé*

46. Semi-desert and desert scrub in Gila Desert, Arizona. To right, large cactus, about 12 feet high, is cholla (*Opuntia imbricata*). In background are saguaros and scrub growth including palo verde (*Cercidium*), mesquite (*Prosopis*) and creosote (*Larrea*)

Prof. F. K. Hare]

[American Geographical Society

47. Close boreal forest of black spruce (*Picea mariana*) on rough ground moraine near Alexander Lake, eastern Canada

Prof. F. K. Hare]

[American Geographical Society

48. Lichen-woodland of boreal forest-tundra transition on sandy plains of Hamilton River delta in eastern Canada. Widely-spaced black spruces rise above a thick carpet of lichens and occasional low bushes of dwarf birch

[*Royal Canadian Navy*

49. Lichen and rock tundra, Ellesmere Island, Canadian Arctic, in summer. In centre, on coastal flat near the inlet, is a summer camp

[*National Film Board of Canada*

50. Tundra, Baffin Island, Canadian Arctic, in summer, showing species of rush, sedge, cotton-grass and saxifrage growing amid lichen-covered boulders

[Dept. of Information, Uganda

51. Savanna woodland at about 6,000 feet in Uganda. In background, surmounted by orographic clouds, is the scarp of an outer spur of Mt. Elgon. On the right, patches of cultivation are visible

[Dr. H. A. Osmaston

52. Dry bush savanna, with baobab, Tanzania

[Uganda Govt. Ministry of Information

53. Giant lobelia at about 7,000 feet on the flanks of Mt. Ruwenzori

[Dr. H. A. Osmaston

54. Giant groundsels (*Senecio johnstonii*) near the snowline on Mt. Ruwenzori

dominant. In Canada between the prairies and the boreal coniferous forest, a deciduous belt, up to 100 miles wide, of dominant balsam poplar, aspen and white spruce (*Picea albertiniana*), forms a gradation similar to that in Siberia.

The variety in these forests was small and extensive stands of the same species were common, especially in Europe. The whole formation, except the secondary pine communities, has largely been cleared for cultivation of its rich brown and grey earths and for charcoal, fuel and building timber.

Intermixed Broad-leaved Deciduous and Coniferous Forest. In the northern continents stands of broad-leaved deciduous and coniferous trees intermixed except near the western Great Lakes where fine stands of evergreen conifers, especially white pine (*Pinus strobus*), completely dominated the vegetation. Elsewhere the gradation consisted of a mosaic of patches and strips composed mainly or entirely of either coniferous or broad-leaved deciduous species. The lack of true mixing, as in the selva, was probably due to the newness of the formation or ecotone, large areas of which had been covered by at least one ice-advance in the Pleistocene and all of which had experienced cold climates each time the ice-sheets expanded. In European Russia, oak associations interdigitated with pine associations; in north America just south of the St. Lawrence and Great Lakes, white (*P. strobus*) and red pines (*P. resinosa*) dominated on sandy tracts, oak and hickory on the heavier soils, and willow-aspen-alder-birch in swampy areas. Often where the landscape was dissected, dominantly deciduous associations covered the sunny slopes while the shady slopes were clothed mainly with conifers. To-day almost all this ecotone has been exploited and arable, pasture and secondary treegrowth reflect the economic value of the forest soils and of the convenient proximity of commercial stands of softwoods and hardwoods.

Boreal Needle-leaved Coniferous Forest. Beyond the intermixed deciduous-coniferous associations in the northern hemisphere there stretches a wide belt of predominantly needle-leaved coniferous forest, which lies largely on areas overrun by Pleistocene ice-sheets and still retaining much marsh, swamp and lake. Where not covered by ice the soils froze deeply and still contain at various depths a permanently frozen, or permafrost, layer. To-day most of these regions are thinly peopled as the winters are very cold and the frost-free period, although the summer daylight is long, rarely exceeds the three warmest months. The annual precipitation equals about 15 to 20 inches of rainfall and comes mainly in summer but winter snowfall is important and the typical conical tree is ideal for shedding snow. Transpiration and growth proceed slowly and most trees are less than 19 inches in diameter. On occasions winds may be strong even in the cold season and then evaporation is reduced by the small leathery needle-shaped leaves that in nearly all species fall a few at a time, the foliage often taking five years or more for complete renewal. The needles form raw humus (mor) deficient in mineral nutrients and discourage a ground flora, although the forest readily regenerates itself. Partly because of the relative newness of the plant colonization, the species are few and relatively pure stands of the same species grow over wide areas. The general aspect is sombre and monochromatic and is little brightened by the poor ground flora of mainly bilberry, crowberry and dwarf birch with abundant mosses and lichens (Plate 47).

The main variations seem controlled by drainage, soil fertility and burning. In Europe, pines predominate on sandy soils and spruce on the richer soils whereas in Siberia, where large areas were never ice-covered, the fir (*Abies sibirica*), larch (which is deciduous), and Siberian stone pine (*Pinus sibirica*) are also common and towards the east various dwarf species of pine and larch. In North America on the Laurentian Shield in Quebec and Ontario the worst and wettest soils carry jack pine, those areas with moderate to poor drainage support black spruce and tamarack or eastern larch (*Larix laricina*) and those of better quality white spruce and balsam fir. Farther west the lodge-pole pine and alpine fir and other western species predominate. Everywhere the muskegs or bogs abound with spagnum moss and cotton grass.

The boreal forest forms the world's chief source of soft timber, particularly for woodpulp, and of furs. If cleared in relatively narrow strips it regenerates itself naturally within a few years but if felled over large areas usually produces birch-aspen woodland that takes a long time to revert to conifers.

Tundra. The northern parts of the boreal forest grade into treeless tundra over a belt often several hundred miles wide (Fig. 159). Usually the boreal formation of close conifers with a shaded, mossy floor degenerates into a lower, more open coniferous stand often with a brightly-lit ground flora of lichen and discontinuous shrub (Plates 47 and 48). Then, probably because radiant energy decreases and wind force increases, this woodland degenerates into an inter-digitation of patches of hardy conifers on the better soils near the watercourses, and of tracts of tundra on the interfluves. Here the stunted conifers often show much dead wood and a veneer of lichen and moss, and usually begin to be replaced by stunted birch and in

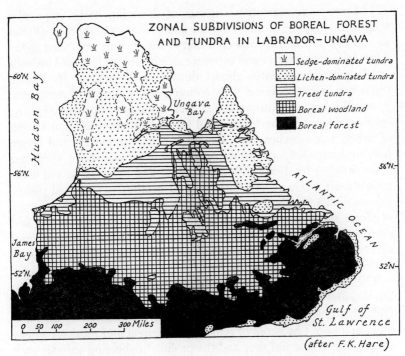

159. Boreal forest–tundra transition in Labrador-Ungava

places also by bushy aspens and larches. These deciduous species seem able to survive better than evergreen conifers in climates prone to strong sub-zero winds.

The tundra has a short summer season of about 2 months and a still shorter frost-free period yet the continuous or almost continuous daylight encourages rapid growth and many plants complete their flowering cycle in three or four weeks. Not infrequently unseasonal frosts cuts off the vegetation when in flower or fruit. But climate also has a severe mechanical effect on vegetation. In late spring or early summer the soil begins to thaw out and tends to liquify above the permafrost, which even in the warmest weather rarely retreats more than a few inches or a foot. In autumn the surface layer expands again on re-freezing. Thus over large areas the root development is very shallow and on slopes is repeatedly disturbed by solifluxion and on flat areas disrupted by the alternate expansion and contraction of freeze and thaw.

The common vegetation consists of flowering herbs, especially sedges and grasses, of dwarf shrubs and of mosses and lichen (Plates 49 and 50). The density and continuity varies mainly according to the depth of surface matter and exposure to wind-blast. Exposed uplands are largely stony wastes almost devoid of vegetation. Ill-drained tracts are mainly covered with cotton grass, sedges and sphagnum moss. Where undulations occur and drainage improves slightly, considerable areas may be carpeted mainly with an intertwined mesh of lichen a few inches high. On thick peat, the mosses and lichen become hummocky and bear tussocks of grass and sedge which are often invaded by dwarf willow and alder and by small-leaved evergreens such as crowberry, bilberry and bearberry. On flood plains, grasses with a wealth of flowering herbs, such as anemone, buttercup, saxifrage and willow-herb, may prevail. On better drained, sunny and sheltered slopes, these flowers may form 'bloom-mats' or heath and shrubs, including dwarf rhododendron, alder, birch, and willow, may coalesce into low-growing thickets.

The economy of the natives of the tundra is based mainly on hunting and fishing in the New World and on reindeer-herding in the Old. Much modern tundra has been strongly influenced by man and other animals. The herds of reindeer or caribou and musk-ox follow annual migrations and they and the prolific fur-bearing animals such as hare and lemming soon deplete the local vegetation. In a recent attempt to introduce reindeer herding in parts of North America, the deer increased so rapidly that they soon outstripped the local natural food supply. Thus it seems highly probable that in the Old World at least, forest would advance farther into the tundra were it not for man's influence through tree-felling, fires and reindeer-herding.

Grasslands

Grasses (*Gramineae*) form one of the most numerous and widespread of the families of flowering plants. Their many thousands of species range in size up to bamboo, which may reach 120 feet in height, and include many grain-bearing plants indispensable to man.

Most grasses have a short flowering-cycle and yield seeds that withstand cold and heat well especially if enmeshed in the humus-mat that accumulates beneath the plant. Many species spread quickly by means of rhizomes and suckers and nearly all avoid the shade. The need for a sunny ripening spell and the relatively short stature of grass species ensure that eventually trees will dominate grass unless conditions are strongly inimical to treegrowth, as happens in the tundra where grasses are often the most abundant plant. To-day the chief grasslands are the savanna of the tropics, and the steppe, prairie and pampa of middle latitudes.

Tropical Savanna. The term savanna, which is derived from an Arawak word current in the Greater Antilles when Columbus arrived and meaning open meadows with scattered trees, is to-day widely applied to tropical landscapes where grasses and sedges abound and trees, if present, are scattered or in strips and patches, particularly in hollows and alongside watercourses. The dominant plants are usually bunch or tussock grasses with dense stiff stems terminating in silvery spikes. Bare soil is often exposed between the tussocks. The species of grass and the amount of treegrowth vary considerably in different regions.

Over two extensive areas on the periphery of the Congo rainforest giant grass-low tree savanna occurs, the dominant elephant grass and other bunch grasses being 5 to 15 feet tall and the numerous deciduous trees about 30 or 40 feet high. This rich growth grades into a more extensive tall grass-low tree savanna in which the tussock grasses seldom exceed 5 feet and grow even beneath the acacia and other umbrella-shaped trees. Similar grassland in South America (campos and llanos) often has a scattering of isolated palms. Plates 51 and 52 show well-wooded savanna and a drier type of savanna in East Africa.

The problem of why grasses predominate in savanna is far from solved. On some flood-plains where waterlogging occurs frequently (as in the llanos of the Orinoco) sedges and marsh plants dominate and palms are the chief trees. These edaphic conditions may favour grasses more than trees but, on the other hand, the palms present are fire-resistant. In the Asiatic tropics and East Indies on poor laterites and on porous soils the trees become stunted and grass is encouraged, which in its turn may encourage fire to the further detriment of the treegrowth. In Australia the open crown nature of eucalypt forests seems to encourage grass beneath them but this continent is notorious for forest fires. There seems little doubt that, particularly in Africa, most of the savanna and tropical grassland is due to burning, accidental or deliberate, and that the surviving trees, such as the water-storing baobab (*Adansonia digitata*), *Bombax* and *Ceiba*, are strongly fire resistant. The baobab develops a wide-spreading root-system, a small branching canopy and a short massive trunk, often 12 to 30 feet in diameter, that stores large quantities of water in its soft tissues (Plates 43 and 52).

Among the reasons why fires favour herbaceous plants at the expense of woody species is the relatively slow growth and longer life-cycle of the woody species. A fire that destroys grasses kills the growth often of a year or a few years only whereas the tree and shrub, unless fire-resistant, loses in a few minutes the growth of decades or even centuries. When growth resumes after the fire, the herbaceous species, with the marked advantage of being adapted to a short-life cycle, recover at a rapid pace. Thereafter, the struggle is like the race between the hare and the tortoise, the slow-growing tree would win in the long run but not if it had to cross a fire-barrier. In this respect it is noticeable that the world's great grasslands occur on relatively flat surfaces over which wind sweeps and fires spread most effectively. Often grasses give way to treegrowth at breaks of slope where valleys incise a plateau or precisely at those locations where the wind-velocity suddenly drops and fires tend to die out.

The direct destructive action of man and other animals must also be considerable. The natives use large quantities of wood for domestic purposes and the larger herbivores, such as antelope, zebra, giraffe and elephant, as well as the minor rodents, are avid destroyers of treegrowth and saplings. Thus the question of climatic influence must be assessed in the light of biotic competition. The dry season heat, the strong winds and tinder-dry grass-stalks favour occasional extensive fires. Were the climate cooler in summer and the regular drought shorter or absent as in the rainforest the treegrowth would re-establish itself. In the dry-season climate, once chance fires and animals have created grassy glades, the process is likely to spread because, with encouragement, the biotic (animal) influence expands at an explosive rate. The total exclusion of animals and fires for long periods would probably convert most tropical grassland into deciduous forest. But the length of time needed must be stressed as much savanna has been created from forest and woodland that dates back to the more pluvial climatic conditions that prevailed in these latitudes during the last main glacial phase. The climate to-day is drier and slightly warmer than it was 20,000 years ago and is definitely less favourable to natural re-afforestation.

Middle-Latitude Grasslands. Large areas of the steppes, prairies and pampa are cultivated to-day and in most continents only the drier fringes remain under pasture. Even in these, however, the natural vegetation has been greatly influenced by man and over-grazing. In North America, when first seen by Europeans, the tall-

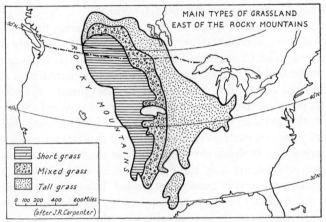

160. Main types of grassland east of the Rocky Mountains
The blank area in the prairies is the Sand Hills of Nebraska

grass prairie nearest the forest edge was a continuous sward that included bunch grasses and numerous herbs that flowered successively throughout the warm season. The dominant types grew to 5 or 6 feet and sometimes to double that height; the annual rainfall averaged more than 20 inches; the soils were pedalfers rather than pedocals; and strips of forest lined the stream banks. Westward on much of the Great Plains where the rainfall is less, trees were rare and the grasses were a mixture of tall and of short species arranged in an upper storey 2 to 4 feet high and a lower storey (of buffalo and grama grass) up to 1 foot in height. Farther west between the mixed grass belt and the Rocky foothills, short grasses prevailed (Fig. 160). These American grasslands have been described as 'a relatively stabilized product of nature, the outgrowth of climate, soil, vegetation, animals, and micro-organisms, all interacting together' (J. C. Malin,

p. 1). But many scholars would place fire and animals before climate as the predominating influences, particularly in the tall-grass prairie.

The steppes of the Old World once showed gradations similar to those in the American prairies. Near the deciduous forest edge the meadow-steppes of mainly sward-forming grasses, often 4 feet or more high, probably occupied the drier interfluves above the ribbons of trees in the valleys. Southward as the dry season strengthened, the meadow-steppe merged into dominantly tussock-grass steppe, 3 to 4 feet high, and then into short-grass steppe dotted with semi-desert shrubs such as wormwood (*Artemisia*).

The prairies and steppes have cold winters whereas the grasslands of the southern hemisphere are relatively mild and relatively humid all the year. The Canterbury Plains in South Island, New Zealand, were dominated by tussock grasses and sparsely dotted with shrubs and cabbage trees when first settled by Europeans. Similarly the western parts of the pampas of Argentina were mainly under tall grass, with forests and thickets of small trees in the moister hollows. The drier eastern parts were dominated by short-grass species with a scattering of xerophytic shrubs. To-day in the wetter pampas, as in the better-watered parts of the South African veld and the Canter-bury Plains, trees will grow if protected from fire and animals. Hence the absence of trees here is apparently largely due to burning and biotic influences. It is noticeable that the aspen, which to-day invades the prairie edge in Canada and Siberia, spreads rapidly by means of underground suckers that normally would not be killed by forest fires. Even scholars who consider short-grass prairie as a climatic climax are faced with the problem that probably the true climax for such a climate would be xerophytic scrub. To-day man's extensive exploitation of the grasslands has made the problem of their origin very difficult to solve.

Vegetation on Mountains

As already discussed in Chapter 7, vegetation on mountains arranges itself in almost horizontal zones that reflect mainly altitude and orientation but which also reflect the general floristic situation of the mountain mass. Normally on mountains vegetation formations occupy the altitudinal zones most similar in climate to that of their main lowland habitat. Thus the needle-leaved boreal forest extends at a steadily rising altitude on mountain ranges all the way from the sub-arctic to the Equator (Fig. 161). The continuity is more complete

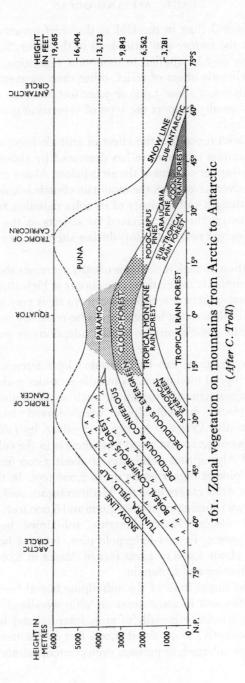

161. Zonal vegetation on mountains from Arctic to Antarctic
(After C. Troll)

DI

in the New World than in the Old as the chief mountain belts are meridional in the former and latitudinal in the latter. Several useful generalizations may be applied to mountain vegetation.

1. As the climatic effect of relief, other than on precipitation increase, is small below about 3,000 or 3,500 feet, the base and foothills of mountains usually support the type of vegetation growing on the adjacent lowlands.

2. Above about 6,000 feet the effect of altitude becomes excessive as the mean annual temperature has decreased by about 20° F. and half the water-vapour content of the air is below. Above these heights, unless cloud cover is frequent, the mountain climate is unique because of the great strength and intensity of its solar radiation and the great difference between the heat received by objects in the sun and in shadow and (except near the poles) during alternate diurnal daylight and darkness.

3. Precipitation on high mountains usually decreases above a certain height and xerophytic associations often occur at high altitudes and in lee-side locations where descending lee-wave air is very dry.

4. Mountain plants may in some areas be relatively isolated relics of ancient plant associations or special adaptions to peculiar conditions.

These generalizations can be illustrated by reference to middle-latitude and tropical mountains. In middle-latitudes with increase of altitude the zonal climax vegetation, such as deciduous forest, is gradually replaced by sub-alpine needle-leaved boreal forest. Commonly in the Pyrenees spruce and silver fir (*Abies excelsa*) dominate between 5,000 and 7,000 feet whereas in the colder winters of the central Alps spruce dominates at about 3,000 feet and larch and cembran pine at higher levels up to 7,000 feet. In the Algerian and Moroccan Atlas ranges near the Mediterranean coast, fine cedars (*Cedrus atlantica*) dominate between 4,000 and 6,000 feet. In the great Cordillera of western North America, sub-alpine boreal forest, especially of spruce, fir and lodgepole pine, forms a belt rising in altitude from about 2,000 to 4,000 feet in Alaska to 8,000 to 12,000 feet in New Mexico and Colorado.

Towards the upper limit of the sub-alpine boreal forest the tree-growth dwindles and in some areas an 'elfin woodland' of gnarled, low-spreading species, especially of pine, interdigitates with patches of grass. Frequently, as at about 7,000 feet in Switzerland, these grassy 'alps' are enlarged, or perhaps even created, mainly by grazing.

However, the strongest influences on the height of the tree-line are probably wind-speed and excessive transpiration or evaporation.

Above the tree-line the vegetation deteriorates into tundra with many species common to the Arctic, as well as endemic (local) species. The plants of the Alpine tundra include bulbs that flower brilliantly in early spring as soon as the snowcover melts and herbs that bloom in summer. Permafrost is absent largely because of good drainage, intensity of solar radiation in summer, rapid surface evaporation and often of a deep snowcover in the cold season. Above the tundra zone, a few isolated plants may survive but bare rock and snow and ice usually mark the upward limit of plant-growth which is virtually the permanent snow-line.

Within the tropics mountains reach the snow-line in central and South America, East Africa and New Guinea. The lower flanks of these lofty ranges and peaks are usually wetter than the adjacent lowlands because of the orographic effect and the middle flanks often show a belt of maximum cloud above which precipitation decreases and sunshine increases rapidly. A typical sequence of vegetation stages would be:

(i) tropical three-tiered rainforest;
(ii) at about 4,500 feet in the Andes and 5,500 feet in East Africa and New Guinea, two-tiered forest with dominants often only 60 to 70 feet high;
(iii) at about 6,000 or 7,000 feet, or the cloud-layer, true montane, 'mossy' forest, a dense growth of stunted trees, seldom above 40 feet tall, usually contorted and richly festooned with mosses and lichens that are moistened frequently by clouds;
(iv) at 10,000 feet, above the usual cloud-layer and where the weather is sunnier and drier, a slightly taller, less-lichened, less dense 'mossy' forest that grades upward as drought and winds increase into dwarfed, gnarled treegrowth;
(v) at about 12,000 feet, alpine tundra of shrubs and grassland of bunch or tussock-species (the punas of the Andes). Many of the plants are common in the Arctic tundra but some, such as the tall groundsel (*Senecio*) and giant lobelias on Ruwenzori and Kilima Njaro in East Africa, are interesting relics or survivals of considerable antiquity (Plates 53 and 54);
(vi) at about 14,800 feet in western New Guinea, 15,000 feet in East Africa and 16,500 feet in the drier climates of the paramos

of Central Peru, the permanent snowline is reached. At and above these altitudes it is rare for a night to pass without frost, and sunny exposures here are the only parts of the world that experience freeze and thaw almost every day of the year.

The Conservation of Plants and Vegetation

To-day, except in primitive societies man depends for his staple food products mainly upon cultivated and improved species of plants. Improvements in plant qualities and in techniques of cultivation are made almost annually. Civilized nations are also busily introducing into their countries foreign species for the sake of either their utility or their beauty. But, on the other hand, the result of millennia of exploitation and interference is that very few countries have preserved appreciable areas of 'natural vegetation' and that thousands of interesting species of plants have been almost or quite exterminated. The preservation of special plants can be aided by artificial conditions in botanical gardens but the conservation of rare plants and of plant associations in 'nature reserves', such as the Sequoia National Park in California, is often difficult. Perhaps the only beech forests left in the northern hemisphere that have not changed since the middle Tertiary period are those on the Elburz mountains in northern Iran and these are now beginning to be exploited.

There is to-day little original or virgin vegetation left to conserve except in the most remote and inaccessible parts of the world. Unfortunately in small nature reserves it is seldom possible to preserve simultaneously both animals *and* plants as animals are great destroyers of vegetation and unless kept in check gradually devastate the surface cover. It is, of course, possible to have a naturally balanced community of wild animals that has evolved into equilibrium with the vegetation, or up to the limit set by available food supplies. But this happens only when all animals, including carnivores, are protected from man, as herbivores, unless kept in check by predators and catastrophes, usually outstrip their local food supply. Man is omnivorous and can acquire additional food supplies by more intensive cultivation of large areas of the continents and by making more use of the abundant life in the oceans. Yet the world population increases at an alarming rate. At present about 60 million more people are added each year and if the existing rate continues the total (nearly 3,300 million in 1965) will be doubled within the next forty years. Already some kind of global planning seems necessary if adequate examples of our great heritage of plants and animals are to be conserved for the enjoyment and enlightenment of posterity.

LIST OF BOOKS

General

Schimper, A. F. W. *Plant Geography upon a Physiological Basis*, 1903; *Pflanzengeographie*, 2nd edition, 1935 (2 vols.; pp. 1,613)
Braun-Blanquet, J. *Plant Sociology*, 1932
Weaver, J. E. & Clements, F. E. *Plant Ecology*, 1938
Van Wijk, W. R. (ed.) *Physics of Plant Environment*, 1963
Cain, A. S. *Foundations of Plant Geography*, 1944
Good, R. *The Geography of Flowering Plants*, 1964
Dansereau, P. *Biogeography: an Ecological Perspective*, 1957
Schmithusen, J. *Allgemeine Vegetationsgeographie*, 1959
Daubenmire, R. F. *Plants and Environment*, 1959
Polunin, N. *Introduction to Plant Geography*, 1960
Eyre, S. R. *Vegetation and Soils*, 1963; 'Determinism and the ecological approach to geography', *Geog.*, 1964, pp. 369–76

Regional

Tropical:
Beard, J. S. 'Climax vegetation in tropical America', *Ecology*, **25**, 1944, pp. 127–58; *The Natural Vegetation of Trinidad*, 1946; 'The classification of tropical American vegetation-types', *Ecology*, **36**, 1955, pp. 89–100
Waibel, L. 'Vegetation and land use in The Planalto Central of Brazil', *Geog. Rev.*, **38**, 1948, pp. 529–54.
Black, G. A. *et al.* 'Some attempts to estimate species diversity and population density of trees in Amazonian forests', *Bot. Gaz.*, **III**, 1950
Richards, P. W. *The Tropical Rain Forest*, 1952
Cole, M. 'Cerrado, Caatinga and Pantanal; . . .', *G.J.*, **126**, 1960, pp. 168–79; 'Vegetation and geomorphology in Northern Rhodesia; . . .', *ibid.*, 1963, pp. 290–310
Harris, D. R. 'The invasion of oceanic islands by alien plants: an example from the Leeward Islands, West Indies', *Inst. Brit. Geog.*, 1962, pp. 67–82
Went, F. W. & Westergaard, M. 'Ecology of desert plants', *Ecology*, **30**, 1949, pp. 26–38
UNESCO *Tropical Soils and Vegetation*, 1961
Langdale-Brown, I., Osmaston, H. A. and Wilson, J. G. *The Vegetation of Uganda*, 1964

Sub-tropical and Mid-latitudes
Schmieder, O. 'The Pampa—A natural or culturally induced grassland?', *Univ. of Calif. Publ. in Geog.*, 2, 1927
Carpenter, J. R. 'The grassland biome', *Ecol. Monogs.*, **10**, 1940, pp. 617–84
Sauer, C. O. 'Grassland climax, fire and man', *Journ. of Range Management*, **3**, 1950, pp. 18–21
Weaver, J. E. & Albertson, F. W. *Grasslands of the Great Plains* (Lincoln, Nebraska), 1957
Coupland, R. T. 'A reconsideration of grassland classification in the northern Great Plains', *Jour. Ecol.*, 1961, pp. 135–67
Malin, J. C. *The Grassland of North America*, 1956
Roseveare, G. M. *The Grasslands of Latin America* (Aberystwyth), 1948
Tansley, A. G. *The British Islands and Their Vegetation*, 1950; *Britain's Green Mantle*, 1949 (shortened version of above)
Kuchler, A. W. 'The broadleaf deciduous forests of the Pacific Northwest', *Annals Assoc. Amer. Geog.*, **36**, 1946, pp. 122–47
Braun, E. L. *Deciduous Forests of Eastern North America*, 1950
Gleason, H. A. & Cronquist, A. *The Natural Geography of Plants*, 1964

Boreal and Tundra
Hare, F. K. Climate and zonal divisions of the boreal forest formation in Eastern Canada, *Amer. Geog. Soc.* Spec. Publ. **32**, 1950; 'New light from Labrador-Ungava', *Ann. Ass. Am. Geog.*, **54**, 1964, pp. 459–76
Sigafoos, R. S. 'Frost action as a primary physical factor in Tundra plant communities', *Ecology*, **33**, 1952, pp. 480–87
Benninghoff, W. S. 'Interaction of vegetation and soil frost phenomena', *Arctic*, **5**, 1952, pp. 34–44
Tikhomirov, B. A. 'The treelessness of the Tundra', *Polar Record*, 1962, pp. 24–30

Mountains
Schweinfurth, U. Die horizontale und vertikale Verbreitung der Vegetation im Himalaya, *Bonner Geogr. Abh.* **20**, 1957
Troll, C. Die tropischen Gebirge, *Bonner Geogr. Abh.*, **25**, 1959
Seifriz, W. 'Vegetation zones in the Caucasus', *Geog. Rev.*, **26**, 1936, pp. 59–66

Deserts
UNESCO *Plant-Water Relationships in Arid and Semi-Arid Conditions* (Arid Zone Research XV & XVI), 1960–1961

Vegetation and Rainfall
Mather, J. R. (ed.) 'The measurement of potential evapo-transpiration', *Publications in Climatology* (Seabrook, N.J.), **7**, 1954
Thornthwaite, C. W. and Mather, J. R. 'The water balance', *ibid*, **8**, 1955
Penman, H. L. *Vegetation and Hydrology*, 1963 (with full bibliography).
U.S. Department of Agriculture. Yearbook. *Water*. 1955

INDEX

ABANDONED ENTRENCHED MEANDERS, 266-69, 273
Adelie Land, 86
Adiabatic cooling, 51; warming, 51, 105
Adret, 99
Adur, river, 261
Agassiz, lake, 325
Ahaggar Mountains, 108, 217, 223, 224
Ailsa Craig, 287
Air masses, 79, 112-14
Aire, river, 236
Akassa, 83
Akyab, 78, 80-81
Alaska, 123, 185, 275
Albedo, 349-50, 352
Aletsch Glacier, 275, 327
Aleutian Islands, 162
Alexandria, 250, 252
Alluvial fans, 248
Alps, climate, 98-106, 108; glaciation, 275, 279; human geography, 109; local winds, 101-6; river régimes, 320-21; structure, 191-93, 199; vegetation, 109, 414, 434, 435
Altdorf, 108
Altitude of sun, 24-25, 42, 69, 85, 419-20
Amazon river, 259, 322; basin, 83-84; régime, 241, 315, 316
America, North, 90, 230, 279, 292, 332
Anabatic winds, 103-4, 344
Andes, 93, 99, 201, 206, 259, 280, 327
Anglesey, 194, 282, 359
Annan, river, 256, 261
Anniviers, Val d', 110
Antarctic Ocean, 167
Antarctica, 42, 200, 274; climate, 62, 85-86, 96-97
Antecedent drainage, 272, 273
Anticyclones, 61, 65, 88, 118-20
Antrim, 183, 209
Apalachicola, river, 329
Aphelion, 22, 278
Appalachians, 190, 191, 192
Arabia, 93, 94, 95, 217
Aral Sea, 325
Ararat, 207
Arctic region, 66, 86, 87; circle, 22, 274; climates, 30, 85-87, 96-97; front, 65, 112, 113, 114; Ocean, 133, 135; soils, 384-85; vegetation, 426-28
Arêtes, 283
Artesian wells, 227, 236
Artifical rain-making, 57-58, 59
Ascension Island, 168
Asia, 24, 31, 75, 90, 91, 92, 221, 312

Assam, 327, 328
Astrakhan, 96, 97
Aswan, 93, 322
Asyut, 94
Athens, 49, 73, 80-81
Atlantic, North, 136, 139-41; South, 137-38
Atlantic Ocean, circulation, 135, 137-41; deposits, 165, 166; relief of bed, 161-62; tides, 152-55
Atmosphere, general circulation, 60-67; composition, 28, 46; density, 28-29; dust in, 29, 42, 43, 51; pollution, 356-58, 359; height, 2, 28, 46; moisture, 29, 42-43, 49-59; pressure, 28-29, 60-64; temperature, 41-48
Atoll, 170-75
Aurora Borealis, 2, 28
Australia, 35, 72, 176, 324; Barrier Reef, 172
Auvergne, 204, 206, 207
Aven, 235
Avernus, lake, 207
Avonmouth, 155, 156
Awe, loch, 195, 285
Aysgarth Falls, 244
Azizia, 94
Azores, 120, 125, 331

BAD LANDS, 218, 223
Baguio, 54
Baikal, lake, 325, 335
Balkash, lake, 325
Baltic Sea, 47, 133, 136, 188; coast, 249, 310; Heights, 292
Baobab, 414, 430
Barchan, 219-20
Barrier reef, 171-72
Barrow Point, 86
Basalt, 183, 203, 208-10
Basic lava, 204-5
Batavia (Jakarta), 49
Batholiths, 211
Bay-bars, 299-303
Beach, 299, 300, 301, 302; ridges, 299
Beachy Head, 297
Beaded esker, 289-90
Bedding planes, 181
Bellegarde, 240
Bellingham, 260, 265
Ben Nevis, 54
Benares, 80-81
Benguela Current, 138, 141
Berg wind, 107
Bergen, 88, 91, 96, 97

Bergschrund, 283
Berlin, 91, 96, 97
Bermuda, 169, 175
Berwick Law, 207
Beult, river, 262
Bevers, 101, 110
Biscay, Bay of, 161
Black Forest, 197
Black Sea, 47
Blackwater, river, 262
Bleawater Tarn, 283
Blizzards, 86, 90
Blyth, river, 260
Bogota, 83
Bolsón, 224
Bombay, 77, 78, 80-81
Bonneville, lake, 337
Bora, 74
Borrowdale, 248, 284, 294
Boulder Clay, 286
Brander, Pass of, 285
Brenets, lake, 328
Brest, 142, 151
Brighton, 346
British Columbia, 308
British Isles, atmosphere, 45; building of, 194-95; climate, 57, 64, 356-57; continental shelf, 157-59; glaciation, 280, 281, 286, 287, 288, 290-91, 292; tides, 153-55; vulcanism, 208, 210, 212
Broads (Norfolk), 304
Bunker Hill, 287
Bunyoni, lake, 330
Buran, 90
Buttermere, 248
Butte, 223

CAIRO, 49
Calcium carbonate, 131, 231, 371, 375
Calcutta, 49, 78
Calgary, 91, 106-7
California, 53, 55, 72, 107, 185, 186, 339
Campos, 80, 81
Canada, 88, 91, 92, 325, 426
Cantal, mountain, 207
Canterbury Plains, 107
Cape Province, 107
Cape Town, 22
Cape Verde Islands, 169
Carbonation, 215, 231, 371
Carbon dioxide, 28, 294, 406
Caribbean Sea, 139
Carlisle, 294
Carse of Gowrie, 107
Casiquiare Channel, 259
Caspian Sea, 312, 324, 325, 355
Cataracts, 243
Catarro, 73
Cauldron Snout, 243

Causses, 235-36, 237
Central Massif of France, 235-36
Ceylon, 76, 78, 175
Chaillexon, lake, 328
Chalk, formation, 179, 180, 182; hydrology, 228, 229; topography, 231, 261
Chandolin, 110
Channels (river), abandoned, 247, 267; buried, 282; overflow, 290, 292; submarine, 158, 159-60
Chebka, 218, 223
Cheddar (Mendips), 232
Chernozem, 380, 391-93
Cherrapunji, 54, 59, 79
Chesil Bank, 301-2
Chile, 72, 423
China, North, 88, 221
Chinook, 106-7
Churn, river, 229, 230, 263
Cirques, 282-83
Cirrus, 55
Clay minerals, 362-65, 369, 370, 372, 395
Cleveland Dike, 212; Hills, 290, 291
Clifton (Avon) gorge, 233
Climate, changes of, 278-79; definition, 82; effect of ocean currents on, 141-43; general factors, 82-83; local factors, 48, 343-58
Climatic statistics, 80-81, 96-97, 110
Climatic types, continental, 88-92, 96-97; cyclonic, 111-28; desert, 92-95, 96-97; equatorial, 61, 83-85, 96-97; map of, 70, 71; Mediterranean, 72-74; monsoon, 75-79; polar, 85-87, 96-97; savanna, 68-72; tropical marine, 68; tundra, 87; West European, 91, 96, 124-27
Clints, 232
Clouds, 55-56, 57-58
Clyde, river, 255, 282
Coastlines, coral, 168-75; deltaic, 248-51, 305; emerged, 304; fault, 310-11; fiord, 308-10; glaciated, 308-10; longitudinal, 304, 305-6; steep or mountainous, 304, 305-11; transverse (ria), 304, 306-8; volcanic, 304, 311
Col (pressure), 121
Coln, river, 262
Colon, 151
Colorado, river, 223, 332
Columbia lava-plateau, 208, 334; river, 272
Como, lake, 285, 334
Condensation, 50, 51
Conduction, 44
Congo, basin, 84, 85; régime, 315; river, 133, 159, 160, 321, 335
Coniston Water, 333

Consequent streams, 258
Continental climates, 88-92, 96, 97; drift, 36-39; islands, 160-61, 175; shelf, 153-55, 157-61
Convection, 44, 52, 56, 62
Coquet, river, 260
Coral, 168; reefs, 168-75, 176
Coriolis force, 25-26, 61, 64
Corn Belt (U.S.A.), 292
Corrasion, by water, 239, 245; by wind and sand, 218-19
Corries, 282
Coruisk, loch, 285
Côte d'Or, 196
Cotopaxi, 206
Cotswold Hills, 102, 182, 228, 229, 230-31; drainage, 229, 262-64, 267-68
Crag and tail, 287
Crater Lake, 207, 331
Cromer Ridge, 286
Crummock Water, 248
Cuckmere, river, 247, 261
Cuesta, 181
Culbin Sands, 222, 224
Cumbria, 280, 283, 284
Cumulus, 55
Cut-offs, 242, 247, 268
Cuyaba, 80, 81
Cycle of erosion, in *karst*, 233-34; river, 242; vulcanism, 213
Cyclone, tropical, 111
Cyclonic depression, 73, 74, 79, 95; formation, 112-17; occluded, 117; secondary, 120-21; speed, 118; tracks, 124, 126; V-shaped, 121

DALMATIA, 305-6, 312
Danube, river, 247
Darent, river, 261, 262
Darjeeling, 78, 110
Dark heat rays, 44
Dartmoor, 183
Davos, 98, 104, 109
D'Aydat, lake, 330, 331
Daylight, 24, 29, 83, 85
Dead Lake, 329
Dead Sea, 133, 198, 324, 335, 355, 360
Death Valley, California, 94
Deccan, 208, 209, 214, 311
Deception Island, 207
Dee, river, 267, 268
Declination of sun, 25
Deferred tributary junctions, 247-48
Deflection, of ocean currents, 25, 137; of rivers, 25; of winds, 25-26, 61, 64
Deltas, 248-51, 252-53; in lakes, 248, 325; on land, 223, 248; in seas, 248-51, 305
Deposition, Æolian, 219-22, 224, 330;

fluvial, 241-42, 247, 248, 329; fluvio-glacial, 278, 286, 288-90; glacial, 278, 285-90; marine, 299-304, 330
Depression or 'Low', *see* Cyclonic depression
Derwent, river, 290, 291
Deserts, climate, 92-95, 96, 97; life in, 95, 224; scenery, 215-24; soils, 393-94; vegetation, 422
Devil's Tower, 207
Dew, 94, 95
Diatom ooze, 163, 166, 167
Dikes, igneous, 211, 212
Dip-slope, 181, 228, 258, 262
Dnieper, river, 320
Dogger Bank, 157, 158
Doldrums, 61, 68, 69, 72, 77
Doline, 233, 235
Doubs, river, 328
Dovedale, 233
Drainage, antecedent, 272, 273; areic, 313; diverted by ice, 290-92; en-doreic, 313; exoreic, 313; super-imposed, 270-72
Drainage of Cotswolds, 262-64; Lake District, 271-72; Northumberland, 259-60, 265; Southern Uplands, 255-57, 260-61; Weald, 261-62; world, 312
Drift, continental, 36-39; glacial, 278, 282, 285-90, 291, 292
Drowned, forests, 188; valleys, 188-89, 306-8
Drumlin, 286, 287, 288, 293
Dry valleys, 228-30
Dumgoyn, 207, 208
Dungeness, 302
Dursley, 231
Dust, atmospheric, 29, 42, 43, 51, 356; storms, 95, 224

EARTH, CURVATURE, 21; interior, 32-37; revolution, 22-23; rotation, 23-27, 61, 63, 64; shape, 21; size, 21; surface, 30-32
Earthquake, 185-87; waves, 32-33, 187
East Africa, 186, 198, 199, 201, 207, 312, 335, 336
East Anglia, coast, 302-4; glacial drift, 286, 292
East Indies, 85, 167, 201, 316, 331
East Lothian, 294
Ebbor gorge, 232
Ebro delta, 250, 251
Eco-climates, *see* Local climates
Eden, river, 261, 262
Edinburgh, 207
Eifel Mountains, 208, 337
Elbow of capture, 258-59, 260
Elbruz Mountains, 207

Engadine valley, 108
English Channel, 155, 157, 159
Entrenched meanders, 266-69
Epeirogenic movement, 194
Equation of Time, 23
Equator, 21, 23, 46, 49, 50, 61, 62
Equatorial air-mass, 79, 112, 113, 114;
  climate, 49, 50, 83-85, 96-97; current,
  136, 137, 138, 139, 141; river régimes,
  315-16; vegetation, 84-85, 415-18
Erg, 217
Erie, lake, 325
Erosion, ice, 276-77, 280-85, 308-9,
  332-34; river, 239-40; wave, 295-99;
  wind, 216-17, 218-19, 332
Erratics, 287
Escarpment, 181, 182, 229, 230, 258,
  261, 262
Esker, 288-90, 293
Étang, 249, 301, 330
Etive, loch, 308
Etna, 199, 214
Euphrates, river, 317-18
Europe, climate, 72-74, 88, 114, 123,
  124-27; ice-sheets, 275, 279, 286, 290
Eustatic movements, 188
Evaporation, 49, 50, 93, 313, 408-12
Evenlode, river, 247, 262, 263, 267-68
Everest, 31
Eyre, lake, 324

FAHLEN, LAKE, 335
Fall Line (U.S.A.), 265, 304
Fault scarps, 196-98, 310-11
Faulting, 195-99
Fault-line scarps, 196-97
Feldspar (felspar), 183, 215, 364
Fetch of waves, 146, 295, 301, 303
Finland, 282, 286, 288, 292, 325, 326,
  338
Fiords, 308-10, 312
Fissure eruptions, 208-10
Flood-plain, 246-48, 252
Floods, river, 241, 316, 319, 320, 321,
  322-23
Florida, 123, 139, 329, 337, 339
Fog, 55, 142-43, 345
Föhn, 105-7, 348
Folding, complex, 190-95; simple,
  189-90
Foraminifera, 166
Forest, boreal, 425-26; broad-leaved
  deciduous, 424-25; mid-latitude ever-
  green, 422-24; monsoon, 418-21; mon-
  tane, 433, 435; selva, 415-21; thorn,
  421-22
Fronts, atmospheric, 65, 66, 67, 114;
  arctic, 65, 114; inter-tropical, 65, 66,
  114; polar, 65, 114, 122-24, 127
Frost drainage, 100-5, 344-45

Fuji Yama, 161, 206
Funafuti atoll, 170-71, 174
Fundy, Bay of, 154, 155, 156

GALICIA, 306, 307
Galilee, sea of, 355, 360
Ganges, river, 316, 321
Gaping Ghyll, 197, 232
Garda, lake, 285, 334, 338
Geneva, 100; lake, 325
Geosynclines, 191
Geysers, 213
Ghardaïa, 223
Giant's Causeway, 183, 209
Gibraltar, 80-81, 347, 348; Strait of,
  133, 134, 135, 136
Giggleswick Scar, 196
Gilbert Islands, 170-71, 175
Glacial deposition, 285-292, 293; ero-
  sion, 280-85, 292, 293; lakes, 283,
  284-85, 286, 290-92, 294
Glaciers, 274, 275, 279
Glaslyn, lake, 283
Glastonbury Tor, 180
Glei soils, 394-95
Glen Tarbet, 285
Globigerina ooze, 163, 165, 166
Gondwanaland, 36
Gower Peninsula, 187-88
Graham Island, 207
Grass: grassland, 428-32, 437
Great Basin, U.S.A., 196, 324, 335, 337
Great Bear Lake, 325
Great Circle, 21-22
Great Lakes (N. America), 48, 90, 290,
  326, 334, 337
Great Slave Lake, 325
Great Whin sill, 212
Greenland, 22, 86, 123, 124, 140, 274,
  326
Ground water, 226-38; position, 226-
  30; scenic effect, 230-36; solvent
  action, 231-36; water-supply, 228,
  236
Grykes, 232
Gstettneralm, *doline*, 345
Guinea, 85; Gulf of, 133, 136
Gulf Stream, 55, 139-40, 142, 169;
  Drift, 124, 140

HAFF, 249, 301, 330
Hail, 57
Hamada, 217, 219, 224
Hanging valleys, 284, 292-93
Harmattan, 69
Hatteras, Cape, 139, 169
Hawaii, 175, 202, 204, 205, 255, 311
Hebrides, 312
Helsinki, 88
Henry Mountains, Utah, 211

High Force, 243
Himalayas, 47, 75, 191, 274, 327
Hog's Back, 182
Holderness, 286, 297, 299
Honolulu, 68
Hood, mountain, 206
Horn, Cape, 22
Horse Latitudes, 61-62, 63, 64, 65
Horst, 197
Hudson, river, 156, 159, 212, 249
Humber, river, 282
Humidity, absolute, 49-50, 355; relative, 50-51, 69, 83, 93, 356
Huron, lake, 325
Hurst Castle Spit, 302
Hwang Ho, river, 221, 247, 316
Hydration, 215-16, 368-69
Hydrolysis, 368-69
Hydrosphere, 30-32

ICE AGE, QUATERNARY, 87, 274, 278-81, 310, 325; Carboniferous, 36-37
Ice deposition, 278, 285-91; erosion, 276-77, 308-10, 332-34; fields, 274, 275; movement, 275-76; transport by, 277-78
Iceland, 124, 201, 204, 209-10, 213, 214, 274
Illite, 363-65, 393
Incised meanders, 266-69
India, 53, 208, 209, 214, 221, 311; climate, 53, 54, 75-79, 80, 81; rivers, 316, 321, 327-28
Indian Ocean, 137, 162, 167
Indus, 316, 321, 327
Ingleborough, 180, 197, 232
Inland drainage, 223, 313
Insalah, 94
Inselberge, 219
Insolation, 41-43, 98-99, 345-47, 355; effect of, 99-100, 109, 345, 347, 349
Insular arcs, 163
Inversion, of relief, 194, 264, 345; of temperature, 101-5
Iquique, 93
Iquitos, 84, 322
Ireland, Antrim, 209; Central Plain, 287, 288, 289, 293, 334; South-west, 91, 272, 306, 307, 308
Irkutsk, 91
Islands, continental, 160-61, 175-76; coral, 168-76; oceanic, 167-76; volcanic, 168, 176
Isostasy, 34-35, 37, 193
Italy, 185, 207, 214, 333; river régimes, 318-19

JACOBABAD, 96, 97
Jaluit, 83
Japan, 162, 184, 186, 201, 206

Java, 84, 214
Jet stream, 48, 61, 77, 128-29
Joints, 181, 183, 184
Jostedalsbreen ice-field, 275
Joux, lake, 335
Jura, mountains, 189, 190, 264, 328, 335

KALAHARI, 93, 141, 225
Kame, 290, 294
Kaolinite, 363-65
Karst, 233-35, 237, 238, 305-6, 330, 334
Katabatic winds, 102-5, 344-45
Kayes, 80, 81
Kent's Cavern, 233
Kenya, mountain, 207
Kermadec Deep, 163, 164
Khamsin, 74
Khartoum, 49, 72, 349
Khasi Hills, 53, 54
Kilauea, 205
Kilima Njaro, mountain, 99, 207
Kingston, West Indies, 68
Kioga, lake, 335
Kirkcaldy, 187
Kivu, lake, 331, 336
Klagenfurt, 101
Klamath, lakes, 335
Kólotta Dyngja, 204
Koum, 217
Krakatoa, 203, 204
Kuroshio, 141

LABRADOR, 188; current, 55, 140-41
Laccolith, 211
Lake District, England, 53, 54, 230, 248, 271-72, 333
Lakes, æolian, 330, 332; beaver, 332, 339; crater, 207-8, 331; glacial, 283, 284-85, 286, 291, 332-34; landslip, 327-29; lava-dammed, 208, 330-31; marine, 300-1, 330; ox-bow, 242, 247, 329; solution, 334; sudd-dammed, 331-32; tectonic, 197, 198, 199, 335-37
Land breezes, 74-75
Landes, 222, 330
Landslips, 230-31, 327-29
Lannemezan, plateau, 248
Lapiés, 235
Lapland, 87
Laterite, 389-90
Latitude, 23
Laurentian Shield, 32, 184, 194, 211, 279, 282, 325-26
Lauterbrunnen valley, 284
Lava, acid, 203, 204, 206; basalt, 183, 203, 208, 214; basic, 204, 208; intrusions, 211-12, 213; plateaux, 208-10
Leh, 98

Len, river, 261
Leningrad, 88, 92
Leveche, 74
Levees, 247, 252
Libya, 217, 219, 220, 221
Lightning, 56
Limestone, 180, 228, 229; Carboniferous, 232-33, 271; chalk, 181, 182, 228, 229; dolomitic, 235; Jurassic, 181, 230, 232, 235; soils, 375, 396; solution of, 231, 233-35, 334; topography, 231-36; vegetation, 412, 424
Lincoln Wolds, 236
Lithosol, 377, 378, 381, 383
Lithosphere, composition, 179-84; density, 32; distribution and extent, 30-32
Llanberis, lake, 326
Llyn Llydau, 54, 287
Local climates, 48, 343-60; influence of, atmospheric pollution, 356-58, 358-59, 360; build-up areas, 355-58, 359, 360; colour and composition, 349-50; crops, 351-53, 360; exposure to winds, 107, 347-49, 359; grass-cover, 351-52; insolation, 345-47; orientation, 99-100, 109, 345-47; shape of ground, 100, 101, 344-45; snow-cover, soils, 343, 350-51; water-bodies, 48, 354-55, 360; woods and forests, 353-54, 360
Lochan Nan Cat, 283
Loe Bar, 300-301
Loess, 221-22, 225, 373
London, 22, 49, 156; basin, 227; climate, 356-57
Long Island, 290
Longitude, 23
Longitudinal coastlines, 305-6
Longshore, drift, 155, 250, 299, 300, 301, 303
Lowther Hills, 255-57, 260-61
Lucerne, 240
Lulworth Cove, 297-98
Lumb Falls, 244

MAARE, 208, 337
Mackenzie valley, 90
Marjelen See, 327
McMurdo Sound, 86, 96
Madagascar, 161, 176
Madeira Islands, 175
Maggiore, lake, 285, 334
Mahabaleshwar, 78
Malaspina Glacier, 275
Malaya, 85
Maldive Islands, 170, 176
Malham Cove, 232-33
Manaus, 84, 96, 97, 322
Manchuria, 88

Marchlyn Mawr, 283
Mariana Deep, 37, 162
Marine, benches, 187-88, 296; erosion, 295-96; life, 143-44; planation, 272; transgressions, 34-35
Marysville Buttes, 207
Mauna Loa, 202, 205, 206
Meanders, 245-46, 247; entrenched, 266-69
Mediterranean climate, 67, 68, 72-74, 80, 81
Mediterranean Sea, 133, 134, 135, 136; deltas, 249, 250-51; tides, 153
Medway, river, 261, 262
Mendip Hills, 232, 237
Mer de Glace, 275
Mesa, 223
Mexico, 224
Michaelmas Cay, 174-75
Michigan, lake, 48, 222, 325, 334, 355
Micro-climates, 344-60
Miles City, U.S.A., 102
Mindanao Deep, 162
Misfit rivers, 262-63
Mississippi, river, 240, 246, 247, 248, 320; basin, 221, 241, 252, 337; delta, 186, 189, 250, 253, 329
Mistral, 74, 107, 347, 348
Moho discontinuity, 34-35
Mole, river, 231, 261, 262
Monsoons, 75, 78-79
Mont Blanc, 100
Mont Pelée, 204
Monte Nuovo, 206
Montmorillonite, 363-65, 396
Moon, tide-raising force, 147-51
Moraines, 285-86, 290, 291, 292
Morar, loch, 285, 333, 334
Moscow, 91
Moses, lake, 334
Mountain, breeze, 103-5, 344; building, 189-95; climates, 98-110, 348, 349; life on, 108-9, 110; river régimes, 320-21; sickness, 60; snow-line, 108, 109
Mull, island, 209, 212
Murmansk, 140
Mutanda, lake, 330

NAIN, 142
Nairn, 54
Nant Ffrancon, 284
Nappe, 192
Narenta, river, 234
Neap tides, 150, 151
Nemi, lake, 207
Nevados, 103
Névé, 277, 282, 283
Nevis, loch, 285
New England, 302, 312

Newfoundland, 55, 124, 141, 142
New York, 22, 142, 155, 156
New Zealand, 54, 161, 176, 213, 214, 308, 311
Niagara Falls, 243
Niger, river, 133, 316-17
Nile, river, 49, 248; delta, 250, 252; régime, 316, 317, 321, 322, 323
Nimbus, 55
Nith, river, 256
Nivation, 230, 275, 279, 282
North Atlantic, Drift, 124, 135, 139-40, 142; Ocean, 139-41
North Downs, 182, 227, 229, 262
North German Plain, 290, 292
North Sea, 154, 157-59, 176
Northumberland, rivers, 259-60, 265
Norway, 275, 308, 309
Nor'-Wester, 107
Nunatak, 278
Nyasa, lake, 198, 199, 325, 335

OBSEQUENT STREAMS, 258, 262
Ocean, area, 30-31; circulation, 135, 137-41; currents, 135-44; deposits, 163-67; depths, 161-63; drifts, 135; salinity, 131-33; temperature, 133-35; tides, 135, 147-56; waves, 145-47
Ocean Island, 96, 97
Okeechobee, lake, 337
Olekminsk, 88
Ontario, lake, 325
Oregon, tectonic lakes, 196, 197, 335
Orfordness, 303
Orientation, 99-100, 345-47
Orogenic movements, 189-99; Alpine, 191-93, 195; Caledonian, 195; Hercynian, 195
Orographic rain, 52-54, 78, 107-8
Outwash plain, 286, 290
Overflow channels, fluvio-glacial, 290-92; glacier, 285; river, 259
Ox-bow lakes, 242, 247, 329
Oxidization, 215
Oymekon, 90
Ozone, 2, 28, 46

PACIFIC OCEAN, CIRCULATION, 141; deposits, 165, 166, 167; relief of floor, 162-63, 164
Pakistan, 186, 220
Palaeomagnetism, 37-38
Palaeowind, 37-38
Palestine, 198, 355
Pampa, 431, 432, 437
Para, 83, 96, 97
Paramos, 109, 435-36
Pedalfer, 372
Pediment, 225

Pedocal, 373
Pen-y-Ghent, 180
Pennines, 232-33, 348
Perihelion, 22
Permafrost, 384
Peru, 93; current, 93, 142, 143
Philippine Islands, 54
Phlegraean Fields, 206, 213
Pickering, Vale of, 290; glacial lake, 290-92
Pike's Peak, 102, 211
Plant, conservation, 436; evolution, 400-1; formations, climax, 403-4; formations, major, 414-36; successions, 404-6; spread of, 402-3
Plateau climate, 99, 101
Plitvicka, lake, 330
Plugs, 207, 208, 287
Po, river, 250, 318, 319
Podzol, 372, 377, 381-83, 385-87
Podzolic soils, 387-88
Polar, climates, 85-88; front, 65, 66-67, 113, 114, 122-24, 127; winds, 66
Polje, 233, 237
Popocatepétl, 206
Porosity, 226, 227, 230
Pot-holes, 239-40
Prairies, 92, 431, 432
Precipitation, causes, 51-55, 56; forms, 55-57; relation to relief, 52-54, 107-8
Precession of equinoxes, 27
Pressure, world distribution, 60-64; gradients, 62-64; migration of belts, 67-74
Pressure systems, Col, 121; High, 118-20; Low, 114-18; secondary depression, 120-21; V-shaped depression, 121, 122; Wedge, 121, 122]
Pteropod ooze, 163, 165, 166
Puna, 99, 109, 435
Puy, 207; de Sarcoui, 204
Pyrenees, 99-100, 248, 274

QUATERNARY ICE AGE, 87, 274; ice-sheets, 278-81, 310, 325
Queensland, 227, 236
Quetta, 186, 200
Quito, 83, 99, 110

RADIATION, FROM EARTH, 43-44, 343
Radio-activity, 34-35
Radiolarian ooze, 163, 166
Rain, 52-56, 57-58
Rainfall, see Precipitation
Rain making, artificial, 57-58
Rainier, mountain, 206
Raised beaches, 187-88
Randkluft, 283
Rannoch Moor, 285

Red Clay, 163, 165, 166, 167
Red Sea, 69, 133, 135
Red Tarn, 283
Ree, lough, 334
Reg, 217, 219
Rejuvenation, 265-70, 273
Relief, of hydrosphere, 31-32, 157, 161-63; inversion of, 264; of lithosphere, 31-32; submarine, 157-60, 161-63; influence on soils, 378-80
Rendzina, 374-75, 396
Reuss valley, 106, 108
Reykjavik, 214
Rhine, river, 106, 319-20; rift-valley, 197, 247
Rhône, delta, 250, 252, 253; glacier, 275; river, 106, 108, 240, 325, 327, 347
Ria, 306-8, 312
Ribble, river, 245
Rickmansworth, 345
Rift valleys, 197-99
Rigikulm, 101
Rio de Janeiro, 141
River action, deposition, 241-42, 246, 247, 248-51; erosion, 239-40; floods, 241; transportation, 240-41
River development, capture, 257-59, 260, 262, 263, 264; headward growth, 254-57; in folded areas, 264; rejuvenation, 264-69; terracing, 269-70; watershed regression, 255-57
River discharge, 313-15
River régimes, continental, 320; equatorial, 315-16; Mediterranean, 317-19; monsoon, 316-17; mountain, 320-21; rainy temperate, 319-20; savanna, 69, 71, 316
River valleys, deepening, 244; effect of climate, 242; effect of geology, 242-44; effect of relief, 242; evolution, 239-48; shallowing, 246-48; terraced, 269-70; widening, 244-48
Rivers, consequent, 258, 261, 262; misfits, 262-63; obsequent, 258, 262; subsequent, 257, 258, 260, 261
Roaring Forties, 66, 138
Roches moutonnées, 287
Rock, lake, 334
Rocks, igneous, 33, 179, 182-84; metamorphic, 183; sedimentary, 33, 34, 179-82, 184
Rocky Mountains, 88, 106, 199
Rome, 73, 207, 231
Roquefort, 237
Rotation, of earth, 23-27, 61, 63, 64
Rother, river, 261, 302

SAGASTYR, 90, 96-97

Sahara, 93-95, 96-97
St. Helena, 161, 168
St. John's, 142, 151
St. Lawrence, river, 124, 334, 337
St. Luc, 110
St. Paul, island, 207
Salisbury Plain, 228, 350-51
Salpausselka, 286, 326
Samun, 107
Sand, coastal, 180, 222, 303; corrasion 218-19; dunes, 217-18, 219-21, 224
Säntis, peak, 102, 108, 110, 335
Santorin Island, 207
Sargasso Sea, 139, 141, 144
Savanna, 69, 80
Scandinavia, 191, 195, 279, 286
Schattenseite, 99, 108, 109
Scilly Isles, 73, 175
Scotland, 54, 197, 208, 209, 287, 308
Scottish Highlands, 186, 191, 195, 210, 280, 285, 308, 333
Screes, 216, 329
Sea cliffs, 295, 296, 297, 299, 311
Sea-level, changes in, 34, 173, 174, 187-89, 265, 279, 285, 304, 306, 309-10
Sea water, composition, 131; movements, 135-41, 145-55; salinity, 131-33; temperature, 133-35
Seasons, cause of, 24
Seif, 220
Seine, river, 155, 252, 319
Selva, 85, 415-18
Sequoia, 423, 426
Severn, river, 262, 291, 319; bore, 155
Seychelle Islands, 176
Shannon, river, 319, 334
Shasta, mountain, 206
Shetland Islands, 295
Shode, river, 262
Shyok, river, 327, 328
Sial, 32-34, 37
Siberia, 90, 91, 92
Sierra Nevada, U.S.A., 53, 54, 56, 196, 248, 329
Sill, 211-12
Sima, 33-35, 37
Sind, 76, 78
Singapore, 83, 96-97, 175
Sion, 108
Sirocco, 74
Skagerrak, 136, 159
Skjer coast, 310
Skye, island, 210, 293
Smoke, 51, 356, 357, 359
Snag, lake, 208
Snow, 56-57; cover, 91-92, 108, 354; line, 109, 274
Snowdon, 57, 194, 264, 280, 283, 284, 287
Sogne Fiord, 308, 309

Soils, 361-99; azonal, 378, 383-84; calcimorphic, 396; chernozem, 380, 391-93; classification, 380-96; desert, 393-94; effect on climate, 48, 349-51; fauna, 375-76; formation of, 372-79; ferralitic, 388-91; ferruginous, 388-89; glei, 394-96; halomorphic, 394; physical content, 361-72; plaggen, 378; podzol, 385-87; podzolic, 377, 387-88; tundra, 384-85
Solana, 100
Solfataras, 213
Solifluxion, 230, 385, 428
Solonchak, 394-95
Solution, 231, 233
Solway Firth, 155, 287
Sonnenseite, 99, 100, 108, 109
Sotch, 235
South Downs, 231, 262
South Wales, 271
Spits, 299-303
Spitzbergen, 86, 96-97, 123
Springs (spring-lines), 228, 236, 254-55
Spurs, truncated, 283-84
Squall-line (front), 121
Stelzing, 101
Steppe, 92, 431-32
Stickle Tarn, 283
Stour, river (Kent), 261, 262
Strandflat, 187, 310
Stratosphere, 2, 45-46, 65
Stratus, 55-56
Stromboli, 203
Stroud (Glos.), 237, 264
Styhead Tarn, 329
Submarine, canyons, 159-60; relief, 157-60, 161-63, 172
Submerged forests, 188
Subsequent streams, 257, 258, 260, 261
Subsidence, of land, 188-89; of sea-level, 188
Sudan, 68-72
Sun, 41; altitude of, 24-25, 42, 69, 85, 346; tidal pull, 150
Sunbaking period, 69, 419-20
Sunshine, 73, 94, 346, 356, 357
Superimposed drainage, 261, 270-72
Superior, lake, 48, 292, 325
Suvadiva, atoll, 170
Swallow holes, 231, 232, 235
Sydney, 80, 81
Symond's Yat, 271
Synclinal ridges, 194, 264

TAÏGA, 92, 425-26, 438
Tanganyika, lake, 198, 199, 325, 335
Tarim Basin, 223
Tarn, river, 235-36
Tay, river, 282
Teak, 420-21

Teise, river, 261, 262
Temperature, atmospheric, 41-48; horizontal distribution, 46-48; inversion, 90, 101-5, 344-45; range, 89, 91, 343, 350-51; vertical distribution, 2, 44-46
Temperature, of earth's interior, 34; of oceans, 133-35; of surface cover, 343-47, 349-55, 357-58
Terraces, river, Avon, 270; Severn, 270; Thames, 269-70
Terra rossa, 375
Tetrahedral theory, 32
Thames, river, 243, 262, 263, 267, 322; régime, 319; terraces, 269-70
Thunderstorms, 56
Tiberias, lake, 208
Tibesti Hills, 108, 217
Tidal currents, 155
Tides, 147-56
Tierra caliente, 109
Timbuktu, 72
Time, equation of, 23; standard, 23-24
Titicaca, lake, 335
Tobolsk, 88, 92, 96, 97
Tokyo, 186
Tonga Deep, 163, 164
Topographies, coastal, 297-311; desert, 216-25; limestone, 231-36; volcanic, 206-14
Topography, climatic control, 98-110
Tornadoes, 69, 72
Torquay headland, 297, 298
Trade Winds, 64-66, 68, 69, 72, 128
Transhumance, 109
Transport, by ice, 277-78, 285-87; river, 240-41, 322; sea-water, 299-304; wind, 219-20, 221-22
Traprain, Law, 211
Trent, river, 155, 246, 247, 248, 252
Tristan da Cunha, 161, 168
Tropical cyclones, 111
Tropopause, 2, 46
Troposphere, 2, 45, 46
Tulare, lake, 329
Tundra, 87, 96, 97, 384-85, 426-28
Turbulence, in atmosphere, 107, 347, 348, 349; in streams, 240-41, 242
Turfan basin, 223
Tuscarora Deep, 163
Tweed, river, 255, 256-57, 287
Twilight, 29-30, 85
Tyne, river, 260, 265, 282, 292
Typhoons, 111

Ubac, 99
Ubaga, 100
Ullswater, 248, 283
Ultra-violet rays, 99
Underground water, 226-31, 236-37
Unst, 295

Upernivik, 86, 96, 97
Upper atmosphere, 2, 45, 46
Upwellings, of cold water, 138, 143
Urban climates, 355-58, 360

VADOSE WATER ZONE, 226
Valentia, 91
Valley, asymmetrical, 243; canoe-shaped, 190; drowned, 188-89, 305-9; dry, 228-30, 238; hanging, 284, 292-93; incised, 265-69; in-valley, 268-70; rift, 197-99; U-shaped, 276; 283-84; V-shaped, 244
Valparaiso, 80, 81
Vapour pressure, 49
Vegetation, 400-38; classifications, 411-36; effect on climate, 350-54; effect on river régimes, 314; equatorial, 85, 415-18; forest, 415-26, 433, 435; grass, 428-32, 437; growth, 406-14; Mediterranean, 73, 74, 378, 423; mountain, 432-36, 438; nature of, 400-2, 404-6, 424-36; savanna, 69, 71, 421, 429-31; spread of, 402-3; tundra, 87, 426-28, 438
Verkhoyansk, 90, 91, 92
Vesuvius, 203, 207, 214
Victoria Falls, 243, 253; lake, 325, 335, 336, 355
Volcanic lakes, 207-8, 330-31, 337; plugs, 207, 208, 287
Volcanoes, cones, 203-6; distribution, 201-2; eruptions, 202-3; topographic effect, 206-8
Volga, river, 320, 322
Vosges, mountains, 197, 280
Vulcanism, 199, 201-14; fissure eruptions, 208-10; lava intrusions, 211-12
Vulcano, 204

WADI HALFA, 49, 95, 317
Wales, North, 54, 293; South, 271
Walfish Bay, 93, 96, 97, 141
Wansbeck, river, 260
Warping, 187, 188, 337
Warsaw, 91, 96, 97
Washington, 208
Waterfalls, 243-44
Watershed regression, 255-57, 261
Water-supply, 228, 236; hardness of, 236-37
Water-table, 226-30

Water vapour, 49-58; pressure, 49; vertical distribution, 46
Waves, erosion, 295-96; fetch, 146, 295, 301, 303; nature, 145-47; propagation, 145
Weald, 182, 191, 194, 195; rivers, 261-62
Wear, river, 266, 292
Weather, 82, 124, 127
Weathering, chemical, 215-16, 218; physical, 216, 218
West Indies, 162, 169, 186
Westerlies, 64, 66, 73, 112, 128
Western Ghats, 53, 78
Wey, river, 261
Whin Sill, 212, 243
Wicklow, mountains, 211
Wight, Isle of, 160, 182, 190, 195
Wind, deflection, 25-26, 61, 64; deposition, 219-22; erosion, 218-19; systems, 64-67
Windrush, river, 262, 263, 267, 268
Winds, anabatic, 103-4; berg, 107; blizzard, 86, 90; bora, 74, 348; buran, 90; chinook, 106-7; föhn, 105-7, 348; harmattan, 69; helm, 348; katabatic, 101-4, 344-45; khamsin, 74, 95; land and lake, 355; land and sea, 74-75, 93; levêche, 74; mistral, 74, 347, 348; monsoon, 75; mountain and valley, 103-4, 344; nevados, 103; nor'-wester, 107; polar, 64, 66-67; Roaring Forties, 66; samun, 107; shamal, 95; sirocco, 74; trades, 64-66, 68, 69; westerlies, 64, 66, 73, 112, 128
Wind-breaks, 348-49
Winnipeg, 22, 96-97, 107; lake, 325, 326
Wookey Hole, 232
Woolhope, Vale of, 182
Wrekin, 207
Wye, river, 266-67, 271

YAKUTSK, 88, 91, 92, 96, 97
Yangtze Kiang, river, 243, 316, 323
Yarmouth, 303, 304
Yenisei, river, 241
York, Vale of, 286-87, 291

ZAMBEZI, 243, 316
Zanzibar, 85
Zeugen, 219
Zurich, 104

Soc
GB
54
B3
1966

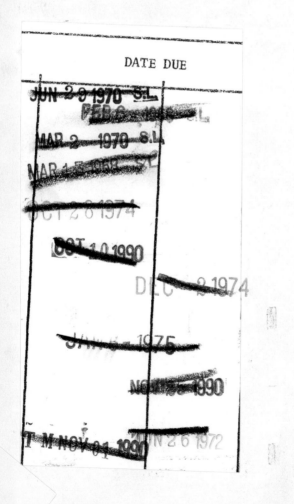

DATE DUE

JUN 29 1970 S.L.

FEB ... S.L.

MAR 2 1970 S.L.

MAR 1 5 1968 S.L.

OCT 26 1974

OCT 10 1990

DEC 2 1974

JAN 1975

NOV 1990

T M NOV 01 1990   JUN 26 1972